MATHEMATICAL PROBABILITY

MATHEMATICAL PROBABILITY

Dr K C Pal MSc (Gold Medallist), PhD

Head of the Dept of Mathematics (UG & PG)
Ramakrishna Mission Vidyamandira
(An Autonomous College with CPE Status)
Chairman of the Board of Studies of Mathematics
Visiting Professor of Calcutta University

New Central Book Agency (P) Ltd

LONDON
HYDERABAD ERNAKULAM BHUBANESWAR
NEW DELHI KOLKATA PUNE GUWAHATI

NCBA

REGD OFFICE
8/1 Chintamoni Das Lane, Kolkata 700 009, India
email: ncbapvtltd@eth.net

OVERSEAS
NCBA (UK) Ltd, 149 Park Avenue North, Northampton, NN3 2HY, UK
email: ncbauk@yahoo.com

EXPORT
NCBA Exports Pvt. Ltd, 212 Shahpur Jat, New Delhi 110 049
email: ncbaexports@ncbapvtltd.com

BRANCHES

212 Shahpur Jat
New Delhi 110 049
email: ncbadel@ncbapvtltd.com

Shop Nos. 3 & 4, Vinayak Towers
681/B Budhwar Peth, Appa Balwant Chowk
Pune 411 002
email: ncbapune@ncbapvtltd.com

Shop No. 15, Bharati Towers, Ground Floor
Forest Park, Bhubaneswar 751 001
email: ncbabub@ncbapvtltd.com

House No. 3-1-315, 1st Floor
Nimboliadda, Kachiguda, Hyderabad 500 027
email: ncbahydb@ncbapvtltd.com

GSS Shopping Complex, 1st Floor
Opposite College Post Office, Convent Road
Ernakulam 682 035
email: ncbaernk@ncbapvtltd.com

Radhamayee Complex (Second Floor)
Opp. SBI, Panbazar, Jaswanta Road
Guwahati 781 001
email: ncbaguwahati@ncbapvtltd.com

MATHEMATICAL PROBABILITY • K C Pal

First Published: July 2012
Reprinted: June 2013

PUBLISHER AND TYPESETTER
New Central Book Agency (P) Ltd
8/1 Chintamoni Das Lane, Kolkata 700 009

PRINTER
New Central Book Agency (P) Ltd
Web-Offset Division, Dhulagarh, Sankrail, Howrah

COVER DESIGNER
Soumen Paul

COVER PRINTER
Techno Process, Kolkata

TECHNICAL EDITOR
Dipan Roy

PROJECT TEAM
Abhijit Sen, Koushik Samanta, and Prabhat Jas

ISBN 978 81 7381 781 6

Price ₹ 325.00

Dedicated to

my mother

Late Puspa Rani Pal

Contents

Preface

In course of teaching the subject in college and university for last 25 years, I have all along been nourishing a desire in my mind to write a comprehensive textbook on **Mathematical Probability** in order to reach the maximum number of students.

Now, as the dream comes true, I earnestly hope the book will enrich the students and will leave them well-equipped for their examinations. A large number of problems from Honours Papers of different leading Indian universities have been solved here in details.

Every attempt has been made in the book for generalisation of theories with the expectation that in some particular situations it will be easier for the students to memorise them. I shall feel contented if the students are benefited from this humble endeavour of mine.

I cordially thank Mr Amitabha Sen, Director of New Central Book Agency (P) Ltd, for offering me such a challenging assignment.

27 March 2012
Kolkata

K C Pal

Chapter 0

Useful Mathematics in Probability Theory

0.1 Mathematical logic

STATEMENT. An information in any form is called a *statement* if it has a truth value, i.e., it is either true (T) or false (F) exclusively; e.g., *The sum of the three angles of a triangle in Euclidean geometry is two right angle* is a statement. *The sum of two sides of any triangle is less than the third side* is also a statement.

The first one is true and the latter is false. The statements are generally denoted by the symbols $p, q, r, \cdots$, etc.

Open statement. It is not a statement as it is true for some cases and false otherwise; e.g., $x^2-5x+6=0$, $x \in \mathbb{R}$ is an open statement, as it is true for $x=2,3$ and is false for any other value of x.

Compound statement. A statement formed by joining two or more statements is called a *compound statement.*

Rules of joining

(a) Conjunction of two statements p and q : Joining two statements p,q by AND is called *conjunction.* $p \wedge q$ is the symbol of expressing the statement p *and* q.

Such a statement is true *only when* both are true. Note that $p \wedge q = q \wedge p$.

(b) Disjunction of two statements p and q : Joining two statements p,q by OR is a disjunction, $p \vee q$ is the symbol of expressing the statement p *or* q.

Such a statement is false *only when* both are false. Note that $p \vee q = q \vee p$.

Negation of a statement. A complementary statement obtained by prefixing a NOT. The negation of a statement p, i.e., NOT p is denoted by $\sim p$.

Such a statement is true when p is false and false when p is true.

Tautology. A statement, which is always true is called a *tautology.* For any statement p the statement $p \vee \sim p$ is a tautology.

De Morgan's laws

For any two statements p,q, the De Morgan's law states

(i) $\sim(p \vee q) = (\sim p) \wedge (\sim q)$ (ii) $\sim(p \wedge q) = (\sim p) \vee (\sim q)$.

Distributive laws

For any three statements p,q,r : (i) $p \vee (q \wedge r) = (p \vee q) \wedge (p \vee r)$ (ii) $p \wedge (q \vee r) = (p \wedge q) \vee (p \wedge r)$.

These are left distributive laws. The right distributive laws also hold as $\vee$ and $\wedge$ are both commutative.

Associative laws

For any three statements p,q,r : (i) $p \wedge (q \wedge r) = (p \wedge q) \wedge r$ (ii) $p \vee (q \vee r) = (p \vee q) \vee r$.

All the above laws may be verified by truth table.

e.g., verification of $p\vee(q\wedge r)=(p\vee q)\wedge(p\vee r)$:

p	q	r	$q\wedge r$	$p\vee(q\wedge r)$	$p\vee q$	$p\vee r$	$(p\vee q)\wedge(p\vee r)$
T	T	T	T	T	T	T	T
T	T	F	F	T	T	T	T
T	F	T	F	T	T	T	T
F	T	T	T	T	T	T	T
F	F	T	F	F	F	T	F
F	T	F	F	F	T	F	F
T	F	F	F	T	T	T	T
F	F	F	F	F	F	F	F

See that the truth values of the statements $p\vee(q\wedge r)$ and $(p\vee q)\wedge(p\vee r)$ are the same in all possible eight situations. Hence, they are the same.

0.2 Conditional statement

Implication between two statements p and q: The statement '*If p, then q*' is called *implication* of q by p and is denoted by $p\Rightarrow q$.

$p\Rightarrow q$ stands for any one of the following expressions. (i) p implies q (ii) p only if q (iii) p is sufficient for q (iv) q is necessary for p. $p\Leftrightarrow q$ is false only when p is true but q is false.

Associated implications

Associated with $p\Rightarrow q$ we have three other implications: (i) Converse of $p\Rightarrow q$ is $q\Rightarrow p$ (ii) Inverse of $p\Rightarrow q$ is $\sim p\Rightarrow\sim q$ (iii) Contrapositive of $p\Rightarrow q$ is $\sim q\Rightarrow\sim p$.

Note: $p\Rightarrow q$ and $\sim q\Rightarrow\sim p$ are logically equivalent.

Biconditional

When $p\Rightarrow q$ and $q\Rightarrow p$ are both true we write $p\Leftrightarrow q$. It stands for p, if and only if q.

$p\Leftrightarrow q$ stands for the following statement. A necessary and sufficient condition for p to be true is that q is true.

0.3 Set

A set is a well-defined collection of distinct objects called *elements*. Well-defined is that property of the set, which guarantees the membership or non-membership of an object in the set.

The concept of a set is too much fundamental and natural. It is so fundamental that even a variable or a constant cannot be defined without it. If x is a variable, then x varies over a set of values. Variable on a singleton set is a constant.

$A, B, C,\cdots$ denote different sets. $a\in A\Rightarrow a$ is a member of A. $b\notin A\Rightarrow b$ is not a member of A.

Notation of some standard sets

$\mathbb{Z}=$ set of all integers

$\mathbb{Z}^+=$ set of all positive integers

$\mathbb{Z}^-=$ set of all negative integers

$\mathbb{N}=$ set of all natural numbers $=\mathbb{Z}^+$

$\mathbb{R}=$ set of all real numbers

$\mathbb{R}^+$ and $\mathbb{R}^-$ are respectively the set of all positive and negative integers

$\mathbb{Q}=$ set of all rational numbers

$\mathbb{Q}^+$ and $\mathbb{Q}^-$ are respectively the set of all positive and negative rational numbers.

$\mathbb{C}=$ set of complex numbers

$\mathbb{Z}_n=$ set of all residue classes modulo $n(n\in\mathbb{N})$

$M_n(\mathbb{R})$ = set of all square matrices of order n over the field of real numbers
S_n = set of all permutations of the set $\{1,2,\cdots,n\}, n \in \mathbb{N}$
$n\mathbb{Z}$ = set of all integers divisible by $n, n \in \mathbb{N}$
ϕ = set of containing no element, called a *null set*
U = set of all elements under discussion, called *universal set.*

Subset

Let A and B be two sets such that $x \in A \Rightarrow x \in B$, then A is called a *subset* of B and is denoted by $A \subset B$.

e.g., $\mathbb{N} \subset \mathbb{Z} \subset \mathbb{Q} \subset \mathbb{R} \subset \mathbb{C}$.

For logical consistency ϕ is taken to be a subset of any set and any set is subset of the universal set. So, for any set A, $\phi \subset A \subset U$.

According to definition, $A \subset A$. We say, A is an improper subset of A. Two sets are said to be comparable if one is a subset of the other. The relation $\subset$ is called *set inclusion.* This relation is reflexive, antisymmetric and transitive, i.e., $A \subset A$, $\forall A$, $A \subset B$ and $B \subset A \Rightarrow A = B$ and $A \subset B$ and $B \subset C \Rightarrow A \subset C$.

0.4 Set algebra

Union. Let A, B be two sets. Their union is denoted by $A \cup B$ and is defined by

$$A \cup B = \{x : x \in A \text{ or } x \in B\}.$$

Obviously, $A \cup B = B \cup A$, $A \cup (B \cup C) = (A \cup B) \cup C$, $A \cup A = A$, $A \cup B = B \Leftrightarrow A \subset B$.

Intersection. Intersection of two sets A and B is denoted by $A \cap B$ and is defined as $A \cap B = \{x : x \in A \text{ and } x \in B\}$.

Obviously, $A \cap B = B \cap A, A \cap A = A, A \cap (B \cap C) = (A \cap B) \cap C, A \cap B = A \Leftrightarrow A \subset B$.

If $A \cap B = \phi$, then A and B are disjoint. Such sets are incomparable.

Complementation

The complement of $A(\subset U)$ is denoted by A^c or A' and is defined by $A^c = \{x \in U : x \notin A\}$.

Obvious properties of complementation is $A \cup A^c = U$ and $A \cap A^c = \phi$. Also, $(A^c)^c = A$.

In this case, we say that $\{A, A^c\}$ is a partition of U.

De Morgan's laws. $(A \cup B)^c = A^c \cap B^c$ and $(A \cap B)^c = A^c \cup B^c$.

Proof. Let $x \in (A \cup B)^c$. $x \in (A \cup B)^c \Leftrightarrow x \notin A \cup B \Leftrightarrow x \notin A$ and $x \notin B$

$\Leftrightarrow x \in A^c$ and $x \in B^c \Leftrightarrow x \in A^c \cap B^c$. $\therefore (A \cup B)^c = A^c \cap B^c$.

This is the proof of first one.

To prove the second, apply the result of first for A^c and B^c.

$\therefore (A^c \cup B^c)^c = (A^c)^c \cap (B^c)^c = (A \cap B)$, or, $\{(A^c \cup B^c)^c\}^c = (A \cap B)^c$, or, $A^c \cup B^c = (A \cap B)^c$.

Relative complement

Let $A, B \subset U$. Then, $A - B = \{x \in A : x \notin B\}$ is called *complement* of B relative to A. Thus complement is the general sense as in particular taking $A = U$, $A - B$ becomes B^c

So, the generalized laws of De Morgan's are $A-(B \cup C) = (A-B) \cap (A-C)$ and $A-(B \cap C) = (A-B) \cup (A-C)$.

Symmetric difference

Let $A, B \subset U$. Then, their symmetric difference is denoted by $A \Delta B$ and is defined by

$$A \Delta B = (A - B) \cup (B - A).$$

$\therefore A \Delta B = (A \cup B) - (A \cap B)$.

Note that $A \Delta B = B \Delta A$. So, Δ is symmetric.

0.5 Family of sets

Collection of sets constitute a family of sets. A family is a set of sets. If I be an index set, then $F = \{A_\alpha, \alpha \in I\}$ is a family of sets, the members of the family are indicated by I. I may be $\mathbb{N}$ or $I = \{1,2,\cdots,n\}$, etc.

Union of the family $= \bigcup_{\alpha\in I} A_\alpha$ and intersection $= \bigcap_{\alpha\in I} A_\alpha$.

De Morgan's laws. $\left\{\bigcup_{\alpha\in I} A_\alpha\right\}^c = \bigcap_{\alpha\in I} A_\alpha^c$ and $\left\{\bigcap_{\alpha\in I} A_\alpha\right\}^c = \bigcup_{\alpha\in I} A_\alpha^c$.

Partition of a set

Let A be a non-empty set. A may be considered as a family of sets, the members are its subsets.

Such a family is said to form a partition of A, if they are pairwise disjoint and their union is the whole set.

Thus a family $\{A_1, A_2, \cdots, A_n\}$ is a partition of A if $A_i \subset A$, $\forall\ i = 1, 2, \cdots, n$, $A_i \cap A_j = \phi$ for $i \neq j$ and $\bigcup_{i=1}^{n} A_i = A$.

Cartesian product

The Cartesian product of two non-empty sets A and B is denoted by $A \times B$ and is defined by

$$A \times B = \{(a, b) : a \in A, b \in B\}.$$

In general, Cartesian product of n non-empty sets is

$$A_1 \times A_2 \times \cdots \times A_n = \{(a_1, a_2, \cdots, a_n) : a_i \in A_i,\ i = 1, 2, \cdots, n\}.$$

$A \times A = \{(a, b) : a, b \in A\}$ is the Cartesian product of A with itself and is denoted by A^2.
Similarly, $A^n = A \times A \times \cdots \times A = \{(a_1, a_2, \cdots, a_n) : a_i \in A,\ i = 1, 2, \cdots, n\}$.
In particular, if $A = \mathbb{R}$, then $\mathbb{R}^2 = \mathbb{R} \times \mathbb{R} = \{(x, y) : x, y \in \mathbb{R}\}$ and

$$\mathbb{R}^n = \mathbb{R} \times \mathbb{R} \times \cdots \times \mathbb{R} = \{(x_1, x_2, \cdots, x_n) : x_i \in \mathbb{R},\ i = 1, 2, \cdots, n\}.$$

Distributive law of $\times$ over $\cup, \cap$ & $-$

If A, B, C be any three subsets of the universal set, then (i) $A \times (B \cup C) = (A \times B) \cup (A \times C)$ (ii) $A \times (B \cap C) = (A \times B) \cap (A \times C)$ (iii) $A \times (B - C) = (A \times B) - (A \times C)$.

Proof. (i) $(a, b) \in A \times (B \cup C) \Leftrightarrow a \in A$ and $b \in B \cup C$
$\Leftrightarrow a \in A$ and $(b \in B$ or $b \in C)$
$\Leftrightarrow (a \in A$ and $b \in B)$ or $(a \in A$ and $b \in C)$
$\Leftrightarrow (a, b) \in A \times B$ or $(a, b) \in A \times C$
$\Leftrightarrow (a, b) \in (A \times B) \cup (A \times C)$.

(ii) $(a, b) \in A \times (B \cap C) \Leftrightarrow a \in A$ and $b \in B \cap C$
$\Leftrightarrow a \in A$ and $(b \in B$ and $b \in C)$
$\Leftrightarrow (a \in A$ and $b \in B)$ and $(a \in A$ and $b \in C)$
$\Leftrightarrow (a, b) \in A \times B$ and $(a, b) \in A \times C$
$\Leftrightarrow (a, b) \in (A \times B) \cap (A \times C)$.

(iii) $(a, b) \in A \times (B - C) \Leftrightarrow a \in A$ and $b \in B - C$
$\Leftrightarrow a \in A$ and $(b \in B$ and $b \notin C)$
$\Leftrightarrow (a \in A$ and $b \in B)$ and $(a \in A$ and $b \notin C)$
$\Leftrightarrow (a, b) \in A \times B$ and $(a, b) \notin A \times C$
$\Leftrightarrow (a, b) \in A \times B - A \times C$.

Note: $(A - B) \times C = (A \times C) - (B \times C)$ [Try yourself].

0.6 Number of elements in a finite set

Let A, B, C be three finite sets and $n(D)$ denote the number of elements in D, then

$$n(A \cup B) = n(A) + n(B) - n(A \cap B),$$

$$n(A \cup B \cup C) = n(A) + n(B) + n(C) - n(A \cap B) - n(B \cap C) - n(C \cap A) + n(A \cap B \cap C),$$

and in general for the finite sets $A_1, A_2, \cdots, A_n$

$$n(A_1 \cup A_2 \cup \cdots \cup A_n) = \sum_{i=1}^{n} n(A_i) - \sum\sum_{1 \le i < j \le n} n(A_i \cap A_j) + \sum\sum\sum_{1 \le i < j < k \le n} n(A_i \cap A_j \cap A_k) + \cdots$$
$$+ (-1)^{n-1} n(A_1 \cap A_2 \cap \cdots \cap A_n).$$

Ex. For any two sets A and B: $A \cup B = (A - B) \cup B$ and $(A - B) \cap B = \phi$.

Solution. $a \in (A - B) \cup B \Leftrightarrow a \in A - B$ or $a \in B \Leftrightarrow a \in B$ or $a \in A - B$
$\Leftrightarrow a \in B$ or ($a \in A$ and $a \notin B$)
$\Leftrightarrow$ ($a \in B$ or $a \in A$) and ($a \in B$ or $a \notin B$)
$\Leftrightarrow a \in A$ or $a \in B$
$\Leftrightarrow a \in A \cup B$ [$a \in B$ or $a \in B^c \Rightarrow a \in B \cup B^c \Rightarrow a \in U$]

$(A - B) \cap B = (A \cap B^c) \cap B = A \cap (B^c \cap B)$, by associative law
$= A \cap \phi = \phi$.

Ex. Union of any two sets may be expressed as the union of three disjoint sets.

Solution. Let A, B be two sets. We, show that $A \cup B = (A - A \cap B) \cup (A \cap B) \cup (B - A \cap B)$ and the sets on RHS are pairwise disjoint.

$(A - A \cap B) \cup (A \cap B) = \{A \cup (A \cap B)\}$ [by the previous example]
$= \{(A \cup A) \cap (A \cup B)\} = \{A \cap (A \cup B)\} = A$ as $A \subset A \cup B$.

Similarly, $(B - A \cap B) \cup (A \cap B) = B$.
$\therefore (A - A \cap B) \cup (A \cap B) \cup (B - A \cap B) = (A - A \cap B) \cup (A \cap B) \cup (B - A \cap B) \cup (A \cap B) = A \cup B$.
Now, $(A - A \cap B) \cap (A \cap B) = \{A \cap (A \cap B)^c\} \cap (A \cap B) = A \cap \{(A \cap B)^c \cap (A \cap B)\} = A \cap \phi = \phi$.
Similarly, $(B - A \cap B) \cap (A \cap B) = \phi$.
$(A - A \cap B) \cap (B - A \cap B) = A \cap (A \cap B)^c \cap B \cap (A \cap B)^c = A \cap B \cap (A \cap B)^c = A \cap B \cap (A^c \cup B^c)$
$= (A \cap B \cap A^c) \cup (A \cap B \cap B^c) = \phi$.

0.7 Function (Mapping)

A function or a mapping f from a non-empty set A into a non-empty set B is a rule that assigns to each element $a \in A$, exactly one element b of B. We than say, f maps A into B and we write, $f: A \to B$.

A is called *domain* and B is the co-domain of the function f. The set $f(A) = \{f(x): x \in A\}$ is called the *range of f*. Obviously, $f(A) \subset B$.

In particular, when $f(A) = B$, then f is an onto function, i.e., $f: A \xrightarrow{\text{onto}} B$.

A function f is called *one-to-one*, if $f(x_1) = f(x_2) \Leftrightarrow x_1 = x_2$.

When a function f is both onto (surjective) and one to-one (injective), then f is called *bijective* or *a bijection.*

When f is a bijection $\exists$ a function denoted by f^{-1}, called the *inverse of the function f* defined by $ff^{-1} = f^{-1}f = I$, where I is the identity mapping [$I(x) = x$, $\forall x \subset$ domain].

That is, $ff^{-1}(x) = f(f^{-1}(x)) = x$ and $f^{-1}f(x) = f^{-1}(f(x)) = x$ and $f^{-1}: B \to A$. When $f(x) = y$, then $f^{-1}(y) = x$.

A function is said to be increasing if for any $x_1, x_2 \in A$ and $x_1 < x_2$, then $f(x_1) \le f(x_2)$. (1)

In case of strictly increasing $x_1 < x_2 \Leftrightarrow f(x_1) < f(x_2)$.

Whenever f is increasing f^{-1} is also so.

In case of decreasing the latter, inequality in (1) will be reserved.

Restriction of a function $f: A \to B$ is $f_D: D \to B$, where $D \subset A$ and $f_D(x) = f(x)$, $\forall x \in D$
Advantage: A discontinuous function may have a continuous restricted function by deleting the points of discontinuities from its domain. A function, which is not differentiable can be made differentiable by taking its restriction, etc.

Step function

Let $[a,b]$ be a closed bounded interval and $P = \{a = x_0, x_1, x_2, \cdots, x_{n-1}, x_n = b\}$ be a partition, i.e., $a = x_0 < x_1 x_2 < \cdots < x_{n-1} < x_n = b$.

A function $f: [a,b] \to \mathbb{R}$ is called a *step function*, if

$$
\begin{aligned}
f(x) &= C_0, \ a \leq x < x_1 \\
&= C_0 + C_1, \ x_1 \leq x < x_2 \\
&\cdots\cdots\cdots\cdots \\
&= C_0 + C_1 + \cdots + C_{n-1}, \ x_{n-1} \leq x < x_n \\
&= C_0 + C_1 + \cdots + C_n, \ x = x_n, \text{ where } C_i \in \mathbb{R}^+, \ i = 0, 1, \cdots, n.
\end{aligned}
$$

Such functions are very common in practice.

e.g., (i) The greatest integer function $f: \mathbb{R} \to \mathbb{R}$ defined by $f(x) = [x]$, $x \in \mathbb{R}$, where $[x]$ = greatest integer $\leq x$.

(ii) The prime number function $\phi: \mathbb{R}^+ \to \mathbb{N} \cup \{0\}$ defined by $\phi(x)$ = number of primes $\leq x$.

$$
[\phi(x) = \begin{cases} 0, & 0 < x < 2 \\ 1, & 2 \leq x < 3 \\ 2, & 3 \leq x < 5, \text{ etc.}] \end{cases}
$$

(iiii) The postage function $p: \mathbb{R}^+ \to \mathbb{R}$ defined by $p(x)$ = postal charge in rupees for posting a letter of x gm.

Suppose,

$$
p(x) = \begin{cases} ₹5, & 0 \leq x < 5 \\ ₹10, & 5 \leq x < 10 \\ ₹15, & 10 \leq x < 15, \text{ etc.} \end{cases}
$$

Note: All step functions are piecewise continuous.

0.8 Functions of several variables

We have two standard formulae $\Delta = \frac{1}{2}ah$ and $V = abc$. In the first case, area Δ of a triangle depends on the base a and the altitude h. Here a and h are two independent variables and Δ is the dependent one. In the second case, volume of a rectangular parallelopiped depends on the length a, breadth b and height c. Here a, b, c are independent variables and V is the dependent variable. In general, if a dependent variable u depends on the independent variables $x_1, x_2, \cdots, x_n$, then the function can be written as $u: \mathbb{R}^n \to \mathbb{R}$ defined by $u = f(x_1, x_2, \cdots, x_n), x_i \in \mathbb{R}$, $\forall i = 1(1)n$.

Partial derivative

Let $u = f(x_1, x_2, \cdots, x_n)$ be a function. So, the value of u depends on the values of $x_1, x_2, \cdots, x_n$. Therefore, any change of the values of $x_1, x_2, \cdots, x_n$ will change the value of u, in general. Now, we are interested to find the change of u rather the rate of change of u caused by the change of x_i only. The rate of change of u w.r.t. some independent variable x_i is obtained by considering the change in u for the change of that particular variable x_i only (i.e., all others remain the same). It is denoted by $\frac{\partial u}{\partial x_i}$ or u_{x_i} or, $\frac{\partial f}{\partial x_i}$ or f_{xi} and is naturally defined as

$$\frac{\partial u}{\partial x_i}=\lim_{h_i\to 0}\frac{u(x_1,x_2,\cdots,x_{i-1},x_i+h_i,x_{i+1},\cdots,x_n)-u(x_1,x_2,\cdots,x_n)}{h_i}.$$

Change of independent variables

Let $u=f(x_1,x_2,\cdots,x_n)$ and

$$x_1=g_1(y_1,y_2,\cdots,y_n),x_2=g_2(y_1,y_2,\cdots,y_n),\cdots,x_n=g_n(y_1,y_2,\cdots,y_n),$$

where f and g_i $(i=1,2,\cdots,n)$ are all differentiable (possesses continuous partial derivatives). Then, u becomes differentiable functions of the independent variables $y_1,y_2,\cdots,y_n$ and

$$\frac{\partial u}{\partial y_i}=\sum_{j=1}^{n}\frac{\partial u}{\partial x_j}\frac{\partial x_j}{\partial y_i},\ i=1,2,\cdots,n$$

i.e.,

$$\frac{\partial x_1}{\partial y_1}\frac{\partial u}{\partial x_1}+\frac{\partial x_2}{\partial y_1}\frac{\partial u}{\partial x_2}+\cdots+\frac{\partial x_n}{\partial y_1}\frac{\partial u}{\partial x_n}=\frac{\partial u}{\partial y_1}$$

$$\frac{\partial x_1}{\partial y_2}\frac{\partial u}{\partial x_1}+\frac{\partial x_2}{\partial y_2}\frac{\partial u}{\partial x_2}+\cdots+\frac{\partial x_n}{\partial y_2}\frac{\partial u}{\partial x_n}=\frac{\partial u}{\partial y_2}$$

....

$$\frac{\partial x_1}{\partial y_n}\frac{\partial u}{\partial x_1}+\frac{\partial x_2}{\partial y_n}\frac{\partial u}{\partial x_2}+\cdots+\frac{\partial x_n}{\partial y_n}\frac{\partial u}{\partial x_n}=\frac{\partial u}{\partial y_n}.$$

Solving for $\frac{\partial u}{\partial x_1},\frac{\partial u}{\partial x_2},\cdots,\frac{\partial u}{\partial x_n}$, we get $\frac{\partial u}{\partial x_i}=\frac{J_i}{J}$, where

$$J=\begin{vmatrix}\frac{\partial x_1}{\partial y_1} & \frac{\partial x_2}{\partial y_1} & \cdots & \frac{\partial x_n}{\partial y_1}\\ \frac{\partial x_1}{\partial y_2} & \frac{\partial x_2}{\partial y_2} & \cdots & \frac{\partial x_n}{\partial y_2}\\ \cdots & \cdots & \cdots & \cdots\\ \cdots & \cdots & \cdots & \cdots\\ \frac{\partial x_1}{\partial y_n} & \frac{\partial x_2}{\partial y_n} & \cdots & \frac{\partial x_n}{\partial y_n}\end{vmatrix}$$

and J_i is the determinant obtained form J by replacing the ith column by $\left[\frac{\partial u}{\partial y_1}\ \frac{\partial u}{\partial y_2}\ \cdots\ \frac{\partial u}{\partial y_n}\right]$, $i=1(1)n$, provided $J\neq 0$. J is called the *Jacobian of transformation* or *simply Jacobian,* named after German Mathematician G. Jacobi (1804–51).

So, Jacobian plays the principal role for the transformation of independent variables.

Notation. The above Jacobian is represented by $\frac{\partial(x_1,x_2,\cdots,x_n)}{\partial(y_1,y_2,\cdots,y_n)}$.

e.g., the Jacobian for the transformation from r,θ to x,y given by $x=r\cos\theta, y=r\sin\theta$ is given by

$$J=\frac{\partial(x,y)}{\partial(r,\theta)}=\begin{vmatrix}\frac{\partial x}{\partial r} & \frac{\partial x}{\partial\theta}\\ \frac{\partial y}{\partial r} & \frac{\partial x}{\partial\theta}\end{vmatrix}=\begin{vmatrix}\cos\theta & -r\sin\theta\\ \sin\theta & r\cos\theta\end{vmatrix}=r$$

and for the transformation r,θ,ϕ to x,y,z when $x=r\sin\theta\cos\phi, y=r\sin\theta\sin\phi, z=r\cos\theta$ is given by

$$J=\frac{\partial(x,y,z)}{\partial(r,\theta,\phi)}=r^2\sin\theta.$$

Chain rule 1. If $x_1, x_2, \cdots, x_n$ are differentiable functions of $y_1, y_2, \cdots, y_n$ and $y_1, y_2, \cdots, y_n$ are differentiable function of $z_1, z_2, \cdots, z_n$, then $x_1, x_2, \cdots, x_n$ are differentiable functions of $z_1, z_2, \cdots, z_n$ and

$$\frac{\partial(x_1, x_2, \cdots, x_n)}{\partial(z_1, z_2, \cdots, z_n)} = \frac{\partial(x_1, x_2, \cdots, x_n)}{\partial(y_1, y_2, \cdots, y_n)} \cdot \frac{\partial(y_1, y_2, \cdots, y_n)}{\partial(z_1, z_2, \cdots, z_n)}.$$

Chain rule 2. If x_1, x_2 are differentiable functions of y_1, y_2, y_3 and y_1, y_2, y_3 are differentiable functions of z_1, z_2, then

$$\frac{\partial(x_1, x_2)}{\partial(z_1, z_2)} = \frac{\partial(x_1, x_2)}{\partial(y_1, y_2)} \cdot \frac{\partial(y_1, y_2)}{\partial(z_1, z_2)} + \frac{\partial(x_1, x_2)}{\partial(y_2, y_3)} \cdot \frac{\partial(y_2, y_3)}{\partial(z_1, z_2)} + \frac{\partial(x_1, x_2)}{\partial(y_3, y_1)} \cdot \frac{\partial(y_3, y_1)}{\partial(z_1, z_2)}.$$

[This rule may be generalized.]

Application of Jacobian in multiple integral

Let $f: A \to \mathbb{R}$, where A is a compact subset of $\mathbb{R}^n$, be a function of the variables $x_1, x_2, \cdots, x_n$.

Let $x_i = f_i(y_1, y_2, \cdots, y_n), i = 1, 2, \cdots, n$ defines the transformation when each f_i possesses continuous first order partial derivatives. Under the transformation let the closed bounded region A of $\mathbb{R}^n$ change to A^* of $\mathbb{R}^n$ and $f(x_1, x_2, \cdots, x_n)$ becomes $f^*(y_1, y_2, \cdots, y_n)$, then the multiple integral

$$\iint_A \cdots \int f(x_1, x_2, \cdots, x_n)\, dx_1\, dx_2 \cdots dx_n$$

becomes
$$\iint_{A^*} \cdots \int f^*(y_1, y_2, \cdots, y_n)|J|\, dy_1\, dy_2 \cdots dy_n, \text{ where } J = \frac{\partial(x_1, x_2, \cdots, x_n)}{\partial(y_1, y_2, \cdots, y_n)}.$$

e.g.,
$$\iint_A f(x, y)\, dx\, dy = \iint_{A^*} f^*(r, \theta)\, r\, dr\, d\theta, \because J = \frac{\partial(x, y)}{\partial(r, \theta)} = r > 0$$

and
$$\iiint_A f(x, y, z)\, dx\, dy\, dz = \iiint_{A^*} f^*(r, \theta, \phi) r^2 |\sin\theta|\, dr\, d\theta\, d\phi, \because J = \frac{\partial(x, y, z)}{\partial(r, \theta, \phi)} = r^2 \sin\theta.$$

A is some closed bounded region of $\mathbb{R}^2$ and A^* is the corresponding transformed region for the first integral and A is a closed bounded region of $\mathbb{R}^3$ for the second integral.

Ex. Find $\iint_A \sqrt{x^2 + y^2}\, dx\, dy$, where $A\{(x, y) \in \mathbb{R}: x^2 + y^2 \le a^2\}$.

Solution. Consider the transformation of independent variables. $x = r\cos\theta, y = r\sin\theta$.

$$\therefore J = \frac{\partial(x, y)}{\partial(r, \theta)} = \begin{vmatrix} \cos\theta & -r\sin\theta \\ \sin\theta & r\cos\theta \end{vmatrix} = r > 0. \therefore |J| = r$$

The transformed region is $A^* = \{(r, \theta) \in \mathbb{R}^2: 0 \le r \le a, 0 \le \theta < 2\pi\}$.

$$\therefore \iint_A \sqrt{x^2 + y^2}\, dx\, dy = \iint_{A^*} r \cdot r\, dr\, d\theta = \int_0^{2\pi} \int_0^a r^2\, dr\, d\theta = \frac{a^3}{3} \int_0^{2\pi} d\theta = \frac{2\pi a^3}{3}.$$

Ex. Find $\iint_A (x - y)\, dx\, dy$, where $A = \{(x, y) \in \mathbb{R}^2: x - y \ge -1, x - y \le 3, x + 2y \le 5, x + 2y \ge 3\}$.

Solution. Consider the transformation. $x - y = u$ and $x + 2y = v$.

$$\therefore x = \frac{2u + v}{3}, \quad y = \frac{v - u}{3}.$$

$$J=\begin{vmatrix}\dfrac{\partial x}{\partial u} & \dfrac{\partial x}{\partial v}\\ \dfrac{\partial y}{\partial u} & \dfrac{\partial y}{\partial v}\end{vmatrix}=\begin{vmatrix}2/3 & 1/3\\ -1/3 & 1/3\end{vmatrix}=\frac{2}{9}+\frac{1}{9}=\frac{1}{3}>0.\ \therefore\ |J|=\frac{1}{3}.$$

The transformed region $A^*=\{(u,v)\in\mathbb{R}^2: -1\le u\le 3, 3\le v\le 5\}$.

$$\therefore \iint\limits_{A}(x-y)\,dx\,dy=\iint\limits_{A^*}u\cdot\frac{1}{3}\,du\,dv=\frac{1}{3}\int_3^5\int_{-1}^3 u\,du\,dv=\frac{1}{3}\Big[v\Big]_3^5\left[\frac{u^2}{2}\right]_{-1}^3=\frac{1}{6}\cdot 2(9-1)=\frac{8}{3}.$$

Note: If $(x,y)\to(u,v)$, then $\iint\limits_{A} dx\,dy=A$ and $\iint\limits_{A^*}|J|\,du\,dv=|J|A^*$, if $|J|=$ constant.

$\therefore$ if A is the area of the closed region A and A^* is the area of the transformed region A^*, then $A=|J|A^*$.

Two more properties of the Jacobians.

1. If $(x_1,x_2,\cdots,x_n)\to(y_1,y_2,\cdots,y_n)$, then $\dfrac{\partial(x_1,x_2,\cdots,x_n)}{\partial(y_1,y_2,\cdots,y_n)}\cdot\dfrac{\partial(y_1,y_2,\cdots,y_n)}{(\partial x_1,x_2,\cdots,x_n)}=1$, provided no factor of the LHS vanishes.

2. $\dfrac{\partial(x_1,x_2,\cdots,x_n)}{\partial(y_1,y_2,\cdots,y_n)}=0\Leftrightarrow x_1,x_2,\cdots,x_n$ are functionally dependent.

0.9 Extrema of a function in two variables

Let $f: A\to\mathbb{R}, A\subset\mathbb{R}^2$ be a function.

Let $(a,b)\in A$. The function f has absolute maximum at (a,b), if $f(a,b)\ge f(x,y)$, $\forall\ (x,y)\in A$. The function f has absolute minimum at (a,b), if $f(a,b)\le f(x,y)$, $\forall\ (x,y)\in A$.

The function f has local maxima at $(a,b)\in A$, if there exists a nbd of (a,b) belonging to A such that $f(a,b)\ge f(x,y)$, $\forall\ (x,y)\in$ nbd of $(a,b)\cap A$.

The reverse inequality is for local minima of f.

In either case, (a,b) is called an *extreme point* and $f(a,b)$ is the corresponding extreme value.

Saddle point

A point $(a,b)\in A$ is said to be a saddle point of f, if every nbd of (a,b) contains point $(x,y)\in A$ such that $f(a,b)\le f(x,y)$; also contains point $(x,y)\in A$ such that $f(a,b)\ge f(x,y)$.

DEFINITION. (For a function of n independent variables). Let $f: A\to\mathbb{R}, A\subset\mathbb{R}^n$ be a function and the point $(a_1,a_2,\cdots,a_n)\in A$. The point $(a_1,a_2,\cdots,a_n)$ is called an *extreme point* of f, if for every point $(x_1,x_2,\cdots,x_n)\in$ some deleted nbd of $(a_1,a_2,\cdots,a_n)\cap A$, the difference $f(x_1,x_2,\cdots,x_n)-f(a_1,a_2,\cdots,a_n)$ maintains the same sign.

If the sign is negative, the point $(a_1,a_2,\cdots,a_n)$ is a point of maximum and

$$f_{\max}=f(a_1,a_2,\cdots,a_n).$$

In case the sign is positive, the point $(a_1,a_2,\cdots,a_n)$ is a point of minimum and

$$f_{\min}=f(a_1,a_2,\cdots,a_n).$$

If the sign of the difference is fluctuating for every deleted nbd of $(a_1,a_2,\cdots,a_n)$, the point $(a_1,a_2,\cdots,a_n)$ is neither a point of maximum nor a point of minimum and it is a saddle point of f.

Critical point

The point $(a_1,a_2,\cdots,a_n)\in A$, defined above, of the function $f: A\to\mathbb{R}, A\subset\mathbb{R}^n$ is said to be a critical point or a stationary point of f.

A critical point of a function is either a point of maxima or minima or a saddle point.

So, every point of maximum is a critical point but the converse is not true as a critical point may be a point of minimum or saddle as well, etc.

So, a point of maxima is necessarily critical. Similar remark for a point of minima or a saddle point is also valid.

RULE. Let $f: A \to \mathbb{R}, A \subset \mathbb{R}^3$ be a function. Let $(a,b,c) \in A$ is a critical point and possesses continuous partial derivatives up to order two in some nbd of the point. Let

$$A = f_{xx}(a,b,c), \quad B = \begin{vmatrix} f_{xx}(a,b,c) & f_{xy}(a,b,c) \\ f_{yx}(a,b,c) & f_{yy}(a,b,c) \end{vmatrix}$$

$$\text{and } C = \begin{vmatrix} f_{xx}(a,b,c) & f_{xy}(a,b,c) & f_{xz}(a,b,c) \\ f_{yx}(a,b,c) & f_{yy}(a,b,c) & f_{yz}(a,b,c) \\ f_{zx}(a,b,c) & f_{zy}(a,b,c) & f_{zz}(a,b,c) \end{vmatrix}.$$

Then, (i) f is maximum at (a,b,c), if $A < 0, B > 0, C < 0$.

(ii) f is minimum at (a,b,c), if $A > 0, B > 0, C > 0$.

(iii) f has neither maxima nor minima at (a,b,c), if $B < 0$ [In this case, (a,b,c) is a saddle point].

0.10 Improper integrals

The definite integral $\int_a^b f(x)\,dx$ is proper if (i) $[a,b]$ is a closed and bounded interval and (ii) the integrand f is bounded on $[a,b]$. Every definite integral must have a definite value.

Lack of either of the conditions (i) and (ii) implies that the integral is improper. Some of the improper integrals are convergent and some of them are divergent.

There are two types of improper integrals, viz., first kind and second kind.

Improper integral of first kind. An improper integral is of first kind, if the first criteria, viz., $[a,b]$ is a closed and bounded interval is violated only.

$$\int_a^\infty f(x)\,dx, \int_{-\infty}^b f(x)\,dx \text{ and } \int_{-\infty}^\infty f(x)\,dx$$

are three types of first kind improper integrals, when $a, b \in \mathbb{R}$.

Convergence of $\int_a^\infty f(x)\,dx$; $\int_{-\infty}^b f(x)\,dx$ and $\int_{-\infty}^\infty f(x)\,dx$:

Let $\int_a^R f(x)\,dx$ exists for $a < R < \infty$. Then, $\int_a^\infty f(x)\,dx = \lim_{R \to \infty} \int_a^R f(x)\,dx$, provided the limit exists finitely. Similarly, $\int_{-\infty}^b f(x)\,dx = \lim_{R \to -\infty} \int_R^b f(x)\,dx$, $-\infty < R < b$, provided both $\int_R^b f(x)\,dx$ exists and the limit exists finitely.

Let $c \in \mathbb{R}$. Then, $\int_{-\infty}^\infty f(x)\,dx = \int_{-\infty}^c f(x)\,dx + \int_c^\infty f(x)\,dx$.

$\therefore \int_{-\infty}^\infty f(x)\,dx$ converges if both the integrals on RHS are convergent.

Ex. Examine the convergence of

(i) $\int_0^\infty \cos x\,dx$ (ii) $\int_0^\infty x^3 e^{-x^2}\,dx$ (iii) $\int_0^\infty x^2 e^{-x}\,dx$ (iv) $\int_{-\infty}^0 x^2 \sin x\,dx$ (v) $\int_{-\infty}^\infty 2x\,e^{-x^2}\,dx$.

Solution. (i) $\int_0^\infty \cos x\, dx = \lim_{R\to\infty}\int_0^R \cos x\, dx = \lim_{R\to\infty} \sin R$, which does not exist.

$\therefore \int_0^\infty \cos x\, dx$ is divergent.

(ii) $\int_0^\infty x^3 e^{-x^2}\, dx = \lim_{R\to\infty}\int_0^R x^3 e^{-x^2}\, dx.$

Let $x^2 = u$. $\therefore\ 2x\,dx = du$.

$$\text{The integral} = \lim_{R\to\infty}\frac{1}{2}\int_0^R u\,e^{-u}\,dx = \frac{1}{2}\lim_{R\to\infty}\left[-u\,e^{-u}\Big|_0^R + \int_0^R e^{-u}\,du\right]$$

$$= \frac{1}{2}\lim_{R\to\infty}\left[-R\,e^{-R} - \left(e^{-R}-1\right)\right] = \frac{1}{2}.$$

$\therefore$ the integral converges and converges to 1/2.

$$\text{(iii)} \int_0^\infty x^2 e^{-x}\,dx = \lim_{R\to\infty}\int_0^R x^2 e^{-x}\,dx = \lim_{R\to\infty}\left[-x^2 e^{-x}\Big|_0^R + \int_0^R 2x\,e^{-x}\,dx\right]$$

$$= 2\lim_{R\to\infty}\int_0^R x\,e^{-x}\,dx, \because \lim_{R\to\infty} x^2 e^{-x} = 0$$

$$= 2\lim_{R\to\infty}\left[-x\,e^{-x}\Big|_0^R + \int_0^R e^{-x}\,dx\right] = -2\lim_{R\to\infty}\left(e^{-R}-1\right) = 2.$$

$\therefore$ the integral converges to 2.

$$\text{(iv)} \int_{-\infty}^0 x^2 \sin x\,dx = \lim_{R\to\infty}\int_{-R}^0 x^2\sin x\,dx.$$

$$\int x^2 \sin x\,dx = -x^2\cos x + 2\int x\cos x\,dx = -x^2\cos x + 2x\sin x - 2\int \sin x\,dx$$

$$= \left(2-x^2\right)\cos x + 2x\sin x.$$

$\therefore \lim_{R\to\infty}\int_{-R}^0 x^2\sin x\,dx = \lim_{R\to\infty}\left[2-\left\{\left(2-R^2\right)\cos R + 2R\sin R\right\}\right]$, which does not exist.

Hence, the integral diverges.

$$\text{(v)} \int_{-\infty}^\infty 2x\,e^{-x^2}\,dx = \int_{\infty}^0 2x\,e^{-x^2}\,dx + \int_0^\infty 2x\,e^{-x^2} = \lim_{R\to\infty}\left\{\int_{-R}^0 2x\,e^{-x^2}\,dx + \int_0^R 2x\,e^{-x^2}\,dx\right\}.$$

Put $x^2 = u$. $\therefore\ 2x\,dx = du$.

$$\text{The integral} = \lim_{R\to\infty}\left\{\int_{R^2}^0 e^{-u}\,du + \int_0^{R^2} e^{-u}\,du\right\} = \lim_{R\to\infty}\left\{-\int_0^{R^2} e^{-u}\,du + \int_0^{R^2} e^{-u}\,du\right\}$$

$$= 0, \text{ as } \int_0^{R^2} e^{-u}\,du = \left[e^{-u}\right]_{R^2}^0 = 1 - e^{-R^2} \to 1 \text{ as } R^2\to\infty, \text{ i.e., } \int_0^\infty e^{-u}\,du \text{ conv.}$$

Note: The integral is zero as the integrand is an odd function.

Absolute convergence

DEFINITION. An improper integral $\int_a^\infty f(x)\,dx$ is said to be absolutely convergent, if $\int_a^\infty |f(x)|\,dx$ converges.

Theorem If $\int_a^\infty |f(x)|\,dx$ is convergent, then $\int_a^\infty f(x)\,dx$ converges. Converse may not be true. When $\int_a^\infty f(x)\,dx$ is convergent but $\int_a^\infty |f(x)|\,dx$ is not, then $\int_a^\infty f(x)\,dx$ is said to converge conditionally.

The sufficient condition for absolute convergence of an improper convergent integral is that the integrand is non-negative throughout the range of integration. It is not necessary.

0.11 Improper integral of the second kind

DEFINITION. An integral $\int_a^b f(x)\,dx$ is said to be an improper integral of the second kind if the integrand $f(x)$ is unbounded for a finite number of points in $[a,b]$.

We consider three cases, viz., (i) f is unbounded at the left end point a only (ii) f is unbounded at the right end point b only and (iii) f is unbounded at some interior point c of $[a,b]$, i.e., $a<c<b$.

Case (i) $\int_a^b f(x)\,dx = \lim\limits_{\delta_1\to 0+} \int_{a+\delta_1}^b f(x)\,dx$, provided the limit exists.

Case (ii) $\int_a^b f(x)\,dx = \lim_{\delta_2\to 0+} \int_a^{b-\delta_2} f(x)\,dx$, provided the limit exists.

Case (iii) $\int_a^b f(x)\,dx = \lim\limits_{\delta_1\to 0+} \int_a^{c-\delta_1} f(x)\,dx + \lim_{\delta_2\to 0+} \int_{c+\delta_2}^b f(x)\,dx$, provided the limits exist.

Sometimes, the integral $\int_a^b f(x)\,dx$ does not converge when $\delta_1 \neq \delta_2$ but converges when $\delta_1=\delta_2=\delta$ (say).

In such a case we say that the principal value of $\int_a^b f(x)\,dx$ exists and we write,

$$P\int_a^b f(x)\,dx = \lim_{\delta\to 0+}\left\{\int_a^{c-\delta} f(x)\,dx + \int_{c+\delta}^b f(x)\,dx\right\}.$$

Absolute convergence

$\int_a^b f(x)\,dx$ is said to converge absolutely, if $\int_a^b |f(x)|\,dx$ converges.

An absolutely convergent integral is convergent but not conversely.

However, if $f(x)\geq 0$, $\forall\, x\in[a,b]$, then every convergent series is absolutely convergent but $f(x)\geq 0$, $\forall\, x\in[a,b]$ is not always necessary for absolute convergence.

Ex. Examine the convergence of (i) $\int_0^1 \frac{1}{x}\,dx$ (ii) $\int_0^1 \frac{dx}{\sqrt{1-x^2}}$ (iii) $\int_{-1}^1 \frac{1}{x^3}\,dx$.

Solution. (i) $\int_0^1 \frac{1}{x}\,dx = \lim\limits_{\delta\to 0+}\int_\delta^1 \frac{1}{x}\,dx = \lim\limits_{\delta\to 0+}\left[\ln|x|\right]_\delta^1 = -\lim\limits_{\delta\to 0+}\ln\delta = \infty.$

$\therefore$ the integral does not converge.

Solution. (i) $\displaystyle\int_0^1 \frac{1}{x}\,dx = \lim_{\delta\to 0+}\int_\delta^1 \frac{1}{x}\,dx = \lim_{\delta\to 0+}\Big[\ln|x|\Big]_\delta^1 = -\lim_{\delta\to 0+}\ln\delta = \infty.$

∴ the integral does not converge.

(ii) $\displaystyle\int_0^1 \frac{dx}{\sqrt{1-x^2}} = \lim_{\delta\to 0+}\int_0^{1-\delta}\frac{dx}{\sqrt{1-x^2}} = \lim_{\delta\to 0+}\sin^{-1}(1-\delta) = \frac{\pi}{2}.$

∴ the integral converges.

(iii) $$\int_{-1}^{1}\frac{1}{x^3}\,dx = \int_{-1}^{0}\frac{dx}{x^3} + \int_0^1\frac{dx}{x^3} = \lim_{\delta_1\to 0+}\int_{-1}^{0-\delta_1}\frac{dx}{x^3} + \lim_{\delta_2\to 0+}\int_{0+\delta_2}^{1}\frac{dx}{x^3}$$

$$= \lim_{\delta_1\to 0+}\left[-\frac{1}{2x^2}\right]_{-1}^{-\delta_1} + \lim_{\delta_2\to 0+}\left[-\frac{1}{2x^2}\right]_{\delta_2}^{1}$$

$$= \lim_{\delta_1\to 0+}\left[\frac{1}{2}-\frac{1}{2\delta_1^2}\right] + \lim_{\delta_2\to 0+}\left[-\frac{1}{2}+\frac{1}{2\delta_2^2}\right] \text{ does not exist,}$$

as $\displaystyle\lim_{\delta_2\to 0+}\frac{1}{2\delta_1^2}$ and $\displaystyle\lim_{\delta_2\to 0+}\frac{1}{2\delta_2^2}$ do not exist finitely.

$$P\int_{-1}^{1}\frac{1}{x^3}\,dx = \lim_{\delta\to 0+}\left[\int_{-1}^{-\delta}\frac{dx}{x^3} + \int_\delta^1\frac{dx}{x^3}\right] = \lim_{\delta\to 0+}\left[\frac{1}{2}-\frac{1}{2\delta^2}-\frac{1}{2}+\frac{1}{2\delta^2}\right] = 0.$$

∴ general value does not exist but principal value exists.

Two important improper integrals

I. $\displaystyle\int_0^\infty e^{-x}x^{l-1}\,dx,\ l>0$ [first type] II. $\displaystyle\int_0^1 x^{l-1}(1-x)^{m-1}\,dx,\ l>0, m>0$ [second type].

These two integrals converge, but do not converge to any elementary functions. They are special functions.

$\displaystyle\int_0^\infty e^{-x}x^{l-1}\,dx = \Gamma(l) =$ value of the gamma function at l.

$\displaystyle\int_0^1 x^{l-1}(1-x)^{m-1}\,dx = B(l,m) =$ value of beta function at (l,m).

Chapter 1

Mathematical Probability

1.1 Introduction

Mathematical probability is concerned with the facts, which occur by chance factor. May be, which includes the meaning of may not be is the theme of the subject. It is the numerical measurement of chance of occurrence of incidents, which are random in nature. The facts or incidents, which are not deterministic are random in nature. In fact, they are probabilistic and their occurrence depends on chance factor. They are happening by chance. Such an apparent randomness is not random in the long run and follows some regularity known as *Statistical regularity.*

1.2 Short history and importance

Before doing an action we are always estimating its effect consciously or unconsciously. As such estimation is unavoidable. Every physical measurement suffers by an error. Such errors are random in nature and follows normal law. e.g., if we are working with the length of a rod we have no option but to estimate its length (actual length is unknown for ever).

For planning, estimation is a must although actual situation will be known later.

Probability theory helps us to estimate in a systematic manner, particularly when we are dealing with a huge data. In this computer age handling a huge data is a matter of programming and so the application of probability, particularly its applied part statistics, is now fascinating in many subjects like: Quantum Mechanics, Physics, Chemistry, Biology, Psychology, Medicine, Economics, Sociology, etc. For the development of an industry, systematic analysis of past data is done by applying probability theory. In actuarial science, highly applied for future planning, probability theory is actively applied. According to my concept: *Every phenomenon on Earth (real world) is probabilistic, whereas every phenomenon in Heaven (ideal world) is deterministic.*

A good number of Mathematicians and Scientists such as Bernoulli (1654–1705), Abraham de Moivre (1667–1754), Laplace (1749–1827), Tsebysheff (1821–1894), A. Markoff (1856–1922), A. Liapounaff (1858–1918), P.L. Chebyshev (1821–1894) and many others contributed a lot for the development of the subject.

1.3 Useful terms

Experiment. It is an act, which can be performed any number of times under uniform or identical conditions.

Random Experiment (RE)

It is an experiment for which the totality of the outcomes or results are known in advance but the outcome for a particular trial (performance of a RE once) is completely unknown. Moreover, in spite of performing the RE with identical conditions the outcomes vary at random from trial to trial.

For example, ⟨I⟩ Tossing a coin ⟨II⟩ Rolling a die from a box ⟨III⟩ Drawing a card from a well-shuffled full pack of playing cards at random ⟨IV⟩ Drawing a ball blindly from an urn (container) containing a white balls and b black balls, etc.

Event

An outcome of a RE is called an *event.* e.g., Head (H), Tail (T), either Head or Tail (H or T), neither head or tail are event associated with the RE⟨I⟩. Face 1 (F(1)), Face 2 (F(2)), Face 3 (F(3)), Face 4 (F(4)), Face 5 (F(5)), Face 6 (F(6)), odd face, even face, either odd face or even face, neither odd face nor even face, face multiple of 3, etc. are events associated with the RE⟨II⟩.

Heart, Spade, Club, Diamond, Queen of Club, Ace of Heart, Ace, Jack, Either Heart or Diamond, Any Card, etc. are the events associated with the RE⟨III⟩.

White ball, black ball, either white ball or black ball, etc. are events associated with the RE⟨IV⟩.

Types of events

Simple events. The most fundamental events, which cannot be decomposed are called *simple events.* They are of atomic structure. Simple events are also called *elementary events.*

For example, H, T of RE⟨I⟩; F(1), F(2), F(3), F(4), F(5) and F(6) of RE⟨II⟩; each card, viz., Jack of Spade, Ace of Heart, Queen of Diamond, etc. of RE⟨III⟩ are simple events.

Compound events. The events, which are not fundamental, and, which are formed by two or more simple events with different connectors, are called compound events.

For example, H or T, odd face, club are respectively compound events associated with the RE ⟨I⟩, ⟨II⟩ and ⟨III⟩.

H or T consisting of H, T with the connector '*or*'. Odd face = $\{F(1), F(3), F(5)\}$ contains three simple events. The event club consisting of all the 13 cards of club. Occurrence of H or T means the occurrence of either Head or Tail, the occurrence of odd face implies the occurrence of either face 1 or face 3 or face 5, the occurrence of club implies the occurrence of any card of Club.

Certain event. The event, which is sure to occur at any performance of the RE is the certain event.

e.g., H or T of RE⟨I⟩, either odd face or even face of RE⟨II⟩, either jack or not jack of RE⟨III⟩, either white ball or black ball of RE⟨IV⟩ are the certain events.

Impossible events. The events, which do not occur in any performance of the RE are called *impossible events.*

e.g., neither H nor T of RE⟨I⟩, neither odd face nor even face of RE⟨II⟩, Neither Heart nor Spade nor diamond nor club of RE⟨III⟩, neither white ball or black ball of RE⟨IV⟩ are the examples of impossible events.

Mutually exclusive events (MEE). Two or more events of a RE are called *mutually* exclusive, if they can never occur simultaneously. Obviously, any two simple events of a RE are MEE. Two compound events may or may not be MEE. e.g., *odd face* and *even face* of RE⟨II⟩ are MEE; whereas *even face* and *multiple of* 3 of RE⟨II⟩ are not MEE.

Complementary events. Two MEE will be complementary, if their union (i.e., connected by either or) is the certain event.

For example, H or T RE⟨I⟩, either odd face or even face of RE⟨II⟩, either Spade or not Spade of RE⟨III⟩, either white ball or black ball of RE ⟨IV⟩ are certain events and the individual events of a certain event are MEE. Hence, the two events H and T of RE⟨I⟩ are complementary, *odd face* and *even face* of RE⟨II⟩ are complementary, *Spade* and *not Spade* of RE⟨III⟩ are complementary, *white ball* and black ball of RE⟨IV⟩ are complementary.

1.4 Event space

The set of all possible simple outcomes of a RE is called the *event space* of that RE. It is generally denoted by S.

Event spaces of the RE ⟨I⟩, ⟨II⟩, ⟨III⟩ and ⟨IV⟩ are respectively

$$S=\{\text{H,T}\},\ S=\{\text{F(1), F(2), F(3), F(4), F(5), F(6)}\},$$

$S=$ the set of all the 52 cards, i.e., $S=\{x\,|\,x \text{ is a playing card}\}$ and $S=\{\text{White, Black}\}$.

Now, we can define *Event* mathematically as follows.

Event

Any subset of the event space S of a RE E is defined as an event of E, so, that if $A\subseteq S$, we say that A is an event of the RE whose event space is S.

As $\phi\subseteq S$ and $S\subseteq S$, $\therefore$ ϕ (the null set) and S (the universal set) corresponds to impossible and certain events respectively.

When A contains only one element (one simple event), A is then simple event and when A contains more than one element of S, it is then a compound event of the RE.

When $A, B\subseteq S$ and $A\cap B=\phi$, the events A and B are MEE.

Further when $A\cap B=\phi$ and $A\cup B=S$, then events A and B are complementary.

When $A\subseteq B\subseteq S$ we say that A implies B.

Note: In our probability theory we represent $A\cup B$ as $A+B$, $A\cap B$ by AB and ϕ by 0.

It is important to note that the union of any two events can always be expressed as the union of three MEE, i.e., $A, B\subseteq S$, then

$$A+B=(A-AB)+AB+(B-AB).$$

The three events on RHS are MEE (pairwise).

Note: $A-B=$ complement of B in A (relative complement of B).

Pairwise mutually exclusive events. A set of events $\{A_1,A_2,\cdots,A_n\}$ are said to be pairwise MEE, if $A_i\,A_j=0,\ i\neq j$.

Exhaustive events. A set of events $\{A_1,A_2,\cdots,A_n\}$ are said to be exhaustive, if $A_1+A_2+\cdots+A_n=S$ (the certain event).

Pairwise ME and exhaustive events. A set of events $\{A_1,A_2,\cdots,A_n\}$ are said to be pairwise ME and exhaustive, if (i) $A_i\,A_j=0,\ i\neq j$ and (ii) $A_1+A_2+\cdots+A_n=S$, i.e., the set $\{A_1,A_2,\cdots,A_n\}$ forms a partition of S.

For example, $A,\bar{A}$ are ME and exhaustive events, i.e., $A\bar{A}=0$ and $A+\bar{A}=S$ [$\bar{A}=$ not A, i.e., negation of A].

$\therefore$ $\{A,\bar{A}\}$ is the simplest example of a partition for any event space S, when $A\subseteq S$.

Example of an infinite event space (discrete)

Consider the RE: Throwing a coin until a head appears. The event space S is

$$S=\{H,\ TH,\ TTH,\ TTTH,\ \cdots\}.$$

1.5 Definitions of probability

There are three definitions of probability, viz., (i) Classical definition (ii) Frequency definition and (iii) Axiomatic definition.

(i) **Classical definition.** (Given by Laplace and Bernoulli in the early 19th century):

Let E be a RE whose event space is S. Let S contains finite number of equally likely elementary events. If an event A of E (i.e., $A\subseteq S$) contains m elementary events, then the probability of A is denoted by $P(A)$ and is defined by $P(A)=m/n$, when $n=n(S)$.

Note 1. Two events are said to be equally like, if no one is more likely than the other.

2. Elementary event = simple event = event point = point.

Criticism of classical definition

(a) as $n(S)=n=$ finite. $\therefore$ it is applicable to a finite discrete event space,

(b) there is no way of testing equally likely without the theory of probability,

(c) equally likely has the same meaning with equally probable. So, defining probability in terms of probability is logically poor.

Immediate consequences

1. Obviously, $0 \leq m \leq n$. $\therefore$ $0 \leq m/n \leq 1 \Rightarrow 0 \leq P(A) \leq 1$. In particular, $P(A)=0 \Rightarrow A=0$ (impossible event) and $P(A)=1 \Rightarrow A=S$ (the certain event).

2. When $A \subseteq B \subseteq S$, $n(A) \leq n(B)$.

$$\therefore \quad \frac{n(A)}{n(S)} \leq \frac{n(B)}{n(S)} \Rightarrow P(A) \leq P(B), \text{ i.e., } A \Rightarrow B \Leftrightarrow P(A) \leq P(B).$$

3. We have $n(A+B)=n(A)+n(B)-n(AB)$.

$$\therefore \quad \frac{n(A+B)}{n(S)}=\frac{n(A)}{n(S)}+\frac{n(B)}{n(S)}-\frac{n(AB)}{n(S)}, \text{ i.e., } P(A+B)=P(A)+P(B)-P(AB).$$

If $AB=0$, then $n(AB)=0$, then $P(A+B)=P(A)+P(B)$. So, when A and B are MEE, the probability of the sum of two events is equal to the sum of their probabilities.

(ii) **Frequency definition of probability.** (Given by Von Mises in 1920):

Suppose, E be a RE with the event space S and $A \subseteq S$ be an event of E. Suppose, the RE be repeated N times under identical conditions and the event A is found to occur $N(A)$ times. Then, $f(A)=\frac{N(A)}{N}$ is called *frequency ratio* or *relative frequency* of A.

It is experimentally found that as N becomes larger and larger the frequency ratio will tend to become more or less a constant. This phenomenon of stability of the frequency ratio is called the *statistical regularity*.

The probability of the event A is defined by $P(A)=\lim\limits_{N\to\infty}\frac{N(A)}{N}$.

Immediate consequences

1. Obviously, $0 \leq N(A) \leq N \Rightarrow 0 \leq \dfrac{N(A)}{N} \leq 1$, i.e., $0 \leq \lim\limits_{N\to\infty}\dfrac{N(A)}{N} \leq 1$, or, $0 \leq P(A) \leq 1$.
   $P(A)=0 \Rightarrow A$ is stochastically impossible and
   $P(A)=1 \Rightarrow A$ is stochastically certain.

2. If A and B are MEE, i.e., $AB=0$, then $N(A+B)=N(A)+N(B)$.

$$\begin{aligned}\therefore\ P(A+B) &= \lim_{N\to\infty}\frac{N(A+B)}{N}=\lim_{N\to\infty}\frac{N(A)+N(B)}{N}\\ &= \lim_{N\to\infty}\frac{N(A)}{N}+\lim_{N\to\infty}\frac{N(B)}{N}, \text{ [as both the limits exist]}\\ &= P(A)+P(B).\end{aligned}$$

3. $N(S)=N$. $\therefore$ $P(S)=\lim\limits_{N\to\infty}\dfrac{N(S)}{N}=1$ and $N(0)=0$. $\therefore$ $P(0)=\lim\limits_{N\to\infty}\dfrac{N(0)}{N}=0$.

Note: (2) Can be extended to any finite number of events, i.e., if $A_1, A_2, \cdots, A_n$ be a set of pairwise ME events, then

$$\begin{aligned}P(A_1+A_2+\cdots+A_n) &= \lim_{N\to\infty}\frac{N(A_1+A_2+\cdots+A_n)}{N}=\lim_{N\to\infty}\frac{N(A_1)+N(A_2)+\cdots N(A_n)}{N}\\ &= \lim_{N\to\infty}\frac{N(A_1)}{N}+\lim_{N\to\infty}\frac{N(A_2)}{N}+\cdots+\lim_{N\to\infty}\frac{N(A_n)}{N}, \text{ [as all the limits exist]}\\ &= P(A_1)+P(A_2)+\cdots+P(A_n).\end{aligned}$$

(iii) **Axiomatic definition of probability.** This is elucidated in next chapter.

1.6 Conditional probability

Suppose, a RE E be repeated under identical condition N times and an event A of E occurred $N(A)(\neq 0)$ times. Along with A the number of times another event B of E also occurs is $N(AB)$. Then, $f(B|A) = \frac{N(AB)}{N(A)}$ is called the conditional frequency ratio B subject to the condition that A has already occurred. The conditional probability of B subject to the condition that the *event* A has already occurred is denoted by $P(B|A)$ and is defined by $P(B|A) = \lim_{N\to\infty} \frac{N(AB)}{N(A)}$, assuming that the limit exists.

$$\therefore\ P(B|A) = \lim_{N\to\infty} \frac{N(AB)/N}{N(A)/N} = \frac{\lim_{N\to\infty} [N(AB)/N]}{\lim_{N\to\infty} [N(A)/N]} = \frac{P(AB)}{P(A)},\ P(A) \neq 0.$$

$\therefore\ P(AB) = P(A)P(B|A)$. Similarly, we can also have $P(AB) = P(B)P(A|B)$, provided $P(B) \neq 0$.

EXERCISE

1. Out of $(2n+1)$ tickets consecutively numbered, three are drawn at random. Find the probability that the numbers on them are in arithmetical progression. $\left(Ans.\ \frac{3n}{4n^2-1}\right)$
2. Two integers m and n are chosen at random with replacement from a set of numbers $\{1,2,\cdots,9\}$. What is the probability that $|m^2 - n^2|$ is divisible by 2. $\left(Ans.\ \frac{41}{81}\right)$ [C.H. 2005]
3. A second order determinant is considered with the elements 0 and 1. What is the probability that the determinant is non-negative? $\left(Ans.\ \frac{13}{16}\right)$
4. Two squares are chosen at random from the small squares drawn on a chessboard. What is the probability that the two chosen squares have exactly one corner in common? $\left(Ans.\ \frac{7}{144}\right)$
5. The sum of the digits of a seven-digit number is 59. Find the probability that this number is divisible by 11. $\left(Ans.\ \frac{4}{21}\right)$
6. If two numbers p and q are chosen at random from the set $\{1,2,3,\cdots,10\}$ with replacement, determine the probability that the roots of the equation $x^2 + px + q = 0$ are real. $\left(Ans.\ \frac{31}{50}\right)$
7. What is the probability that a leap year selected at random will contain 53 Sundays? $\left(Ans.\ \frac{2}{7}\right)$ [C.H. 2006]
8. What is the probability that all the four students of a family must have different birthdays? (Assume 1 year = 365 days) (*Ans.* 0.98 (approx.))
9. An integer is chosen at random from the set $\{1,2,\cdots,200\}$. What is the probability that the integer is divisible by 2 or 3? $\left(Ans.\ \frac{133}{200}\right)$ [C.H. 2007]
10. If 15 persons be placed at random (i) on a bench (ii) in a ring, find the probability that three particular persons will be next to each other.
    [**Hints.** (i) Possible number of placements of 15 person on a bench is 15!.
    Possible number of placements of 15 person on a bench with the restriction that particular 3 persons will be next to one another is $13! \times 3!$. $\therefore$ required probability $= \frac{13!\,3!}{15!} = \frac{6}{15\times 14} = \frac{1}{35}$.
    (ii) 15 persons may be placed in a ring in 14! ways.
    With restriction they are placed in $12! \times 3!$ ways. $\therefore$ required probability $= \frac{12!\,3!}{14!} = \frac{6}{13\times 14} = \frac{3}{91}$.]
11. 12 persons, amongst whom are A and B, are seated at random round a round table. What is the probability that there are 3 persons between A and B.
    [**Hints.** A can occupy seat in 12 ways and for each seating of A, B can occupy seat in 11 ways. Hence, A and B can occupy seats in 12×11, i.e., 132 ways. Let the seats are marked $1,2,\cdots,12$. 3 persons between A and B will be seated if A, B occupy the pair of seats $(1,5)$, $(2,6)$, $(3,7)$, $(4,8)$, $(5,9)$, $(6,10)$, $(7,11)$, $(8,12)$, $(9,1)$, $(10,2)$, $(11,3)$, $(12,4)$. As A and B may interchange their seats. $\therefore$ the total number of favourable cases is $12 \times 2 = 24$. $\therefore$ required probability $= \frac{24}{132} = \frac{2}{11}$.]

12. Find the probability of obtaining a sum of 10 in a single throw with 5 fair dice. $\left(Ans.\ \frac{7}{432}\right)$
Note: Any problem of permutation and combination may be arranged as a problem of probability (all of which may be solved by classical rule).

13. Out of 14400 tickets 3 are drawn at random, what is the probability that they are in geometrical progression? $(Ans.\ 1.34 \times 10^{-8})$

14. 7 Mathematics and 8 Physics books are placed at random on a bookshelf. Find the probability that none of the Physics books placed consecutively. $\left(Ans.\ \frac{7!\times 8!}{15!}\right)$
[C.H. 2003]

15. In a Bernoullian sequence of 4 trials it is given that the probability of success in a trial is p and in the first three trials success occurred. What is the probability that the fourth one is also a success?
$(Ans.\ p)$

16. A missile was fired on a plane at which there are two targets T_1 and T_2. The probability of hitting T_1 is p_1 and that of hitting T_2 is p_2. It is known that T_2 was not hit. Find the probability that T_1 was hit.
$\left(Ans.\ \frac{p_1}{1-p_2}\right)$
[C.H. 1992]

Chapter 2

Fundamental Axioms

2.1 Axiomatic definition of probability (Given by Kolmogorov in 1933)

Let E be a RE with the event space S. Let $A \subseteq S$ be an event of E, then the probability of A is denoted by $P(A)$ and is defined by the following axioms: $\langle \text{I} \rangle$ $P(A) \geq 0$ $\langle \text{II} \rangle$ $P(S) = 1$ $\langle \text{III} \rangle$ If $\{A_n\}_n$ be a sequence of pairwise mutually exclusive events (MEE) of the RE E, i.e., $A_i A_j = 0, \ \forall\, i, j, i \neq j$, then

$$P\left(\sum_{i=1}^{\infty} A_i\right) = \sum_{i=1}^{\infty} P(A_i), \text{ i.e., } P(A_1 + A_2 + \cdots + A_n + \cdots) = P(A_1) + P(A_2) + \cdots + P(A_n) + \cdots$$

Some important deductions

1. $P\left(\bar{A}\right) = 1 - P(A)$. We have $A + \bar{A} = S$ and $A\bar{A} = 0$.
   $\therefore$ by axioms $\langle \text{II} \rangle$ and $\langle \text{III} \rangle$, $P\left(A + \bar{A}\right) = P(S) \Rightarrow P(A) + P\left(\bar{A}\right) = 1$. $\therefore$ $P\left(\bar{A}\right) = 1 - P(A)$.
   $\therefore$ $P(A) = 1 - P\left(\bar{A}\right) \leq 1$ as $P\left(\bar{A}\right) \geq 0$ by axiom $\langle \text{I} \rangle$.
2. $P(0) = P\left(\bar{S}\right) = 1 - P(S) = 1 - 1$, [by axiom $\langle \text{II} \rangle$]
   $= 0$
3. Let $A \Rightarrow B$, i.e., $A \subseteq B$. $\therefore$ $B = A + (B - A)$, where A and $B - A$ are MEE.
   $\therefore$ $P(B) = P\{A + (B - A)\} = P(A) + P(B - A)$, [by axiom $\langle \text{III} \rangle$].
   So, $P(B) \geq P(A)$ as $P(B - A) \geq 0$, [by axiom $\langle \text{I} \rangle$] or, $P(A) \leq P(B)$.
4. **Deduction of classical rule**
   Let the event space S of a RE E contains n event points (n is finite), say, $U_1, U_2, \cdots, U_n$.
   $\therefore$ $S = U_1 + U_2 + \cdots + U_n$. Since, elementary events (event points) are MEE, so $\{U_1\}, \{U_2\}, \cdots, \{U_n\}$ form a partition of S.
   $\therefore$ $P(U_1 + U_2 + \cdots + U_n) = P(U_1) + P(U_2) + \cdots + P(U_n)$, [by axiom $\langle \text{III} \rangle$]
   or, $P(S) = P(U_1) + P(U_2) + \cdots + P(U_n)$, or, $P(U_1) + P(U_2) + \cdots + P(U_n) = 1$, [by axiom $\langle \text{II} \rangle$].
   Now, if the events are assumed to be equally likely, then

$$P(U_1) = P(U_2) = \cdots = P(U_n), \text{ i.e., } \frac{P(U_1)}{1} = \frac{P(U_2)}{1} = \cdots = \frac{P(U_n)}{1}.$$

$$\therefore \text{ each ratio} = \frac{P(U_1) + P(U_2) + \cdots + P(U_n)}{1 + 1 + \cdots + 1 \ (n \text{ terms})} = \frac{1}{n}.$$

   Let A be an event of E containing $m (\leq n)$ event points. As the event points are equally likely, $\therefore$ we can take $A = U_1 + U_2 + \cdots + U_m$ without any loss of generality.

$$\therefore \ P(A) = P(U_1 + U_2 + \cdots + U_m) = P(U_1) + P(U_2) + \cdots + P(U_m), \text{ [by axiom } \langle \text{III} \rangle]$$
$$= \tfrac{1}{n} + \tfrac{1}{n} + \cdots \text{ to } m \text{ terms} = \tfrac{m}{n}, \text{ which is the classical rule.}$$

5. **General addition rule**
Let E be a RE and A, B be two events of E.
Obviously, $A+B=(A-AB)+AB+(B-AB)$, where the events on RHS are MEE.
$\therefore\ P(A+B)=P(A-AB)+P(AB)+P(B-AB)$, [by axiom ⟨III⟩] (1)
Now, $A=(A-AB)+AB$ and $B=(B-AB)+AB$.
$\therefore\ P(A)=P(A-AB)+P(AB)$ and $P(B)=P(B-AB)+P(AB)$.
$\therefore$ (1) $\Rightarrow P(A+B)=P(A)-P(AB)+P(AB)+P(B)-P(AB)=P(A)+P(B)-P(AB)$. (2)
[In particular, when A and B are MEE, then $P(AB)=P(0)=0$ and $P(A+B)=P(A)+P(B)$]
For any three events A, B, C of E,

$$\begin{aligned}
P(A+B+C) &= P\{(A+B)+C\}=P(A+B)+P(C)-P\{(A+B)C\} \\
&= P(A+B)+P(C)-P(AC+BC), \text{ [by distributive law]} \\
&= P(A)+P(B)-P(AB)+P(C)-\{P(AC)+P(BC)-P(ABC)\}, \\
&\qquad\qquad \text{[by (2) and } AB\cdot AC=ABC] \\
&= P(A)+P(B)+P(C)-P(BC)-P(CA)-P(AB)+P(ABC).
\end{aligned}$$

$\therefore$ for any n events $A_1, A_2, \cdots, A_n$ of E we have by inference:

$$\begin{aligned}
&P(A_1+A_2+\cdots+A_n) \\
&=\sum_{i=1}^{n} P(A_i)-\sum_{\substack{i,j=1\\ i<j}}^{n} P\left(A_iA_j\right)+\sum_{\substack{i,j,k=1\\ i<j<k}}^{n} P\left(A_iA_jA_k\right)-\cdots+(-1)^{n-1}P(A_1A_2\cdots A_n).
\end{aligned}$$

Note: The result can be deduced easily by Mathematical Induction.
[**Hints.** First deduce $P(A_1+A_2)=P(A_1)+P(A_2)-P(A_1A_2)$. (1)
Hence, the result is true for $n=2$. Let it be true for $n=m(\geq 2)$,

i.e., let $P(A_1+A_2+\cdots+A_m)$

$$=\sum_{i=1}^{m} P(A_i)-\sum_{\substack{i,j=1\\ i<j}}^{m} P\left(A_iA_j\right)+\sum_{\substack{i,j,k=1\\ i<j<k}}^{m} P\left(A_iA_jA_k\right)-\cdots+(-1)^{m-1}P(A_1A_2\cdots A_m). \quad (2)$$

$$\begin{aligned}
&P(A_1+A_2+\cdots+A_m+A_{m+1}) \\
&=P\{(A_1+A_2+\cdots+A_m)+P(A_{m+1})\} \\
&=P(A_1+A_2+\cdots+A_m)+P(A_{m+1})-P\{(A_1+A_2+\cdots+A_m)A_{m+1}\}, \text{ [by (1)]} \\
&=P(A_1+A_2+\cdots+A_m)+P(A_{m+1})-P(A_1A_{m+1}+A_2A_{m+1}+\cdots+A_mA_{m+1}) \\
&=\sum_{i=1}^{m} P(A_i)-\sum_{\substack{i,j=1\\ i<j}}^{m} P\left(A_iA_j\right)+\sum_{\substack{i,j,k=1\\ i<j<k}}^{m} P\left(A_iA_jA_k\right)-\cdots+(-1)^{m-1}P(A_1A_2\cdots A_m) \\
&\quad +P(A_{m+1})-\left\{\sum_{i=1}^{m} P(A_iA_{m+1})-\sum_{\substack{i,j=1\\ i<j}}^{m} P\left(A_iA_jA_{m+1}\right)+\cdots+(-1)^{m-1}P(A_1A_2\cdots A_mA_{m+1})\right\}
\end{aligned}$$

$$=\sum_{i=1}^{m+1}P(A_i)-\left\{\sum_{\substack{i,j=1\\i<j}}^{m}P\left(A_iA_j\right)+\sum_{i=1}^{m}P(A_iA_{m+1})\right\}$$

$$+\left\{\sum_{\substack{i,j,k=1\\i<j<k}}^{m}P\left(A_iA_jA_k\right)+\sum_{\substack{i,j=1\\i<j}}^{m}P\left(A_iA_jA_{m+1}\right)\right\}-\cdots$$

$$+(-1)^{m-1}\left\{P(A_1A_2\cdots A_m)+\sum_{\substack{i,j,\cdots,u=1\\i<j<\cdots<u}}^{m}P\left(A_iA_j\cdots A_uA_{m+1}\right)\right\}$$

$\left[A_iA_j\cdots A_u \text{ is a product of } (m-1) \text{ elements}\right]$

$$+(-1)^m P(A_1A_2\cdots A_mA_{m+1})$$

$$=\sum_{i=1}^{m+1}P(A_i)-\sum_{\substack{i,j=1\\i<j}}^{m+1}P\left(A_iA_j\right)+\sum_{\substack{i,j,k=1\\i<j<k}}^{m+1}P\left(A_iA_jA_k\right)-\cdots+(-1)^m P(A_1A_2\cdots A_{m+1}).]$$

Hence, whenever the result is true for $m(\geq 2)$ events, it is true for $(m+1)$ events. Hence, by the principle of Mathematical induction, the result is true for all natural number $n(\geq 2)$.

6. Let $\{A_n\}_n$ be a sequence of events of a RE E and is monotonic, i.e., either

$$A_1\subseteq A_2\subseteq\cdots\subseteq A_n\subseteq\cdots \text{ [expanding]}$$
$$\text{or, } A_1\supseteq A_2\supseteq\cdots\supseteq A_n\supseteq\cdots \text{ [contracting]}.$$

Then, $P\left(\lim\limits_{n\to\infty}A_n\right)=\lim\limits_{n\to\infty}P(A_n)$, where $\lim\limits_{n\to\infty}A_n=\sum\limits_{k=1}^{\infty}A_k$. [C.H. 2006]

Proof. First, consider that $\{A_n\}_n$ is expanding.

$$\therefore\ A_1+A_2+\cdots+A_n=A_n, \text{ i.e., } \sum_{k=1}^{n}A_k=A_n\ [A_kA_s\neq 0,\ 1\leq k,\ s\leq n].$$

Let $B_1=A_1, B_2=A_2-A_1,\cdots,B_k=A_k-A_{k-1},\cdots$.

Then, $B_kB_s=0, k\neq s$ and $\sum\limits_{k=1}^{n}A_k=\sum\limits_{k=1}^{n}B_k$, $n\in\mathbb{N}$ (set of natural numbers).

$$\therefore\ A_n=\sum_{k=1}^{n}B_k. \tag{1}$$

$$\therefore\ P\left(\lim_{n\to\infty}A_n\right)=P\left(\lim_{n\to\infty}\sum_{k=1}^{n}B_k\right)=P\left(\sum_{k=1}^{\infty}B_k\right)=\sum_{k=1}^{\infty}P(B_k), \text{ [by axiom (III)]}$$
$$=\lim_{n\to\infty}\sum_{k=1}^{n}P(B_k)=\lim_{n\to\infty}P\left(\sum_{k=1}^{n}B_k\right), \text{ [by axiom (III)]}$$
$$=\lim_{n\to\infty}P(A_n), \text{ [by (1)]}.$$

In case $\{A_n\}_n$ is contracting, $\left\{\bar{A}_n\right\}_n$ is expanding.

Hence, $P\left(\lim\limits_{n\to\infty}\bar{A}_n\right)=\lim\limits_{n\to\infty}P\left(\bar{A}_n\right)$. (2)

We know that for any monotone sequence $\{A_n\}_n$, $\overline{\lim_{n\to\infty} A_n} = \lim_{n\to\infty} \bar{A}_n$.

Hence (2) $\Rightarrow P\left(\overline{\lim_{n\to\infty} A_n}\right) = \lim_{n\to\infty} P\left(\bar{A}_n\right)$ or, $1 - P\left(\lim_{n\to\infty} A_n\right) = \lim_{n\to\infty}(1 - P(A_n))$

or, $P\left(\lim_{n\to\infty} A_n\right) = \lim_{n\to\infty} P(A_n)$. Hence proved.

2.2 Conditional probability

Let A and B be two events of a RE E. Let $P(B) \neq 0$. Then, the conditional probability of the event A subject to the condition B has occurred is denoted by $P(A\,|\,B)$ and is defined as $P(A\,|\,B) = \frac{P(AB)}{P(B)}$.

Similarly, $P(B\,|\,A) = \frac{P(AB)}{P(A)}$, provided $P(A) \neq 0$.

$\therefore\ P(AB) = P(A)P(B\,|\,A) = P(B)P(A\,|\,B)$.

Ex. For any three events A, B, C, show that $P(ABC) = P(A)P(B|A)P(C|AB)$, provided the conditional probabilities are defined.

Solution. RHS $= P(A)\frac{P(AB)}{P(A)}\frac{P(ABC)}{P(AB)} = P(ABC)$.

Ex. For any n events $A_1, A_2, \cdots, A_n$, show that $P(A_1A_2\cdots A_n) = P(A_1)\,P(A_2|A_1)\,P(A_3|A_1A_2)\cdots P(A_n|A_1A_2\cdots A_{n-1})$, provided all the conditional probabilities are defined.

Ex. Show that conditional probabilities satisfy all the three axioms of probability. [C.H. 2006]

Proof. (i) Let A, B be two events such that $P(B) \neq 0$, then $P(A\,|\,B) = \frac{P(AB)}{P(B)} \geq 0$ as $P(AB) \geq 0$ and $P(B) > 0$. Thus axiom ⟨I⟩ is satisfied by the conditional probabilities.

(ii) Let S be the certain events and B be an event such that $P(B) \neq 0$.

$$\therefore\ P(S\,|\,B) = \frac{P(SB)}{P(B)} = \frac{P(B)}{P(B)}\ [\because\ B \subseteq S.\ \therefore\ SB = B]$$
$$= 1.$$

Thus axiom ⟨II⟩ is satisfied by the conditional probabilities.

(iii) Let $\{A_n\}_n$ be a sequence events such that $A_iA_j = 0, i \neq j$, i.e., $\{A_n\}_n$ be a sequence of pairwise exclusive events and B be an event such that $P(B) \neq 0$.

Then $P\{(A_1 + A_2 + \cdots + A_n + \cdots)\,|\,B\}$

$$= \frac{P\{(A_1 + A_2 + \cdots + A_n + \cdots)B\}}{P(B)} = \frac{P(A_1B + A_2B + \cdots + A_nB + \cdots)}{P(B)}$$
$$= \frac{P(A_1B) + P(A_2B) + \cdots + P(A_nB) + \cdots}{P(B)}\ [\text{by axiom ⟨III⟩ as } A_iBA_jB = A_iA_jB = 0, i \neq j]$$
$$= \frac{P(A_1B)}{P(B)} + \frac{P(A_2B)}{P(B)} + \cdots + \frac{P(A_nB)}{P(B)} + \cdots$$
$$= P(A_1|B) + P(A_2|B) + \cdots + P(A_n|B) + \cdots,$$

which is the third axiom for conditional probabilities. Hence, conditional probabilities satisfy all the three axioms of probability.

2.2.1 Bayes' theorem

[C.H. 2005]

Let $\{A_1, A_2, \cdots, A_n\}$ be a set of mutually exclusive and exhaustive events of a RE E (i.e., $A_iA_j = 0$, $\forall\ i, j$ when $i \neq j$ and $A_1 + A_2 + \cdots + A_n = S$, the certain event) and X be an arbitrary event following any one of them.

Then, $P(X) = P(A_1)\,P(X|A_1) + P(A_2)\,P(X|A_2) + \cdots + P(A_n)\,P(X|A_n)$.

Further if $P(X) \neq 0$,
$$P(A_i|X) = \frac{P(A_i)\,P(X|A_i)}{\sum_{k=1}^{n} P(A_k)\,P(X|A_k)}, \; i = 1, 2, \cdots, n$$

Proof. $A_1 + A_2 + \cdots + A_n = S$.

$\therefore A_1X + A_2X + \cdots + A_nX = SX = X$ as $X \subseteq S$. (1)

Now, when $i \neq j, A_iXA_jX = A_iA_jX = 0X = 0, \forall\, i, j$.

$\therefore A_1X, A_2X, \cdots, A_nX$ are pairwise MEE, whenever $A_1, A_2, \cdots, A_n$ are so.

Hence, from (1), $P(A_1X + A_2X + \cdots + A_nX) = P(X)$, i.e.,
$$\begin{aligned} P(X) &= P(A_1X) + P(A_2X) + \cdots + P(A_nX), \text{ [by axiom } \langle \text{III} \rangle] \\ &= P(A_1)\,P(X|A_1) + P(A_2)\,P(X|A_2) + \cdots + P(A_n)\,P(X|A_n) \\ &= \sum_{k=1}^{n} P(A_k)\,P(X|A_k). \end{aligned} \quad (2)$$
$$\therefore P(A_iX) = P(X)P(A_i|X) = P(A_i)P(X|A_i), \; i = 1, 2, \cdots, n$$
$$\Rightarrow P(A_i|X) = \frac{P(A_i)\,P(X|A_i)}{P(X)} = \frac{P(A_i)\,P(X|A_i)}{\sum_{k=1}^{n} P(A_k)P(X|A_k)}, \; i = 1, 2, \cdots, n \textbf{ (Proved)}.$$

2.2.2 Stochastic independence

Two events A and B of a RE E are said to be stochastically independent or simply independent, if
$$P(A|B) = P(A), \text{ when } P(B) \neq 0$$
$$\text{and } P(B|A) = P(B), \text{ when } P(A) \neq 0.$$

$\therefore$ A and B will be independent if $P(AB) = P(A)P(B)$.

$A_1, A_2, \cdots, A_n$ will be mutually independent if $P(A_1, A_2, \cdots, A_n) = P(A_1)\,P(A_2) \cdots P(A_n)$.

Note: Pairwise independence does not imply mutual independence [Ex. 2(iii)].

Ex. 1 If two events A and B are independent, then show that (i) their complementary events $\bar{A}$ and $\bar{B}$ are also independent (ii) $A, \bar{B}$ are independent (iii) $\bar{A}, B$ are independent. [C.H. 2007]

Solution. (i) Given that A and B are independent, so $P(AB) = P(A)\,P(B)$. (1)
$$\begin{aligned} \therefore P\left(\bar{A}\bar{B}\right) &= P\left(\overline{A+B}\right), \text{ [by De Morgan's law]} \\ &= 1 - P(A+B) = 1 - \{P(A) + P(B) - P(AB)\} = 1 - P(A) - P(B) + P(A)P(B), \text{ [by (1)]} \\ &= 1 - P(A) - P(B)\,(1 - P(A)) = (1 - P(A))\,(1 - P(B)) = P\left(\bar{A}\right)\,P\left(\bar{B}\right). \end{aligned}$$

Hence, $\bar{A}$ and $\bar{B}$ are independent.

(ii) We have $A = AS = A\left(B + \bar{B}\right) = AB + A\bar{B}$.

As $B\bar{B} = 0$, $\therefore ABA\bar{B} = AB\bar{B} = A0 = 0$.

$\therefore$ by axiom $\langle$III$\rangle$, $\quad P(A) = P(AB) + P\left(A\bar{B}\right)$
$$\begin{aligned} \text{or, } P\left(A\bar{B}\right) &= P(A) - P(A)P(B), \text{ [as } A \text{ and } B \text{ are independent]} \\ &= P(A)\,(1 - P(B)) = P(A)\,P\left(\bar{B}\right). \end{aligned}$$

Hence, A and $\bar{B}$ are independent.

(iii) Exactly similar to (ii).

Ex. 2 (i) Show that two proper events cannot be simultaneously mutually exclusive and independent. [C.H. 1997]

Solution. Let A and B be two proper events of a RE E, i.e., $P(A) \neq 0$ and $P(B) \neq 0$.

If A and B are independent and mutually exclusive, then $P(AB) = P(A)P(B)$ and $AB = 0$.

These two cannot hold simultaneously, as $P(AB) = P(0) = 0$ but $P(A)P(B) \neq 0 \because P(A) \neq 0$ and $P(B) \neq 0$.

Ex. 2 (ii) Suppose that A, B is an independent pair and that B, C is an independent pair. Does it follow that A, C is an independent pair? [C.H. 1997]

Solution. Let the RE be $E =$ throw of two dice simultaneously.

Let A be the event that first die shows odd face.

B be the event that second die shows odd face.

C be the event that first die shows even face.

Then, A, B are independent as they are the outcomes of two different dice thrown simultaneously.

Similarly, B, C are independent.

But, A, C are dependent as $P(AC) = P(0) = 0$ and $P(A)P(C) = \frac{1}{2} \cdot \frac{1}{2} = \frac{1}{4}$, i.e., $P(AC) \neq P(A)P(C)$.

Ex. 2 (iii) Give an example to, show that if an event C is independent of the events A and B respectively, it will not necessarily be independent of AB.

Solution. Let the RE be $E =$ throw of a pair of dice simultaneously (which is the same as throwing one after another).

Let the events $A =$ first die shows odd face, $B =$ second die shows odd face and $C =$ sum of the faces of the two dice is odd.

$\therefore$ $P(A) = \frac{1}{2}, P(B) = \frac{1}{2}$ and $P(C) = \frac{1}{2}$ [for C, out of 36 points 18 is in favour].

$\therefore$ $P(AB) = P(A)P(B) = \frac{1}{2} \cdot \frac{1}{2} = \frac{1}{4}$ as A, B are independent.

$P(AC) = P(A)P(C|A) = P(A)P\left(\bar{B}\right) = \frac{1}{2} \cdot \frac{1}{2} = \frac{1}{4} = P(A)P(C)$ $\left[P(C) = \frac{18}{36} = \frac{1}{2}\right]$ and

$P(BC) = P(B)P(C|B) = P(B)P\left(\bar{A}\right) = \frac{1}{4} = P(B)P(C)$.

$P(ABC) = P(0) = 0 \neq P(A)P(B)P(C)$ [as $P(A)P(B)P(C) = \frac{1}{2} \cdot \frac{1}{2} \cdot \frac{18}{36} = \frac{1}{8}$].

This shows that A, B, C are pairwise independent but not mutually independent. So, pairwise independence does not necessarily imply their mutual independence.

Now, $P(C|AB) = 0 \neq P(C)$.

$\therefore$ C is independent of both A and B but C is not necessarily be independent of AB.

2.2.3 Problems based on Bayes' theorem

Ex. 3 In a bolt factory, machines A, B, C manufacture respectively 25%, 35% and 40% of the total. Of their output 5%, 4% and 2% are defective bolts. A bolt is drawn at random from the production and is found defective. What is the probability that it was produced by the machine A?

Solution. Let A_1, B_1, C_1 be respectively the events that a bolt was produced by machines A, B and C.

$\therefore$ $P(A_1) = \frac{25}{100}$, $P(B_1) = \frac{35}{100}$ and $P(C_1) = \frac{40}{100}$.

Let E be the event of drawing a defective bolt from the production.

$\therefore$ $P(E|A_1) = \frac{5}{100}$, $P(E|B_1) = \frac{4}{100}$ and $P(E|C_1) = \frac{2}{100}$.

Using Bayes' formula, $P(A_1|E) = \dfrac{P(A_1)\,P(E|A_1)}{P(A_1)\,P(E|A_1) + P(B_1)\,P(E|B_1) + P(C_1)\,P(E|C_1)}$, as the events A_1, B_1, C_1 are MEE and exhaustive

$$= \frac{\frac{25}{100} \cdot \frac{5}{100}}{\frac{25}{100} \cdot \frac{5}{100} + \frac{35}{100} \cdot \frac{4}{100} + \frac{40}{100} \cdot \frac{2}{100}} = \frac{125}{345} = \frac{25}{69}$$

$\left[\text{Similarly, } P(B_1|E) = \frac{28}{69},\ P(C_1|E) = \frac{16}{69}.\right]$

Ex. 4 There are two identical urns containing 4 white, 3 red balls and 3 white, 7 red balls respectively. An urn is chosen at random and a ball is drawn from it. Find the probability that the ball drawn is white. If the ball drawn is white, what is the probability that it was drawn from the first urn?

Solution. Let U_1, U_2 and W be respectively the events of choosing first urn, second urn and the ball drawn is white.

$\therefore\ P(U_1)=P(U_2)=\frac{1}{2}$, as the urns are identical.

Events U_1 and U_2 are MEE and exhaustive, so far any event W, $P(W)=P(U_1)\,P(W\,|\,U_1)+P(U_2)\,P(W\,|\,U_2)$

$$=\frac{1}{2}\cdot\frac{4}{7}+\frac{1}{2}\cdot\frac{3}{10}=\frac{61}{140}.$$

By Bayes' theorem $P(U_1|W)=\dfrac{P(U_1)P(W|U_1)}{P(U_1)P(W|U_1)+P(U_2)P(W|U_2)}=\dfrac{\frac{1}{2}\cdot\frac{4}{7}}{\frac{61}{140}}=\dfrac{2}{7}\cdot\dfrac{140}{61}=\dfrac{40}{61}.$

Ex. 5 There are three identical urns each containing 5 white and 5 black balls. One ball is transferred blindly from the first to the second urn and then one ball is transferred from the second to the third urn. Finally, a ball is drawn from the third urn. What is the probability that this ball is white?

Solution. Let $W_i\ (i=1,2,3)$ be the event of drawing a white ball from the ith urn and $B_i\ (i=1,2,3)$ be the event of drawing a black ball from the ith urn.

$\therefore\ P(W_1)=\frac{5}{10}=\frac{1}{2},\ P(B_1)=\frac{5}{10}=\frac{1}{2}.$

W_1 and B_1 are MEE and exhaustive, so the probability of W_2 following either of them is given by

$$P(W_2)=P(W_1)\,P(W_2\,|\,W_1)+P(B_1)\,P(W_2\,|\,B_1)=\frac{1}{2}\cdot\frac{6}{11}+\frac{1}{2}\cdot\frac{5}{11}=\frac{1}{2}.$$

Similarly, $P(B_2)=P(W_1)\,P(B_2\,|\,W_1)+P(B_1)\,P(B_2\,|\,B_1)=\frac{1}{2}\cdot\frac{5}{11}+\frac{1}{2}\cdot\frac{6}{11}=\frac{1}{2}.$

Events W_2 and B_2 are MEE and exhaustive.

$\therefore\ P(W_3)=P(W_2)\,P(W_3\,|\,W_2)+P(B_2)\,P(W_3\,|\,B_2)=\frac{1}{2}\cdot\frac{6}{11}+\frac{1}{2}\cdot\frac{5}{11}=\frac{1}{2}.$

Ex. 6 The probability of detecting tuberculosis in X-ray examination of a person suffering from the disease is $1-b$. The probability of diagnosing a healthy person as tubercular is a. If the ratio of the tubercular patients to the whole population is c, find the probability that a person is healthy if after examination he is diagnosed as tubercular. [C.H. 1994]

Solution. The RE in the problem contains two parts, viz., selecting a person at random who is either tubercular (T) or healthy (H) and then he is subjected to X-ray examination when the result is either diagnosed tubercular (t) or diagnosed healthy (h).

So, the event space $S=\{(T,t),\ (T,h),\ (H,t),\ (H,h)\}$.

Let A be the event that a person is tubercular and B be the event that person is diagnosed tubercular.

$\therefore\ A=\{(T,t),\ (T,h)\}$ and $B=\{(T,t),\ (H,t)\}$.

$\therefore\ P(B\,|\,A)=1-b, P\left(B\,|\,\bar{A}\right)=a, P(A)=c.$

As $A,\bar{A}$ are MEE and exhaustive.

$\therefore\ P(B)=P(A)P(B\,|\,A)+P\left(\bar{A}\right)\,P\left(B\,|\,\bar{A}\right)=c(1-b)+(1-c)a.$

$\therefore$ the required probability is $P\left(\bar{A}\,|\,B\right)=\frac{P(\bar{A})P(B|\bar{A})}{P(B)}=\frac{(1-c)a}{c(1-b)+(1-c)a}.$

Alternative. Let $T=$ the event that a person is tubercular patient.

$H=$ the event that a person is healthy.

$D=$ the event that a person is diagnosed as tubercular.

$\therefore\ P(T)=c, P(H)=1-c, P(D\,|\,T)=1-b$ and $P(D\,|\,H)=a.$

As T and H are MEE and exhaustive, $\therefore$ for any event D,

$$P(D)=P(T)P(D\,|\,T)+P(H)P(D\,|\,H)=c(1-b)+(1-c)a.$$

$\therefore$ by Bayes' theorem $P(H\,|\,D)=\frac{P(H)P(D|H)}{P(D)}=\frac{(1-c)a}{c(1-b)+(1-c)a}.$

Ex. 7 A and B are two independent witness in a particular case. The probability that A will speak the truth is x and the probability that B will speak the truth is y. A and B agree in a certain statement. Show that the probability that the statement is true is $\frac{xy}{1-x-y+2xy}$. [C.H. 2009]

Solution. Since, A and B agree in a certain statement, so both of them either speak the truth or tell a lie.

Let the event that both will speak the truth be T, the event that both will speak a lie be L and the event that both agree in a certain statement be X.

$\therefore$ $P(T)=xy$, as the witnesses are independent.

$P(L)=(1-x)(1-y)$.

$$\therefore\ P(X)=P(T+L)=P(T)+P(L) \text{ as } T \text{ and } L \text{ are MEE}$$
$$=xy+(1-x)(1-y)=1-x-y+2xy.$$

$\therefore$ the required probability $P(T|X)$ is given by

$$P(T|X)=\frac{P(TX)}{P(X)}=\frac{P\{T(T+L)\}}{P(X)}=\frac{P(T+0)}{P(X)} \text{ as } TL=0$$
$$=\frac{P(T)}{P(X)}=\frac{xy}{1-x-y+2xy}.$$

Ex. 8 The chance that a doctor will diagnose a certain disease correctly is 60%. The chance that the patient will die after correct diagnosis is 40% and the chance of death by wrong diagnosis is 70%. *A patient of the doctor having the disease dies.* What is the probability that the disease was diagnosed correctly?

Solution. Let A be the event that the doctor diagnoses correctly and B be the event that the patient having the disease dies.

$\therefore$ $P(A)=\frac{60}{100}=\frac{3}{5}$, $P(B|A)=\frac{40}{100}=\frac{2}{5}$, $P\left(B|\bar{A}\right)=\frac{70}{100}=\frac{7}{10}$. We are to find $P(A|B)$.

As A and $\bar{A}$ are MEE and exhaustive. $\therefore$ for any event B following A or $\bar{A}$ is given by

$$P(B)=P(A)P(B|A)+P\left(\bar{A}\right)P\left(B|\bar{A}\right)=\tfrac{3}{5}\cdot\tfrac{2}{5}+\left(1-\tfrac{3}{5}\right)\cdot\tfrac{7}{10}=\tfrac{6}{25}+\tfrac{7}{25}=\tfrac{13}{25}.$$

$\therefore$ the required probability $P(A|B)$ is given by

$$P(A|B)=\frac{P(A)P(B|A)}{P(B)}=\frac{\frac{3}{5}\cdot\frac{2}{5}}{\frac{13}{25}}=\frac{6}{13}.$$

2.3 Two important inequalities

(i) Boole's inequality and (ii) Bonferroni's inequalities.

2.3.1 Boole's inequality

For any n events $A_1, A_2, \cdots, A_n$ of a RE E,

$$P(A_1+A_2+\cdots+A_n)\leq P(A_1)+P(A_2)+\cdots+P(A_n).$$ [C.H. 2005, 2007]

Proof. We proceed to prove by Mathematical Induction. For any two events A_1 and A_2,

$P(A_1+A_2)=P(A_1)+P(A_2)-P(A_1A_2)\leq P(A_1)+P(A_2)$ as $P(A_1A_2)\geq 0$ [by axiom (I)].

$\therefore$ the inequality hold for $n=2$.

Assume that the inequality is true for $n=m(\geq 2)$,

i.e., assume $P(A_1+A_2+\cdots+A_m)\leq P(A_1)+P(A_2)+\cdots+P(A_m)$. (1)

$$\therefore\ P(A_1+A_2+\cdots+A_m+A_{m+1})$$
$$=P\{(A_1+A_2+\cdots+A_m)+A_{m+1}\}$$
$$\leq P(A_1+A_2+\cdots+A_m)+P(A_{m+1}), \text{ [as the inequality holds for two events]}$$
$$\leq P(A_1)+P(A_2)+\cdots+P(A_m)+P(A_{m+1}) \text{ [by (1)]}.$$

Hence, the inequality is true for $n=m+1$, whenever it is true for $n=m$.

Hence, by the principle of Mathematical Induction the inequality is true for all natural number $n\geq 2$. Also, the inequality is valid for a single event A_1 as equality. So, for all natural number n the inequality (known as *Boole's inequality*) is established.

2.3.2 ...qualities

Fo... ...RE E, $P(A_1A_2\cdots A_n)\geq 1-\sum\limits_{i=1}^{n}P\left(\bar{A}_i\right)$

and al... $\ldots_1A_2\cdots A_n)\geq \sum\limits_{i=1}^{n}P(A_i)-(n-1).$ [C.H. 2007]

Proof. We have $P\left(\bar{A}_1+\bar{A}_2+\cdots+\bar{A}_n\right)\leq P\left(\bar{A}_1\right)+P\left(\bar{A}_2\right)+\cdots+P\left(\bar{A}_n\right)$ [by Boole's inequality]

or, $P\left(\overline{A_1A_2\cdots A_n}\right)\leq P\left(\bar{A}_1\right)+P\left(\bar{A}_2\right)+\cdots+P\left(\bar{A}_n\right)$, [by De Morgan's law]

or, $1-P(A_1A_2\cdots A_n)\leq \sum\limits_{i=1}^{n}P\left(\bar{A}_i\right)$.

$\therefore P(A_1A_2\cdots A_n)\geq 1-\sum\limits_{i=1}^{n}P\left(\bar{A}_i\right)$, which proves the first part. (1)

Second Part. We have $P\left(\bar{A}\right)=1-P(A)$.

$$\therefore \text{RHS of (1)}=1-\sum_{i=1}^{n}\{1-P(A_i)\}=1-\left\{n-\sum_{i=1}^{n}P(A_i)\right\}=\sum_{i=1}^{n}P(A_i)-(n-1).$$

$\therefore (1)\Rightarrow P(A_1A_2\cdots A_n)\geq \sum\limits_{i=1}^{n}P(A_i)-(n-1)$ **(Proved).**

2.3.3 An interesting problem

Suppose, an urn contains a white balls and b black balls.

Then, $P(W)=\frac{a}{a+b}$ and $P(B)=\frac{b}{a+b}$.

If we draw a ball and it happens to be white, then probability of white ball in the second draw (without replacement) is $\frac{a-1}{a+b-1}$.

But, if we do not know the result of the first draw, then the probability of while ball in the second draw (without replacement) is again $\frac{a}{a+b}$. How?

Let W_1 = event of drawing a white ball in the first draw, B_1 = event of drawing a black ball in the first draw and W_2 = event of drawing a white ball in the second draw without knowing the result of first draw and without replacement.

Then, W_1 and B_1 are MEE and exhaustive.

$$\therefore P(W_2)=P(W_1)\,P(W_2\,|\,W_1)+P(B_1)\,P(W_2\,|\,B_1)=\frac{a}{a+b}\cdot\frac{a-1}{a+b-1}+\frac{b}{a+b}\cdot\frac{a}{a+b-1}$$

$$=\frac{a^2-a+ab}{(a+b)\,(a+b-1)}=\frac{a(a+b-1)}{(a+b)\,(a+b-1)}=\frac{a}{a+b}.$$

EXERCISE

1. Show that the probability that exactly one of the events A ad B occurs is $P(A)+P(B)-2P(AB)$.
2. If A and B be two mutually exclusive events and $P(A\cup B)\neq 0$, prove that $P(A\,|\,A\cup B)=\frac{P(A)}{P(A)+P(B)}$. [B.H. 2005]
3. If $P(A)=\frac{3}{4}$ and $P(B)=\frac{5}{8}$, show that $\frac{3}{8}\leq P(A\cap B)\leq\frac{5}{8}$. [B.H. 2006]
4. A problem in Mathematics is given to $(n-1)$ students whose chances of solving it are respectively $\frac{1}{2},\frac{1}{3},\cdots,\frac{1}{n}$. What is the probability that the problem will be solved?
[**Hints.** E_i = the event that ith student can solve the problem, $i=1,2,\cdots,n-1$.
$\therefore\ P(E_1)=\frac{1}{2},P(E_2)=\frac{1}{3},\cdots,P(E_{n-1})=\frac{1}{n}$.

$E_1+E_2+\cdots+E_{n-1}$ is the event that the problem will be solved.

$$\therefore\ P(E_1+E_2+\cdots+E_{n-1})=1-P\left(\overline{E_1+E_2+\cdots+E_{n-1}}\right)$$
$$=1-P\left(\bar{E}_1\bar{E}_2\cdots\bar{E}_{n-1}\right) \text{ [by De Morgan's law]}$$
$$=1-P\left(\bar{E}_1\right)P\left(\bar{E}_2\right)\cdots P\left(\bar{E}_{n-1}\right) \text{ [as the events are independent]}$$
$$=1-\frac{1}{2}\cdot\frac{2}{3}\cdot\frac{3}{4}\cdots\frac{n-2}{n-1}\cdot\frac{n-1}{n}=1-\frac{1}{n}.]$$

5. Two players A and B throw a pair of dice. A wins if he throws a sum of 5 before B throws 8. (i) What is the probability of winning of A if A starts and (ii) what is the probability of winning of A if B starts? (*Ans.* (i) 9/19 (ii) 31/76)
6. Two urns contain respectively 5 white, 6 black, 4 red balls and 3 white, 7 black, 5 red balls. One ball from each urn is drawn randomly. What is the probability that the two balls must be of same colour?
[**Hints.** $P(W_1W_2+B_1B_2+R_1R_2)=P(W_1W_2)+P(B_1B_2)+P(R_1R_2)$, events are MEE
$=P(W_1)P(W_2)+P(B_1)P(B_2)+P(R_1)P(R_2)$, as events are independent
$=\frac{5}{15}\cdot\frac{3}{15}+\frac{6}{15}\cdot\frac{7}{15}+\frac{4}{15}\cdot\frac{5}{15}=\frac{77}{225}$.]
7. From an urn containing 3 white and 5 black balls, four balls are transferred into an empty urn. From this urn a ball is drawn and is found to be whi... ..nat is the probability that out of four balls transferred 3 are white and 1 is black? [C.H. 2007]
(*Ans.* 1/7)
[**Hints.** $E_i=i$ white balls are transferred to the empty ve... $=0,1,2,3$.
$B=$ one white ball is drawn from second urn.
$P(B\mid E_0)=0,\ P(B\mid E_1)=1/4,\ P(B\mid E_2)=2/4,\ P(B\mid E_3)=3/4.\ \therefore\ P(E_3\mid B)=\cdots=1/7$.]
8. For n mutually independent events $A_1,A_2,\cdots,A_n$, show that $P(A_1+A_2+\cdots+A_n)=1-(1-P(A_1))(1-P(A_2))\cdots(1-P(A_n))$. [C.H. 2007]
9. For any two events A and B with $P(A)>0$, prove that $P(B\mid A)\geq 1-\frac{P(\bar{B})}{P(A)}$. [C.H. 2007]
10. State precisely axiomatic definition of probability. Show that for any two events A and B belonging to the event space S, $P\left(\bar{A}\bar{B}\right)=P\left(\bar{A}\right)-P(B)+P(AB)$, where $\bar{A}=S-A$. [C.H. 2005]
11. An event A is known to be independent of the events B, $B+C$ and BC. Show that it is also independent of C.
[**Hints.** Given: $P(AB)=P(A)P(B), P(A(B+C))=P(A)P(B+C)$ and $P\{A(BC)\}=P(A)P(BC)$. From second condition, $P(AB+AC)=P(A)P(B+C)$, or, $P(AB)+P(AC)-P(ABC)=P(A)\,(P(B)+P(C)-P(BC))$. Using first and third conditions, $P(AC)=P(A)P(C)$. $\therefore$ A and C are independent.]
12. A problem is given to three students whose chances of solving it are respectively 1/2, 1/3 and 1/4. What is the probability that exactly two of them can solve the problem? Also, find the probability that the problem will not be solved.
[**Hints.** Let A, B, C be respectively the events that the students can solve the problem.
$\therefore\ P(A)=1/2,\ P(B)=1/3,\ P(C)=1/4;\ P\left(\bar{A}\right)=1/2,\ P\left(\bar{B}\right)=2/3,\ P\left(\bar{C}\right)=3/4$.
$\therefore\ AB\bar{C}+A\bar{B}C+\bar{A}BC$ and $\bar{A}\bar{B}\bar{C}$ be respectively the events that exactly two of them can solve the problem and no one can solve the problem, i.e., the problem will not be solved.
$P\left(AB\bar{C}+A\bar{B}C+\bar{A}BC\right)=P\left(AB\bar{C}\right)+P\left(A\bar{B}C\right)+P\left(\bar{A}BC\right)$, as the events are MEE.
$=P(A)P(B)P\left(\bar{C}\right)+P(A)P\left(\bar{B}\right)P(C)+P\left(\bar{A}\right)P(B)P(C)$, as the events A, B, C are independent.
$=\frac{1}{2}\cdot\frac{1}{3}\cdot\frac{3}{4}+\frac{1}{2}\cdot\frac{2}{3}\cdot\frac{1}{4}+\frac{1}{2}\cdot\frac{1}{3}\cdot\frac{1}{4}=\frac{6}{2\cdot3\cdot4}=\frac{1}{4}$.
$P\left(\bar{A}\bar{B}\bar{C}\right)=P\left(\bar{A}\right)P\left(\bar{B}\right)P\left(\bar{C}\right)=\frac{1}{2}\cdot\frac{2}{3}\cdot\frac{3}{4}=\frac{1}{4}$.]
13. A, B, C play a game and the chances of their winning are respectively $\frac{1}{2}$, $\frac{1}{3}$ and $\frac{1}{4}$ ... followed by B and then C, find the chances of their winning. ...3,2/9,1/9)

[Hints. $A+\bar{A}\bar{B}\bar{C}A+\bar{A}\bar{B}\bar{C}\bar{A}\bar{B}\bar{C}A+\cdots\infty$ is the event of A's winning, etc.

14. Three events A, B, C of RE be such that they are pairwise independent and A is independent of BC, then show that A, B, C are mutually independent.
15. If an event A is independent of the events B, BC and $B+C$, then show that it is also independent of C.
16. If n objects are distributed among a men and b women $(b < a)$, show that probability that the women get an odd number of objects is $\frac{1}{2}[(a+b)^n-(a-b)^n]/(a+b)^n$. [C.H. 2008]
17. An urn contains r red and b black balls. Three random drawings are made; such that the ball drawn is always replaced and in addition c balls of the same colour drawn are added to the urn. Find the probability of a run of 3 black balls. [C.H. 2003]

    [Hints. $P(B_1 B_2 B_3)=P(B_1)\,P(B_2|B_1)\,P(B_3|B_1 B_2)=\frac{b}{b+r}\cdot\frac{b+c}{b+c+r}\cdot\frac{b+2c}{b+2c+r}$.]
18. For two arbitrary events A and B defined on the event space W, show that $P(B|A)\geq 1-\frac{P(\bar{B})}{P(A)}$, $P(A)\neq 0$.

    [Hints. As $P(A)\geq 0$ and given $P(A)\neq 0$. $\therefore$ $P(A)>0$.
    Obviously, $P(A+B)\leq 1$. $\therefore$ $P(A)+P(B)-P(AB)\leq 1$,
    or, $1-P(B)\geq P(A)-P(AB)$, or, $P\left(\bar{B}\right)\geq P(A)-P(AB)$],
    or, $\frac{P(\bar{B})}{P(A)}\geq 1-\frac{P(AB)}{P(A)}$ [$\because$ $P(A)>0$]
    or, $\frac{P(AB)}{P(A)}\geq 1-\frac{P(\bar{B})}{P(A)}$, i.e., $P(B|A)\geq 1-\frac{P(\bar{B})}{P(A)}$.]

Chapter 3

Compound Experiment/Repeated Trials

3.1 Introduction

If a random experiment be repeated n times and if the sequence of all trials is considered as a single trial of a RE, then the latter one is called a *compound experiment* or *repeated trials*. If S be the event space of the basic RE, then the even space of the compound RE is S^n $(= S \times S \times \cdots n$ times) and the compound experiment is denoted by E^n, when E is the basic RE.

In this chapter we consider two types of compound RE (i) Bernoullian (ii) Poissonian.

3.2 Bernoullian trials

A sequence of independent trials is called *Bernoullian*, if

(i) the event space of each trial consists of only two event points called *success* (S) and *failure* (F)

(ii) the probability of success remains the same for every trial of the sequence.

Binomial law. Let E^n be a compound experiment of Bernoullian trials and A_i $(0 \le i \le n)$ the event of i-successes, then $P(A_i) = \binom{n}{i} p^i q^{n-i}$, where $P(S) = p, P(F) = q, p+q = 1$ and $0 < p < 1$. [C.H. 2005]

Proof. S^n, the event space E^n, consists of 2^n event points of the type $(S, F, S, S, \cdots, S, F)$. The first tuple corresponds the result of first trial, the second tuple, corresponds to the result of the second trial, etc.

Now, the number of event points in favour of A_i is $\binom{n}{i}$ out of 2^n. The probability of an event point of A_i is $p^i q^{n-i}$ as the trials are independent.

Further the elementary events, i.e., event points are always mutually exclusive.

$\therefore\ P(A_i) = p^i q^{n-i} + p^i q^{n-i} + \cdots$ to $\binom{n}{i}$ times $= \binom{n}{i} p^i q^{n-i}$.

3.2.1 Some important deductions

1. $P(A_0 + A_1 + \cdots + A_n) = 1$.

   LHS $= P(A_0) + P(A_1) + \cdots + P(A_n)$, as the events are MEE

   $$= \sum_{r=0}^{n} P(A_r) = \sum_{r=0}^{n} \binom{n}{r} p^r q^{n-r} = (p+q)^n = 1\ [\because\ p+q = 1].$$

   Alternative. $A_0 + A_1 + \cdots + A_n = S^n$. $\therefore\ P(A_0 + A_1 + \cdots + A_n) = P(S^n) = 1$.

2. The probability of at most k successes is given by $P(A_0 + A_1 + \cdots + A_k) = \sum_{r=0}^{k} P(A_r) = \sum_{r=0}^{k} \binom{n}{k} p^k q^{n-k}$.

3. The probability of at least k successes is given by $P(A_k + A_{k+1} + \cdots + A_n) = \sum_{r=k}^{n} P(A_r) = \sum_{r=k}^{n} \binom{n}{k} p^k q^{n-k}$.

4. The probability of an event whose number of successes lying between r_1 to r_2 is given by

   $$P(A_r:\ r_1 \le r \le r_2) = \sum_{k=r_1}^{r_2} \binom{n}{k} p^k q^{n-k}.$$

5. Now, we want to find $\max\{P(A_0), P(A_1), \cdots, P(A_n)\}$.
$P(A_r)$ will be maximum, if $P(A_{r-1}) \le P(A_r) \ge P(A_{r+1})$.

$$P(A_{r-1}) \le P(A_r) \Rightarrow \binom{n}{r-1} p^{r-1} q^{n-r+1} \le \binom{n}{r} p^r q^{n-r}$$

$$\text{or, } \frac{n!}{(r-1)!\,(n-r+1)!} p^{r-1} q^{n-r+1} \le \frac{n!}{r!\,(n-r)!} p^r q^{n-r}$$

$$\text{or, } \frac{q}{n-r+1} \le \frac{p}{r} \text{ as } p>0 \text{ and } q>0$$

$$\text{or, } qr \le np - pr + p, \text{ or, } r \le np + p \text{ as } p+q=1. \tag{1}$$

$$\text{Again, } P(A_r) \ge P(A_{r+1}) \Rightarrow \frac{n!}{r!\,(n-r)!} p^r q^{n-r} \ge \frac{n!}{(r+1)!\,(n-r-1)!} p^{r+1} q^{n-r-1}, \text{ or, } \frac{q}{n-r} \ge \frac{p}{r+1}$$

$$\text{or, } qr + q \ge np - pr, \text{ or, } r \ge np - q \text{ as } p+q=1. \tag{2}$$

$\therefore$ $P(A_r)$ will be maximum, when r satisfies $np - q \le r \le np + p$ [by (1) and (2)].

3.3 Poissonian trials

A sequence of independent repeated trials is called *Poissonian*, if

(i) the event space of each trial consists of only two event points called *success* (S) and failure (F)

(ii) the probability of success varies from trial to trial.

If p_i is the probability of success at the ith trial and q_i is the probability of failure at the ith trial, then $p_i + q_i = 1$, $i = 1, 2, \cdots, n$.

In this case, we cannot have precise formula for $P(A_i)$ as in the case of Bernoullian trials.

Consider the case of the RE E^3. Then, $P(A_i)$, $i = 0,1,2,3$ are given by $P(A_0) = q_1\, q_2\, q_3$, $P(A_1) = p_1\, q_2\, q_3 + q_1\, p_2\, q_3 + q_1\, q_2\, p_3$, $P(A_2) = p_1\, p_2\, q_3 + p_1\, q_2\, p_3 + q_1\, p_2\, p_3$, $P(A_3) = p_1\, p_2\, p_3$.

In particular, when $p_1 = p_2 = p_3 = p$ and $q_1 = q_2 = q_3 = q$, say, then we get $P(A_0) = q^3$, $P(A_1) = 3pq^2 = \binom{3}{1} p^1 q^{3-1}$, $P(A_2) = 3p^2 q = \binom{3}{2} p^2 q^{3-2}$ and $P(A_3) = p^3$, which are the same as obtained by binomial law.

3.4 Multinomial law

Here the sequence of trials is no longer Bernoullian. In present situation, we consider a more general case. Here the event space for each trial consists of more than two event points, viz., $s_1, s_2, \cdots, s_m$ such that $P(s_i) = p_i$ $(i = 1, 2, \cdots, m)$ and $\sum_{i=1}^{m} p_i = 1$.

If $A_{i_1 i_2 \cdots i_m}$ be an event of E^n in which s_1 occurs i_1 times, s_2 occurs i_2 times, etc. and $i_1 + i_2 + \cdots + i_m = n$, then $P\left(A_{i_1 i_2 \cdots i_m}\right) = \frac{n!}{i_1!\, i_2! \cdots i_m!} p_1^{i_1}\, p_2^{i_2} \cdots p_m^{i_m}$, which is the multinational law.

Each event point of E^n in favour of $A_{i_1\, i_2 \cdots i_m}$ consists of i_1 number of s_1, i_2 number of s_2, $\cdots$ i_m number of s_m and $0 \le i_1, i_2, \cdots, i_m \le n$.

If S^n be the event space of E^n, then S^n contains m^n event points. Out of m^n points, the number of event points in favour of $A_{i_1\, i_2 \cdots i_m}$ is $\frac{n!}{i_1!\, i_2! \cdots i_m!}$, because out of n-tuples of an event point in favour of $A_{i_1\, i_2 \cdots i_m}$ contains i_1 identical entity (viz., s_1), i_2 different identical entity (viz., s_2), $\cdots$, i_m number of s_m.

The probability of each such event point is $p_1^{i_1}\, p_2^{i_2} \cdots p_m^{i_m}$ as the trials are independent.

Hence, $P\left(A_{i_1\, i_2 \cdots i_m}\right) = \frac{n!}{i_1!\, i_2! \cdots i_m!} p_1^{i_1}\, p_2^{i_2} \cdots p_m^{i_m}$, as the event points in favour of $A_{i_1\, i_2 \cdots i_m}$ are MEE.

WORKED-OUT EXAMPLES

Ex. 1. A man takes a step forward with probability 0.7 and backward with probability 0.3. Find the probability that at the end of 13 steps he is 3 step forward from the starting point.

Solution. Let a step forward is taken as success and a step backward as failure. $\therefore$ $p =$ probability of success $= 0.7$ and $q =$ probability of failure $= 0.3$.
$p+q = 0.7+0.3 = 1$. So, at every attempt he is at a step forward or a step backward from the present state. At the end of 13 steps he is three steps forward from the initial position. This can happen when he faces 8 successes and 5 failures.
Hence, the required probability $= \binom{13}{8}(0.7)^8(0.3)^{13-8} = \binom{13}{5}(0.7)^8(0.3)^5$.
Note: If instead of 3 steps forward it is given that 3 steps away from the starting point, then the event is either 8 successes, 5 failures or 5 successes, 8 failures.
$\therefore$ required probability $= \binom{13}{8}(0.7)^8(0.3)^5 + \binom{13}{5}(0.7)^5(0.3)^8$.

Ex. 2. An even number of cards are drawn from a well-shuffled full pack of playing cards. Find the probability that half of them are black. [C.H. 2006]
Solution. Let S be the event space of the RE 'an even number of cards are drawn'.
Then, $n(S) = {}^{52}C_2 + {}^{52}C_4 + \cdots + {}^{52}C_{52} = \frac{1}{2}\{(1-1)^{52} + (1+1)^{52}\} - ({}^{50}C_0) = 2^{51} - 1$.
$n(A) =$ number of ways of drawing $2n$ cards among, which n are black and n are white ($n = 1, 2, \cdots, 26$).

$$= \sum_{n=1}^{26} {}^{26}C_{2n} \times {}^{26}C_{2n} = \binom{26}{1}^2 + \binom{26}{2}^2 + \cdots + \binom{26}{26}^2, \ \binom{n}{r} = {}^nC_r$$

$$= \text{term independent of } x \text{ in } (1+x)^{26}\left(1+\frac{1}{x}\right)^{26} - \binom{26}{0}^2$$

$$= \text{term independent of } x \text{ in } \frac{(1+x)^{52}}{x^{26}} - 1 = \binom{52}{26} - 1 = \frac{52!}{(26!)^2} - 1.$$

$$\therefore \text{ required probability} = \frac{n(A)}{n(S)} \text{ [by classical rule]} = \frac{\frac{52!}{(26!)^2} - 1}{2^{51} - 1}.$$

Ex. 3. If n coins are distributed among m beggars at random, what is the probability that a particular beggar will get exactly k coins?
Solution. We apply binomial law to solve the problem. A coin distributed to the particular beggar is taken as success, otherwise failure. So, $p =$ probality of success $= 1/m$.
So, we are to find the probability of exactly k successes out of n trials.
$\therefore$ required probability $= \binom{n}{k}\left(\frac{1}{m}\right)^k\left(1-\frac{1}{m}\right)^{n-k} = \binom{n}{k}\frac{(m-1)^{n-k}}{m^n}$.
Note: If we are to find the probability that one of the beggars will get exactly k coins, the required probability $= \binom{m}{1}\binom{n}{k}\frac{(m-1)^{n-k}}{m^n} = \binom{n}{k}\frac{(m-1)^{n-k}}{m^{n-1}}$.

Ex. 4. Show that the probability of obtaining no head in an infinite sequence of independent tosses of a coin is zero. [C.H. 1991]
Solution. First we consider n independent tosses of a fair coin. If A be the event that all the outcomes are tails, then

$$P(A) = \tfrac{1}{2}\cdot\tfrac{1}{2}\cdots \text{ to } n \text{ factors} = \tfrac{1}{2^n}.$$

Now, as $n \to \infty$, $\frac{1}{2^n} \to 0$. $\therefore$ $\lim_{n\to\infty} P(A) = 0$.

Ex. 5. The probability that a man hitting a target is 1/3. How many times must he fire so that the probability of hitting the target at least once is more than 0.9.
Solution. The trials are Bernoullian. If hitting the target is a success (S), then $P(S) = 1/3 = p$, say.
$\therefore$ if A_i be the event of i successes out of n trials, then by binomial law

$$P(A_i) = \binom{n}{i}p^i(1-p)^{n-i} = \binom{n}{i}\left(\tfrac{1}{3}\right)^i\left(\tfrac{2}{3}\right)^{n-i}$$

$\therefore$ probability of no success $= P(A_0) = \binom{n}{0}\left(\frac{1}{3}\right)^0\left(\frac{2}{3}\right)^n = \left(\frac{2}{3}\right)^n$.

$\therefore$ probability of at least one success (hitting) in n trials $= 1 - \left(\frac{2}{3}\right)^n$.

By the given condition, $1 - \left(\frac{2}{3}\right)^n > 0.9$, or, $\left(\frac{2}{3}\right)^n < 0.1$,

or, $\left(\frac{3}{2}\right)^n > 10$ $\therefore$ $n \log\frac{3}{2} > \log 10$, as log is an increasing function.

or, $n > \frac{1}{\log 3/2} \simeq 5.68$.

$\therefore$ the number of times, he has to fire for at least one hit having probability more than 0.9 is 6.

Ex. 6. A fair die is rolled 23 times in succession. Find the probability of obtaining face 6 three times, face 5 eleven times, face 4 twice, face 3 four times, face 2 twice and face 1 once.

Solution. Since, die is fair. $\therefore$ probability of each face is 1/6. Hence, the required probability by multinomial law is

$$\frac{23!}{3!\,11!\,2!\,4!\,2!\,1!}\left(\frac{1}{6}\right)^3\left(\frac{1}{6}\right)^{11}\left(\frac{1}{6}\right)^2\left(\frac{1}{6}\right)^4\left(\frac{1}{6}\right)^2\left(\frac{1}{6}\right)^1 = \frac{23!}{24\times 24\times 11!}\left(\frac{1}{6}\right)^{23} = 0.0000014.$$

Ex. 7. The probability that a college student will be a graduate is 0.4. Determine the probability that out of 15 college students (i) none (ii) one (iii) at least one (iv) exactly 5 will be a graduate.

Solution. If being a graduate is success, probability of success for each student is $p = 0.4$.

(i) Probability that none will be a graduate [i.e., no success] $= \binom{15}{0}(0.4)^0(1-0.4)^{15}$, by binomial law $= (0.6)^{15} = 0.00047$.

(ii) Probability that one will be a graduate [i.e., one success] $= \binom{15}{1}(0.4)^1(1-0.4)^{15-1} = 15 \times 0.4 \times (0.6)^{14} = 0.0047$.

(iii) Probability that at least one will be a graduate $= 1 - 0.00047 = 0.9995$.

(iv) Probability that exactly 5 will be a graduate $= \binom{15}{5}(0.4)^5(0.6)^{10} = 0.186$.

Ex. 8. What is the probability of throwing a sum of at least 8 exactly 4 times in 7 throws of 2 dice?

Solution. This a compound RE. Total number of points in the event space $= 6 \times 6 = 36$.

At least 8 means *either* 8 *or* 9 *or* 10 *or* 11 *or* 12.

Event	Event points	Number of points	Total number of favourable points
8	(2,6),(3,5),(4,4),(5,3),(6,2)	5	
9	(3,6),(4,5),(5,4),(6,3)	4	
10	(4,6),(5,5),(6,4)	3	
11	(5,6),(6,5)	2	15
12	(6,6)	1	

$\therefore$ $p =$ probability of success $= \frac{15}{36} = \frac{5}{12}$, by classical rule.

Required probability $= P(A_4) = \binom{7}{4}\left(\frac{5}{12}\right)^4\left(1-\frac{5}{12}\right)^{7-4} = \binom{7}{3}\left(\frac{5}{12}\right)^4\left(\frac{7}{12}\right)^3$.

Ex. 9. A coin is tossed repeatedly, show that the probability of getting m heads before n tails is

$$\frac{1}{2^{m+n-1}}\sum_{i=m}^{m+n-1}\binom{m+n-1}{i}.$$

Solution. The event will occur, if at least m heads occur in $(m+n-1)$ trials.

$$\therefore \text{ required probability} = \binom{m+n-1}{m}\left(\frac{1}{2}\right)^m\left(\frac{1}{2}\right)^{n-1} + \binom{m+n-1}{m+1}\left(\frac{1}{2}\right)^{m+1}\left(\frac{1}{2}\right)^{n-2} + \cdots + \binom{m+n-1}{m+n-1}\left(\frac{1}{2}\right)^{m+n-1}$$

$$= \frac{1}{2^{m+n-1}}\sum_{i=m}^{m+n-1}\binom{m+n-1}{i}.$$

Ex. 10. A and B throw a pair of fair dice. A wins, if he throws 6 before B throws 7 and B wins, if he throws 7 before A throws 6. If A begins, what is the chance of B's winning? Also, find the chance of B's winning when B begins.

Solution. p_i = probability of throwing 6 with a pair of dice $= \frac{5}{36}$ as the 5 elementary events (1,5), (2,4), (3,3), (4,2) and (5,1) are in favour out of 36 elementary events.

p_2 = probability of throwing 7 with a pair of dice $= \frac{6}{36}$ as the 6 elementary events (1,6), (2,5), (3,4), (4,3), (5,2) and (6,1) are in favour.

$\therefore$ B's chance of winning when A starts $= \bar{p}_1 p_2 + \bar{p}_1\bar{p}_2\bar{p}_1 p_2 + \bar{p}_1\bar{p}_2\bar{p}_1\bar{p}_2\bar{p}_1 p_2 + \cdots \infty$

$$= \frac{\bar{p}_1 p_2}{1-\bar{p}_1\bar{p}_2} = \frac{\frac{31}{36}\times\frac{6}{36}}{1-\frac{31}{36}\times\frac{30}{36}} = \frac{186}{366} = \frac{31}{61}.$$

B's chance of winning, when B starts $= p_2 + \bar{p}_2\bar{p}_1 p_2 + \bar{p}_2\bar{p}_1\bar{p}_2\bar{p}_1 p_2 + \cdots \infty$

$$= \frac{p_2}{1-\bar{p}_1\bar{p}_2} = \frac{\frac{6}{36}}{1-\frac{31}{36}\times\frac{30}{36}} = \frac{6\times 36}{366} = \frac{36}{61}.$$

Ex. 11. A and B play a game, which must be either win or loss. If the probability that A will win any game is p, show that the probability that A wins m games before B wins n games is

$$\sum_{i=m}^{m+n-1} \binom{m+n-1}{i} p^i (1-p)^{m+n-1-i} = \frac{\int_0^p x^{m-1}(1-x)^{n-1}\,dx}{\int_0^1 x^{m-1}(1-x)^{n-1}\,dx}.$$ [C.H. 2004]

[**Hints.** Follow worked-out ex. 9 for first part and for second part integrate numerator by parts and use $B(m,n) = \int_0^1 x^{m-1}(1-x)^{n-1}\,dx = \frac{\Gamma(m)\Gamma(n)}{\Gamma(m+n)}$, etc.]

Ex. 12. A and B play a game in which A's chance of winning is 2/3. In a series of 8 games what is the chance that A will win at least 6 games. [C.H. 2004]

[**Hints.** Probability of A's winning = 2/3. $\therefore$ B's chance of winning is $1-\frac{2}{3} = 1/3$.

$\therefore$ required probability $= \binom{8}{6}\left(\frac{2}{6}\right)^6\left(\frac{1}{3}\right)^2 + \binom{8}{7}\left(\frac{2}{3}\right)^7\left(\frac{1}{3}\right) + \binom{8}{8}\left(\frac{2}{3}\right)^8$.]

EXERCISE

1. There are n letters and to each letter there is a corresponding envelope. Letters are placed at random in the envelopes. What is the probability that (i) no letter is placed in the right envelope (ii) exactly r letters are placed in the right envelopes.
$\left(\text{Ans. (i) } \frac{1}{2!} - \frac{1}{3!} + \frac{1}{4!} - \cdots + (-1)^n \frac{1}{n!} \text{ (ii) } \frac{1}{r!}\left(\frac{1}{2!} - \frac{1}{3!} + \frac{1}{4!}\right) - \cdots + (-1)^{n-r}\frac{1}{(n-r)!}\right)$
2. A coin is tossed $(m+n)$ times when $m > n$. Show that (i) the probability of getting at least m consecutive tails is $\frac{n+2}{2^{m+1}}$ (ii) the probability of getting exactly m consecutive tails is $\frac{n+3}{2^{m+1}}$.
3. A and B throw a pair of unbiased dice. A wins if he throws a sum of 8 before B throws 9. What is the chance of B's winning, when A starts the game. (*Ans.* 0.41)
4. If n biscuits are distributed at random among N beggars, what is the chance that a particular beggar receives $r(< n)$ biscuits? $\left(\text{Ans. } \binom{n}{r}\frac{(N-1)^{n-r}}{N^n}\right)$
[C.H. 2005]
5. A bag contains 5 red, 3 white, 6 black and 4 green balls. A ball is drawn and replaced after noting its colour. Such an operation is done 10 times. Show that among the 10 noted colours, 2 will be red, 2 white, 3 black and 3 green has the probability $\frac{10!}{2!\,2!\,3!\,3!}\left(\frac{5}{18}\right)^2\left(\frac{1}{6}\right)^2\left(\frac{1}{3}\right)^3\left(\frac{2}{9}\right)^3$.
6. Out of 101 tickets consecutively numbered, three are drawn at random. Find the probability that the numbers on them are in geometrical progression. (*Ans.* 0.0032)
7. A machine produces 5% defective product. From the productions 3 articles are chosen at random. What is the probability that one of them is defective. $\left(\text{Ans. } \binom{3}{1}\left(\frac{1}{20}\right)^1\cdot\left(\frac{19}{20}\right)^2\right)$

8. In a Bernoullian sequence of n trials with probability of success p in each trial, show that the probability of at most k successes is given by $\dfrac{\displaystyle\int_0^{1-p} x^{n-k-1}(1-x)^k\,dx}{\displaystyle\int_0^1 x^{n-k-1}(1-x)^k\,dx}$.

9. k numbers are selected at random from the set $\{1,2,\cdots,10\}$, repetitions being allowed, show that the probability that the product of the numbers is of the form $3n-1$ or $3n+1$ (n is a positive integer or zero) is $\left(\frac{7}{10}\right)^k$.
[Hints. The product of k numbers must be one of the forms $3n-1$ or $3n$ or $3n+1$ (n being a non-negative integer).
Then, the product is of the form $3n$; it contains a factor multiple of 3 (i.e., 3, 6 or 9). Let a selection of multiple of 3 be success (S), then $P(S)=\frac{3}{10}=p$, say. The product is either of the form $3n-1$ or $3n+1$, there is no success in the selection of k numbers.
$\therefore$ required probability $=\binom{k}{0}p^0(1-p)^{k-0}=\left(1-\frac{3}{10}\right)^k=\left(\frac{7}{10}\right)^k$.
Note: Product multiple of 3, i.e., of the form $3n$ is $1-\left(\frac{7}{10}\right)^k$]

10. Two players A and B have n fair coins each. They toss their coins simultaneously. What is the probability that both of them face equal numbers of tails.
[Hints. Both may face either no tail or one tail or two tails, ... or all the n tails. He is tossing independently. Hence, the required probability, by binomial law, is given by

$$\binom{n}{0}\left(\frac{1}{2}\right)^n\cdot\binom{n}{0}\left(\frac{1}{2}\right)^n+\binom{n}{1}\left(\frac{1}{2}\right)^n\cdot\binom{n}{1}\left(\frac{1}{2}\right)^n+\cdots+\binom{n}{n}\left(\frac{1}{2}\right)^n\binom{n}{n}\left(\frac{1}{2}\right)^n$$

$$=\frac{1}{2^{2n}}\left\{\binom{n}{0}^2+\binom{n}{1}^2+\cdots+\binom{n}{n}^2\right\}$$

$$=\frac{1}{2^{2n}}\times\text{term independent of } x \text{ in } (1+x)^n\left(1+\frac{1}{x}\right)^n$$

$$=\frac{1}{2^{2n}}\times\text{term independent of } x \text{ in } \frac{(1+x)^n}{x^n}$$

$$=\frac{1}{2^{2n}}\binom{2n}{n}=\frac{2n!}{2^{2n}(n!)^2}]$$

11. Suppose, a special unbiased die is made with 10 faces. Let it be thrown repeatedly up to 7 times. What is the probability that face 6 appears all the times? (*Ans.* $1/10^7$)

12. In a Poissonian 10 repeated trials, let p_i and q_i are respectively the probability of success and failure in the ith trial ($i=1,2,\cdots,10$); show that the probability of r failures ($0\le r\le 10$) is the coefficient of x^r in the product $(p_1+q_1x)(p_2+q_2x)\cdots(p_{10}+q_{10}x)$.
Note: A large number of interesting problems may be solved by binomial law after suitably defining a success.

13. A total of n shells are fired at a target. The probability of the jth shell hitting the target is p_j, $j=1,2,\cdots,n$. Assuming that the n firings are n mutually independent events, find the probability that at least two shells out of n find the target.
[Hints. It is a Poissonian sequence of n trials.
$\therefore$ probability of no success (hitting the target) out of n independent trials = coefficient of x^0 in the product $(p_1x+q_1)(p_2x+q_2)\cdots(p_nx+q_n)=q_1q_2\cdots q_n$.

Probability of one success = coefficient of x^1 in the product $(p_1x+q_1)(p_2x+q_2)\cdots(p_nx+q_n)$ $= p_1q_2q_3\cdots q_n + q_1p_2q_3\cdots q_n + q_2q_2p_3q_4\cdots q_n + \cdots + q_1q_2\cdots q_{n-1}p_n$.

$\therefore$ probability of at least two successes $= 1 - [q_1q_2\cdots q_n + p_1q_2q_3\cdots q_n + q_1p_2q_3\cdots q_n + \cdots + q_1q_2\cdots q_{n-1}p_n.]]$

14. If a die is thrown k times, show that the probability of an even number of sixes is $\frac{1}{2}\left\{1+\left(\frac{2}{3}\right)^k\right\}$.

[C.H. 2002, 2006, 2008]

15. A player tosses a coin and he scores one point for every head turned up and two for every tail. He will play until his score reaches or crosses n.

If p_n is the probability of attaining exactly n points, then show that $p_n = \frac{1}{2}\{p_{n-1}+p_{n-2}\}$.

Hence, find p_n and the limit of p_n as $n \to \infty$.

[**Hints.** Let H = the event that a head turns up, T = the event that a tail turns up and S_n = the event that the total score is n.

Then, S_n occurs exclusively in two ways, viz., (i) In a certain stage score attains $(n-1)$ points and H appears immediately next or (ii) In a certain stage score attains $(n-2)$ points and T appears immediately next. $\therefore S_n = S_{n-1}H + S_{n-2}T$.

$\therefore p_n = P(S_n) = P(S_{n-1}H + S_{n-2}T) = P(S_{n-1})P(H) + P(S_{n-2})P(T)$, as $S_{n-1}H$ and $S_{n-2}T$ are MEE and the trials are independent.

or, $p_n = p_{n-1}\cdot\frac{1}{2} + p_{n-2}\cdot\frac{1}{2} = \frac{1}{2}\{p_{n-1}+p_{n-2}\}$.

$\therefore p_n + \frac{1}{2}p_{n-1} = p_{n-1} + \frac{1}{2}p_{n-2}, p_{n-1} + \frac{1}{2}p_{n-2} = p_{n-2} + \frac{1}{2}p_{n-3}$, etc.

So, $p_n + \frac{1}{2}p_{n-1} = p_2 + \frac{1}{2}p_1 = P(T+HH) + \frac{1}{2}P(H) = \frac{1}{2} + \frac{1}{4} + \frac{1}{2}\cdot\frac{1}{2} = 1$.

$$\therefore p_n = 1 - \tfrac{1}{2}p_{n-1} = 1 - \tfrac{1}{2}\left(1 - \tfrac{1}{2}p_{n-2}\right) = 1 - \tfrac{1}{2} + (-1)^2\cdot\tfrac{1}{2^2}\left(1 - \tfrac{1}{2}p_{n-3}\right)$$
$$= 1 - \tfrac{1}{2} + \tfrac{1}{2^2} + (-1)^3\cdot\tfrac{1}{2^3}\left(1 - \tfrac{1}{2}p_{n-4}\right) = 1 - \tfrac{1}{2} + \tfrac{1}{2^2} - \tfrac{1}{2^3} + (-1)^4\cdot\tfrac{1}{2^4}p_{n-4}$$
$$= \cdots = 1 - \tfrac{1}{2} + \tfrac{1}{2^2} - \tfrac{1}{2^3} + \cdots + (-1)^{n-2}\cdot\tfrac{1}{2^{n-2}} + (-1)^{n-1}\cdot\tfrac{1}{2^{n-1}}p_{n-(n-1)}$$
$$= \frac{1-\left(-\frac{1}{2}\right)^{n-1}}{1+\frac{1}{2}} + (-1)^{n-1}\cdot\tfrac{1}{2^{n-1}}p_1 = \tfrac{2}{3}\left\{1 + (-1)^n\cdot\tfrac{1}{2^{n-1}} + \tfrac{3}{2}\cdot(-1)^{n-1}\cdot\tfrac{1}{2^n}\right\}\ \left[\because p_1 = P(H) = \tfrac{1}{2}\right]$$
$$= \tfrac{1}{3}\left\{2 + (-1)^n\cdot\tfrac{1}{2^{n-2}} + 3(-1)^{n-1}\cdot\tfrac{1}{2^n}\right\} = \tfrac{1}{3}\left\{2 + \tfrac{(-1)^n}{2^n}(4-3)\right\} = \tfrac{1}{3}\left\{2 + (-1)^n\cdot\tfrac{1}{2^n}\right\},$$

$\therefore \lim_{n\to\infty} p_n = \frac{2}{3}$]

16. What is the probability of obtaining a multiple of 3 twice in a throw of 6 dice? $\left(\textit{Ans. } \binom{6}{2}\left(\frac{1}{3}\right)^2\left(\frac{2}{3}\right)^4\right)$

[C.H. 2006]

17. From a pack of 52 cards, an even number of cards are drawn. Find the probability that these consist half of red and half of black.

[C.H. 2006]

[**Hints.** Number of ways of drawing even number of cards $= \binom{52}{2} + \binom{52}{4} + \cdots + \binom{52}{52} = 2^{51} - 1$.

Number of favourable cases $= \binom{26}{1}\binom{26}{1} + \binom{26}{2}\binom{26}{2} + \cdots + \binom{26}{26}\binom{26}{26} = \binom{26}{1}^2 + \binom{26}{2}^2 + \cdots + \binom{26}{26}^2 = \frac{(2\times 26)!}{(26!)^2} - 1$ $= \binom{52}{26} - 1$.

$\therefore$ required probability $= \frac{\binom{52}{26}-1}{2^{51}-1}$.]

Chapter 4

One-dimensional Probability Distributions

4.1 Mapping or function

A function f from a non-empty set A into a non-empty set B (which may be A as well) is a rule of correspondence that assigns to each element $x \in A$, a uniquely determined element $f(x)$ in B. We also call it a mapping from A into B and write $f: A \to B$ (read: f maps A into B).

More mathematically a function f from A into B is a subset of their Cartesian product $A \times B$, i.e., $f \subset A \times B$ such that each element x of A appears exactly once in f as the first element.

A is called the *domain* and B is the co-domain of the function. The set $\{f(x) : x \in A\}$ is called the *range* of the function and is denoted by $f(A)$ and obviously $f(A) \subseteq B$.

Whenever $f(A) = B$, the function is surjective or onto. Whenever $x_1, x_2 \in A$ and $x_1 \neq x_2 \Rightarrow f(x_1) \neq f(x_2)$, i.e., distinct element maps into distinct element, the function is then called *injective* or *into*.

f is bijective, if it is both injective and surjective.

When f is surjective, each element of B has at least one pre-image.

When f is injective, each element of B has at most one pre-image.

In case of bijection each element of B has exactly one pre-image.

Sum function

Let $f_1 : A \to B, f_2 : A \to B, \cdots, f_n : A \to B$ be n function, then their sum $f_1 + f_2 + \cdots + f_n : A \to B$ is also a function defined by $(f_1 + f_2 + \cdots + f_n)(x) = f_1(x) + f_2(x) + \cdots + f_n(x), x \in A$ and $f_1 + f_2 + \cdots + f_n$ is a sum function, i.e., if $f_i : A \to B, i = 1, 2, \cdots, n$, then $\sum_{i=1}^{n} f_i : A \to B$ is defined by $\left(\sum_{i=1}^{n} f_i\right)(x) = \sum_{i=1}^{n} f_i(x), x \in A$.

Product function

Product of the functions $f_i : A \to B, i = 1, 2, \cdots, n$ is $\prod_{i=1}^{n} f_i : A \to B$ defined by

$$\left(\prod_{i=1}^{n} f_i\right)(x) = \prod_{i=1}^{n} f_i(x), x \in A, \text{ may be called } \textit{product function.}$$

Composite function

Let $f : A \to B$ and $g : B \to C$ be two functions, then $f \circ g : A \to C$ defined by $f \circ g(x) = f(g(x))$ is known as *composite function*. It can be easily generalised.

Piecewise continuous function

Let $f : A \to B$ be a function when A is a closed and bounded interval, i.e., $f : [a, b] \to B$.

Let $P = \{a = x_0, x_1, x_2, \cdots, x_{r-1}, x_r, \cdots, x_n = b\}$ be a partition, i.e., $a = x_0 < x_1 < x_2 \cdots < x_{r-1} < x_r \cdots < x_n = b$.

Let f be continuous on $[a, b]$ except at the point $x_1, x_2, \cdots, x_{n-1}$, then f is said to be piecewise continuous on $[a, b]$, i.e., f is continuous on $[a, x_1) \cup (x_1, x_2) \cdots \cup (x_{r-1}, x_r) \cdots \cup (x_{n-1}, b]$.

Step function

Let $f:[a,b]\to\mathbb{R}$, where $[a,b]$ is a closed bounded interval of the real line, be a piecewise continuous function and possible points of discontinuities are $x_1, x_2, \cdots, x_{n-1}$, where $a < x_1 < \cdots < x_{n-1} < x_n = b$.

Let $k_0, k_1, \cdots, k_r, \cdots, k_n$ be all positive constants.

Suppose, f is defined as follows:

$$\begin{aligned} f(x) &= k_0 && \text{when } a \le x < x_1 \\ &= k_0 + k_1 && \text{when } x_1 \le x < x_2 \\ &\cdots\cdots\cdots\cdots \\ &= k_0 + k_1 + \cdots + k_{r-1} && \text{when } x_{r-1} \le x < x_r \\ &\cdots\cdots\cdots\cdots \\ &= k_0 + k_1 + \cdots + k_{n-1} && \text{when } x_{n-1} \le x < x_n \\ &= k_0 + k_1 + \cdots + k_{n-1} + k_n && \text{when } x = b. \end{aligned}$$

Such a piecewise continuous function is called a *step function.* A graph of such a function looks like as shown in the right-hand side.

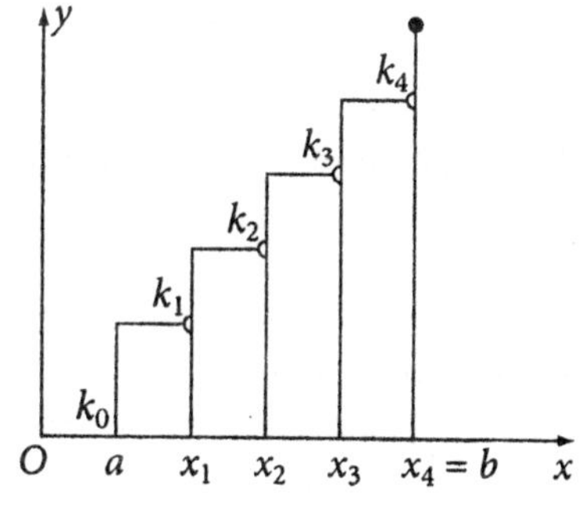

Figure for $n = 4$

4.2 Random variable (RV)

Let E be a Random Experiment (RE) and S be its event space. Let U be an arbitrary event point of S. Let $X: S \to \mathbb{R}$ be a function, i.e., X assigns each event point of S to a definite real number by some rule of correspondence. Then, this real valued function X defined on S is called a *random variable* or *variate.* The range of the function $X(S)$ is called the *spectrum* of X. If $X(S)$ consists of discrete set of values, then the RV is discrete RV and if $X(S)$ takes up all values from a set of intervals, then the RV X is continuous.

For example, let us consider a compound RE E^4, where $E^4 \equiv$ Bernoulli sequence of 4 independent trials with success (s) and failure (f).

Then, its event space (S) consists of 2^4, i.e., 16 event points, viz.,

$$\begin{aligned} S = \{ & U_1 = (s,s,s,s),\ U_2 = (s,s,s,f),\ U_3 = (s,s,f,s),\ U_4 = (s,f,s,s), \\ & U_5 = (f,s,s,s),\ U_6 = (f,f,s,s),\ U_7 = (f,s,f,s),\ U_8 = (f,s,s,f), \\ & U_9 = (s,f,f,s),\ U_{10} = (s,f,s,f),\ U_{11} = (s,s,f,f),\ U_{12} = (f,f,f,s), \\ & U_{13} = (f,f,s,f),\ U_{14} = (f,s,f,f),\ U_{15} = (s,f,f,f),\ U_{16} = (f,f,f,f)\} \end{aligned}$$

Let a RV X be defined on S by $X = X(U_i) =$ number of success in U_i, $i = 1, 2, \cdots, 16$.

$\therefore$ $(X=4) \equiv U_1$, $(X=3) \equiv U_2 + U_3 + U_4 + U_5$, $(X=2) \equiv U_6 + U_7 + U_8 + U_9 + U_{10} + U_{11}$, $X(=1) \equiv U_{12} + U_{13} + U_{14} + U_{15}$ and $(X=0) \equiv U_{16}$.

$\therefore$ the spectrum of X has 5 values, viz., $0, 1, 2, 3, 4$. By the classical rule $P(X=0) = 1/16$, $P(X=1) = 4/16$, $P(X=2) = 6/16$, $P(X=3) = 4/16$ and $P(X=4) = 1/16$.

Note that the events $X=0$, $X=1$, $X=2$, $X=3$ and $X=4$ are mutually exclusive and exhaustive.

$P(X=0) + P(X=1) + P(X=2) + P(X=3) + P(X=4) = 1$.

So, the total probability 1 is distributed among the mutually exclusive and exhaustive events $X = i$, $i = 0, 1, 2, 3, 4$.

$\therefore$ the probability distribution of the discrete RV X is given by $X = 0, 1, 2, 3, 4$ and $f_0 = P(X=0) = 1/4$, $f_1 = P(X=1) = 4/16$, $f_2 = P(X=2) = 6/16$, $f_3 = P(X=3) = 4/16$ and $f_4 = P(X=4) = 1/16$.

Let a set A be $\{0, 1\}$.

$\therefore$ $X \in A$ is an event $(X=0) + (X=1)$, i.e., $X \in A \equiv U_{12} + U_{13} + U_{14} + U_{15} + U_{16}$.

When $A = [0, 1]$, then also $X \in A = U_{12} + U_{13} + U_{14} + U_{15} + U_{16}$.

In fact, $X \in A$ consists of all those event points U_i for which $X(U_i) \in A$.

$X \in (a,b]$, i.e., $a < x \le b$ is the event, which consists of all those event points U_i for which $a < X(U_i) \le b$.

Usefulness of half open interval

If we consider closed interval only to represent an event and $[a,b], [b,c]$ be two events, then $[a,b] + [b,c] = [a,c]$.

Observe that the events on the LHS are not mutually exclusive as the $X = b$ in common, if there be any such event.

If we consider open interval only, then events (a,b) and (b,c) are MEE but their sum (union) is not the event (a,c) as the event $X = b$ belongs to (a,c), whereas $X = b$ neither belongs to (a,b) nor to (b,c).

When we consider half open interval (which is the same half closed interval or clo-open interval) like $(a,b]$ and $(b,c]$, then they are not only MEE but also $(a,b] + (b,c] = (a,c]$.

The event $(-\infty, a] = -\infty < X \le a$ and the event $[a, \infty) = a \le X < \infty$.

The event $(-\infty, \infty)$, i.e., $-\infty < X < \infty$ is always the certain event for every RV X.

So, $P(-\infty < X < \infty) = 1$.

4.3 Distribution function

Distribution function of a RV X completely determines the probability distribution of X. Probability distribution of RV X is the distribution of the total probability 1 to the possible events associated to X.

DEFINITION. The distribution function of a RV X is a function of a real variable x, which X takes up. It is denoted by $F_x(x)$ and is defined by $F_x(x) = P(-\infty < X \le x)$, $-\infty < x < \infty$.

The use of the subscript x in the distribution function F_x is to represent that F_x is the distribution function of X.

Such a representation of a distribution function is helpful and essential when we are dealing with two or more RV at the same time.

Suppose, we are dealing with RV X and Y. The notations of the distribution functions F_x and F_y help us in two ways: (i) we can easily understand, which one is the distribution function for X, and, which one for Y (ii) F_x and F_y represent different functions in general. Different symbols like F, G need not be used.

Important properties of a distribution function

(a) F_x is non-decreasing, i.e., if $a, b \in \mathbb{R}$ and $a < b, F_x(b) \ge F_x(a)$.

Proof. Clearly, $-\infty < X \le a + a < X \le b = -\infty < X \le b$. The events on the LHS are MEE.

$\therefore$ $P(-\infty < X \le a + a < X \le b) = P(-\infty < X \le b)$ or, $P(-\infty < X \le a) + P(a < X \le b) = P(-\infty < X \le b)$ or, $F_x(a) + P(a < x \le b) = F_x(b)$.

$\therefore$ $F_x(b) - F_x(a) = P(a < X \le b) \ge 0$, as the probability of any event is greater or equal to zero by the axiom of probability.

$\therefore$ $F_x(b) \ge F_x(a)$. So, F_x is non-decreasing. Note that $P(a < X \le b) = F_x(b) - F_x(a)$.

(b) As $F_x(x) = P(-\infty < X \le x)$, $\forall\, x \in (-\infty, \infty)$ and $0 \le P(-\infty < X \le x) \le 1$, $\forall\, x \in (-\infty, \infty)$.

$\therefore$ $0 \le F_x(x) \le 1$, $\forall\, x \in (-\infty, \infty)$.

(c) $F_x(-\infty) = 0$ and $F_x(\infty) = 0$. Consider the sequence of events $\{A_n\}_n$ defined by $A_n = (-\infty < X \le -n)$.

$\therefore$ $\lim A_n = 0$, an impossible event.

$\therefore$ $P(\lim A_n) = P(0) = 0$ or, $\lim(P(A_n)) = 0$, as $\{A_n\}_n$ is monotone contracting

or, $\lim\{P(-\infty < X \le -n)\} = 0$ or, $\lim F_x(-n) = 0$ or, $F_x(-\infty) = 0$.

Again, consider the sequence of events $\{B_n\}_n$ defined by $B_n = (-\infty < X \le n)$. $\therefore$ $\lim B_n = S$, the certain event.

$\therefore$ $P(\lim B_n) = P(S) = 1$ or, $\lim P(B_n) = 1$, as $\{B_n\}_n$ is monotone expanding

or, $\lim P(-\infty < X \le n) = 1$ or, $\lim F_x(n) = 1$ or, $F_x(\infty) = 1$.

(d) $F_x(a) - F_x(a-0) = P(X = a)$, $\forall\, a \in \mathbb{R}$, i.e., F_x is discontinuous from left at every step point or spectrum point a. [A point a is a step point, **when $P(X = a) > 0$**]

To prove we consider the sequence of contracting events $\{A_n\}_n$ defined by $A_n = \left(a - \frac{1}{n} < X \le a\right)$.
$\therefore \lim A_n = (X = a)$.
$\therefore P(A_n) = P\left(a - \frac{1}{n} < X \le a\right) = F_x(a) - F_x\left(a - \frac{1}{n}\right)$.
$\therefore \lim P(A_n) = F_x(a) - F_x(a-0)$ or, $P(\lim A_n) = F_x(a) - F_x(a-0)$, as $\{A_n\}_n$ is monotone contracting
or, $P(X = a) = F_x(a) - F_x(a-0)$.

(e) F_x is continuous from right at every point, i.e., $F_x(a+0) = F_x(a)$, $\forall\, a \in \mathbb{R}$.
To show this property we consider the monotone contracting sequence of events $\{A_n\}_n$ defined by

$$A_n = \left(a < X \le a + \tfrac{1}{n}\right).$$

$\lim A_n = 0$, an impossible event.
$\therefore P(\lim A_n) = P(0) = 0$ or, $\lim P(A_n) = 0$, as $\{A_n\}_n$ is monotone sequence
or, $\lim P\left(a < X \le a + \frac{1}{n}\right) = 0$ or, $\lim \left\{F_x\left(a + \frac{1}{n}\right) - F_n(a)\right\} = 0$ or, $F_x(a+0) - F_x(a) = 0$,
i.e., $F_x(a+0) = F_x(a)$.
Thus about the distribution function F_x we have the following:
(i) $F_x : [-\infty, \infty] \to [0,1]$ with $F_x(-\infty) = 0$, $F_x(\infty) = 1$,
(ii) it is monotonically non-decreasing,
(iii) it is continuous from the right at every point,
(iv) it is discontinuous from the left at a point $x = a$, when $P(X = a) > 0$.
It is a jump discontinuity and height of jump is $P(X = a)$. Conversely, a function F_x satisfying the above properties is a probability distribution function.

Distribution of a RV X is uniquely determined by the distribution function F_x of X.

Distribution curve

The curve $y = F_x(x)$ is called the *distribution curve* of a RV X.

It is always confined between $y = 0$ and $y = 1$. When X is a continuous RV, the distribution curve is continuous throughout its domain $(-\infty, \infty)$. The lines $y = 0$ and $y = 1$ are asymptotes of the curve $y = F_n(x)$.

Discrete distribution

When the spectrum of X is a at most countable (i.e., either finite or countably infinite) the probability distribution of X is discrete type.

The discrete distribution is defined by stating the spectrum of X and the probability at each spectrum point. It is necessary that the sum of the probabilities at each such spectrum point is 1.

Let the spectrum of X consists of the points $x_i, i = 0, \pm 1, \pm 2, \cdots$.

Let $f_i = P(X = x_i) > 0$, $\forall\, i$, when $\sum_{i=-\infty}^{\infty} f_i = 1$.

Then, the distribution function is defined as $F_x(x) = P(-\infty < X \le x) = P\left(\sum_{\alpha=-\infty}^{i} (X = x_\alpha)\right)$, when $x_i \le x < x_{i+1}$, i.e., $F_x(x) - \sum_{\alpha=-\infty}^{i} P(X = x_\alpha)$, as the events are MEE

$$= \sum_{\alpha=-\infty}^{i} f_\alpha,\ i \in \mathbb{Z}.$$

Thus F_x is a step function with steps of height f_α at x_α. Conversely, when the distribution function F_x of a RV X is a step function, we conclude that the distribution of X is a discrete distribution.

$f_i (i = 0, \pm 1, \pm 2, \cdots)$ is called the *pmf*.

Note 1. At every non-spectrum point the probability is zero. Let b be a non-spectrum point. Then $P(X = b) = F_x(b) - F_x(b-0) = 0$, as F_x is continuous at b.

2. $P(c < X \le d) = \sum_{c < x_\alpha \le d} F_\alpha =$ sum of the probabilities at every step points $x_\alpha \in (c, d]$.

Explanation (Analytically).
Let $x_i \le c < x_{i+1}$ and $x_j \le d < x_{j+1}$.

Then, $P(c < X \le d) = F_x(d) - F_n(c) = \sum_{\alpha-\infty}^{j} f_\alpha - \sum_{\alpha=-\infty}^{i} f_\alpha = \sum_{\alpha=i+1}^{j} f_\alpha = \sum_{c<x_\alpha \le d} f_\alpha \; [f_\alpha = P(X = x_\alpha)]$

Explanation (Graphically).

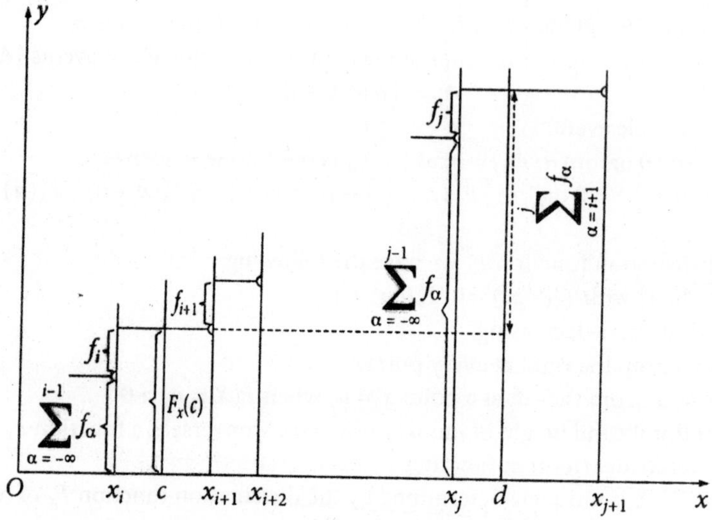

Graph: $y = F_x(x)$

Probability diagram for a discrete distribution

Probability distribution of a discrete RV X may be visualised and may be understood at a glance by the graphical representation.

First we locate the spectrum points $x_i (i = 0, \pm1, \pm2, \cdots)$ on x-axis and at each point x_i we draw an ordinate equal to f_i in some suitable scale and the resulting diagram is the visual representation of a discrete probability distribution and is known as probability diagram.

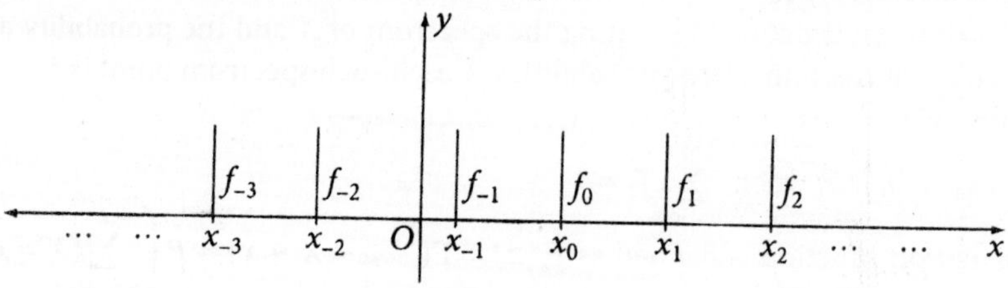

Probability diagram

Illustration. Determine the distribution function and draw the probability diagram of the probability distribution of a discrete RV X whose spectrum is the set $\{0, 1, 2, 3\}$ and $f_0 = 0.2$, $f_1 = 0.3$, $f_2 = 0.3$, $f_3 = 0.2$.

The distribution function F_x of X is given by

$$F_x(x) = \begin{cases} 0, & -\infty < x < 0 \\ 0.2, & 0 \le x < 1 \\ 0.5, & 1 \le x < 2 \\ 0.8, & 2 \le x < 3 \\ 1, & 3 \le x < \infty \end{cases}$$

The probability diagram is given below:

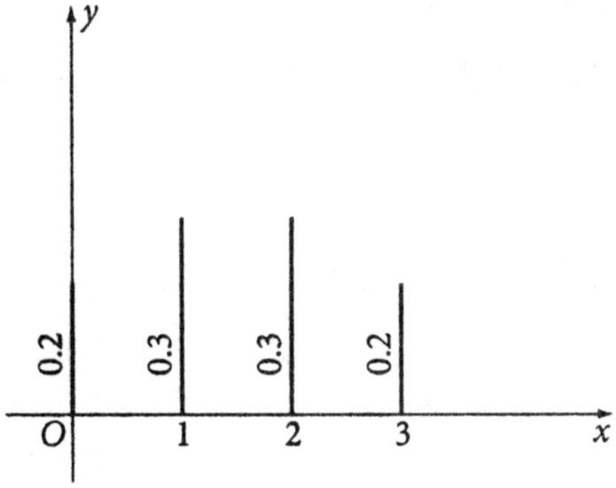

4.4 Useful discrete distributions

I. Binomial distribution

A discrete distribution of a RV X is said to be binomial if the spectrum is the set $\{0,1,2,\cdots,n\}$, i.e., $X=x_i=i$, $i=0(1)n$ and $f_i=P(X=x_i)=P(X=i)=\binom{n}{i}p^i(1-p)^{n-i}$, $i=0(1)n$.

Here $n(n\in\mathbb{N})$ and $p(0<p<1)$ are the two parameters of the distribution.

This is known as *binomial* (n,p) *distribution*. We note that $f_i>0$, $\forall\ i$ and

$$\sum_{i=0}^{n} f_i=\sum_{i=0}^{n}\binom{n}{i}p^i(1-p)^{n-i}=\{p+(1-p)\}^n=1.$$

This is the most useful distribution among the discrete distributions.

The number of success X in a Bernoullian sequence of n independent trials follows binomial distribution with parameters n and p, when p is the probability of success in each trial.

So, a large number of distributions belong to this category.

We draw the figures of distribution function and the probability diagram for binomial (4, 1/2) distribution below.

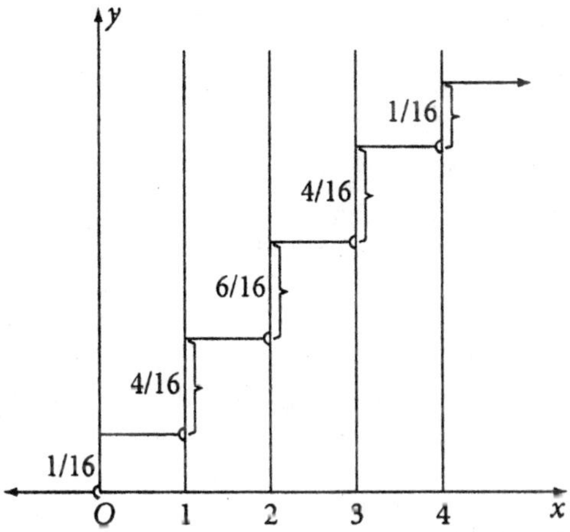

Fig $y=F_x(x)$

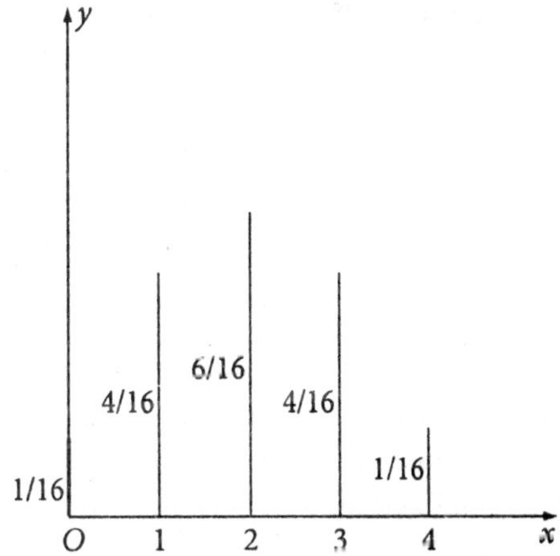

Probability diagram

Observations. 1. The graph of the distribution function F_X shows that it is a step function having steps of height 1/16, 4/16, 6/16, 4/16 and 1/16 at the spectrum points $x=0, x=1, x=2, x=3$ and $x=4$

respectively.

$$F_X(x)=\begin{cases}0, & -\infty<x<0\\ 1/16, & 0\le x<1\\ 5/16, & 1\le x<2\\ 11/16, & 2\le x<3\\ 15/16, & 3\le x<4\\ 1, & 4\le x<\infty\end{cases}$$

2. The probability diagram shows that the greatest probability 6/16 occurs at the step point (spectrum point) $x = 2$. It may be looked upon as the probability of 2 successes out of 4 Bernoullian sequence of independent trials with probability of success 1/2 at each trial. We know that the probability of r successes will be maximum, when r satisfies $np-q\le r\le np+p$, $p+q=1$, i.e., $4\times\frac{1}{2}-\frac{1}{2}\le r\le 4\times\frac{1}{2}+\frac{1}{2}$, i.e., $\frac{3}{2}\le r\le\frac{5}{2}$, where r is a nonnegative integer.

$\therefore$ $r=2$, i.e., probability of 2 successes is the maximum.

II. Negative binomial distribution

Consider an infinite sequence of independent Bernoullian trials with probability of success p at each trial. Let the RV X denotes the number of failures preceding the rth success ($r\ge 1$), then the distribution of X is given by the pmf

$$f_i=P(X=i)=\binom{i+r-1}{i-1}(1-p)^i p^r,\ i=0,1,2,\cdots$$

The distribution of X is known as *negative binomial distribution* with parameters r and $p(0<p<1)$.

Explanation. rth trial is a success and in the remaining $i+r-1$ trials the number of success is $r-1$ and failure is i. The probability of $(r-1)$ successes out of $i+r-1$ independent trials as given by the binomial law $\binom{i+r-1}{r-1}p^{r-1}(1-p)^i$.

$\therefore$ the required probability $f_i=\binom{i+r-1}{r-1}(1-p)^i p^{r-1}\cdot p=\binom{i+r-1}{r-1}(1-p)^i p^r$.

III. Geometric distribution

This is a very particular case of the negative binomial distribution, when $r = 1$. Hence, here X is the number of failures preceding the first success. So, the geometric distribution is given by the pmf $f_i = (1-p)^i p$, $i=0,1,2,\cdots$, or, $f_i=q^i p$, when $p+q=1$ ($q=$ probability of failure at each trial).

IV. Hypergeometric distribution

The probability distribution of a discrete RV X, given by the pmf

$$f_i=P(X=i)=\frac{\binom{M}{i}\binom{N-M}{n-i}}{\binom{N}{n}},\ i=0,1,2,\cdots,\min\{n,M\}.$$

Here N,M and n are three parameters. N,M,n are positive integers with the restrictions $M\le N$ and $n\le N$.

Type of the situations when it will be applied is illustrated below

Consider a basket containing N balls of which M are white and the remaining $N-M$ are black. Consider the RE of drawing n balls one by one from the basket without replacement and X is the RV denoting the number of white balls drawn. Then, the distribution of X is given by the pmf

$$f_i=P(X=i)=\frac{\binom{M}{i}\binom{N-M}{n-i}}{\binom{N}{n}},\ i=0,1,2,\cdots,\min\{n,M\}.$$

Explanation. i white balls out of M may occur in $\binom{M}{i}$ ways and $(n-i)$ black balls out of $(N-M)$ in $\binom{N-M}{n-i}$ ways. As the draws are independent the number of favourable cases is $\binom{M}{i}\binom{N-M}{n-i}$. The total number of ways of drawing n balls out of N balls is $\binom{N}{n}$.

So, by classical rule $f_i = \frac{\binom{M}{i}\binom{N-M}{n-i}}{\binom{N}{n}}$, $i = 0,1,2,\cdots,\min\{n,M\}$.

V. Poisson distribution

Here the spectrum is $\{0,1,2,\cdots,\infty\}$. Poisson distribution is defined by the pmf $f_i = P(X = i) = e^{-\mu}\frac{\mu^i}{i!}$, $i = 0,1,2,\cdots$.

Here $\mu(>0)$ is the only parameter of the distribution.

VI. Uniform distribution

Discrete uniform distribution is very simple and highly applied in Statistics. The spectrum of a RV X having uniform distribution is given by $X = x_1, x_2, \cdots, x_n$ and the probability of $X = x_i$ is denoted by f_{xi} as usual and is defined as

$$f_{xi} = P(X = x_i) = \frac{1}{n},\ i = 1,2,\cdots,n.$$

So, there are n step points, viz., $x_1, x_2, \cdots, x_n$ with equal probability. That is, why it is called *uniform*. The basic need for f_{xi} to be a pmf, viz., $f_{xi} > 0$ and $\Sigma f_{xi} = 1$ are satisfied.

If $x_m \le x \le x_{m+1}$, $m < n$, then $P(X \le x) = m/n$ and $P(X \le x_n) = 1$, following classical law.

$\therefore$ the distribution function is given by

$$F_x(x) = \begin{cases} m/n, & x_m \le x < x_{m+1},\ m < n \\ 1, & x = x_n. \end{cases}$$

Let us consider an unbiased n-faced die and X be a discrete RV denoting the number of a face, then X follows uniform distribution. The distribution of X is given by

$$X = x_i = i \text{ and } f_{xi} = P(X = i) = 1/n,\ i = 1,2,\cdots,n.$$

Following are the graphs for pmf and $y = F_x(x)$ for discrete uniform distribution.

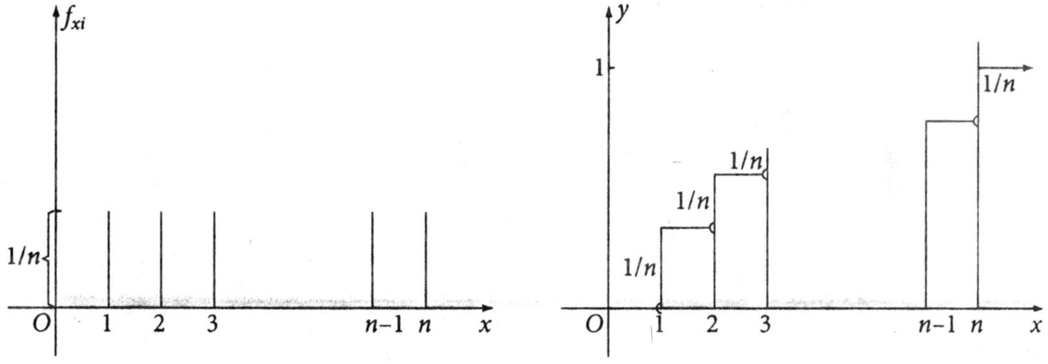

Graph of pmf   Graph of $y = F_x(x)$

Area of application with suitable conditions of Poisson law

Stochastic process. A time dependent RV $X(t)$ is a stochastic process. Let $X(t)$ denotes the number of incidents occurring in a time interval t. Let us assume the following conditions: (i) the probability of occurring one incident in a small interval of time $(t, t+\Delta t)$ is $\lambda\Delta t, \lambda \in \mathbb{R}^+$; (ii) the probability of occurring more than one incident is of order $(\Delta t)^2$, i.e., $0\left[(\Delta t)^2\right]$, which means $\lim_{\Delta t \to 0} \frac{0(\Delta t)^2}{\Delta t} = 0$; (iii) the probability of occurrence of an incident in the time interval $(t, t+\Delta t)$ is independent of the actual time t and also independent of previously occurred incidents.

A stochastic process satisfying the above conditions is called a *Poisson process*.

We propose to prove that the probability distribution of the Poisson process $X(t)$ has a Poisson distribution. That is, the distribution of the RV $X(t)$ denoting the number of incidents in a time interval t satisfying the conditions (i), (ii) and (iii) is a Poisson distribution.

Proof. Let $P_k(t)$ denote the probability of occurring k incidents during the time interval t.
Let the probability of at least one incident during the interval $(t, t+\Delta t) = \mu\Delta t + 0\left[(\Delta t)^2\right]$.
Again, let the probability of more than one incident in the same interval $(t, t+\Delta t) = 0\left[(\Delta t)^2\right]$.
The event of occurring exactly k incidents in time $t+\Delta t$ happens in two mutually exclusive ways, viz.,
1. k incidents occurred during $(0, t)$ and no incident during $(t, t+\Delta t)$.
2. $(k-1)$ incidents during $(0, t)$ and one incident during $(t, t+\Delta t)$.
The probability for first case is $P_k(t)\left[1-\left(\mu\Delta t + 0(\Delta t)^2\right)\right]$, by condition (iii).
The probability for second case is $P_{k-1}(t)(\mu\Delta t)$.

$$\therefore\ P_k(t+\Delta t) = P_k(t)\left[1-\mu\Delta t - 0(\Delta t)^2\right] + P_{k-1}(t)\mu\Delta t$$

$$= P_k(t) - \mu P_k(t)\Delta t - 0(\Delta t)^2 P_k(t) + P_{k-1}(t)\mu\Delta t. \quad (1)$$

$$\therefore\ \lim_{\Delta t\to 0}\frac{P_k(t+\Delta t)-P_k(t)}{\Delta t} = -\mu P_k(t) + \mu P_{k-1}(t) \text{ as } \lim_{\Delta t\to 0}\frac{0(\Delta t)^2}{\Delta t} = 0.$$

$$\therefore\ P_k'(t) = -\mu P_k(t) + \mu P_{k-1}(t),\ k \geq 1. \quad (2)$$

For $k = 0$ eq. (1) becomes

$$P_0(t+\Delta t) = P_0(t)\left\{1-\mu\Delta t - 0(\Delta t)^2\right\} \text{ [as } P_{-1}(t) = \text{probability of } (-1) \text{ incidents in time } t \text{ is } 0]$$

or, $P_0(t+\Delta t) - P_0(t) = -P_0(t)\left\{\mu\Delta t + 0(\Delta t)^2\right\}$

or, $\lim_{\Delta t\to 0}\frac{P_0(t+\Delta t)-P_0(t)}{\Delta t} = -\mu P_0(t)\ \left[\text{as } \lim_{\Delta t\to 0}\frac{0(\Delta t)^2}{\Delta t} = 0\right]$

or, $\frac{P_0'(t)}{P_0(t)} = -\mu.$

Integrating w.r.t. t, $\ln|P_0(t)| = -\mu t + C$, C = constant of integration.
Initial condition is $P_0(0) = 1$, as no incident in no time is the certain event.
$\therefore\ C = 0.\ \therefore\ \ln|P_0(t)| = -\mu t$, or, $|P_0(t)|\, e^{-\mu t}$, or, $P_0(t) = e^{-\mu t}$ as $P_0(t) \geq 0$.
$\therefore$ for $k = 1$, the eq. (2) becomes

$$P_1'(t) = -\mu P_1(t) + \mu e^{-\mu t} \text{ as } P_0(t) = e^{-\mu t} \text{ or, } P_1'(t) + \mu P_1(t) = \mu e^{-\mu t}.$$

This is a linear ODE (ordinary differential equation). $\therefore$ integrating factor (IF) $= e^{\mu\int dt} = e^{\mu t}$.
Multiplying, by integrating factor, $e^{\mu t}P_1'(t) + \mu P_1(t)e^{\mu t} = \mu$ or, $\frac{d}{dt}(e^{\mu t}P_1(t)) = \mu$.
Integrating w.r.t. t, $e^{\mu t}P_1(t) = \mu t + C_1$, C_1 = constant of integration.
As $P_1(0)$ = probability of 1 incident in no time $= 0$,
$\therefore$ initial condition gives $C_1 = 0$. $\therefore\ e^{\mu t}P_1(t) = \mu t$ or, $P_1(t) = \mu t e^{-\mu t}$.
For $k = 2$, the eq. (2) becomes $P_2'(t) + \mu P_2(t) = \mu^2 t e^{-\mu t}$ as $P_1(t) = \mu t e^{-\mu t}$.
It is again a linear ODE with IF $= e^{\mu t}$. Multiplying by IF and integrating,

$P_2(t)e^{\mu t} = \mu^2\int t\, dt + C_2$, [$C_2$ = constant of integration] or, $P_2(t)e^{\mu t} = \frac{\mu^2 t^2}{2} + C_2$. As $P_2(0) = 0$, $\therefore\ C_2 = 0$.

$$\therefore\ P_2(t) = e^{-\mu t} = \frac{(\mu t)^2}{2!}.$$

Again, for $k = 3$, eq. (2) becomes $P_3'(t) = -\mu P_3(t) + \mu e^{-\mu t}\frac{(\mu t)^2}{2!}$

or, $P_3'(t) + \mu P_3(t) = \frac{\mu^3}{2!}t^2 e^{-\mu t}$.

This is linear ODE with IF $= e^{\mu t}$. Multiplying by IF and integrating,

$$P_3(t)e^{\mu t} = \frac{\mu^3}{2!}\int t^2\,dt + C_3, \; [C_3 = \text{constant of integration}]$$
$$= \frac{\mu^3 t^3}{3!} + C_3.$$

$P_3(0) = 0 \Rightarrow C_3 = 0.$ $\therefore$ $P_3(t) = e^{-\mu t}\frac{(\mu t)^3}{3!}.$

$\therefore$ proceeding in the same way by inference, we get $P_k(t) = e^{-\mu t}\frac{(\mu t)^k}{k!}$, $k = 0, 1, 2, \cdots$, which is the pmf of a Poisson distribution with parameter μt.

Poisson distribution as limiting binomial distribution

Now, we study that binomial (n, p) distribution will be approximated by **Poisson (μ) distribution**, when n is very large and p is very small but $np (= \mu)$ is of moderate magnitude.

Let us start with binomial (n, p) distribution. It is given by the pmf

$f_i = \binom{n}{i} p^i q^{n-i}$, $i = 0, 1, 2, \cdots, n$ and $p + q = 1$,

$$\text{i.e., } f_i = \frac{n(n-1)\cdots(n-i+1)}{i!}\left(\frac{\mu}{n}\right)^i\left(1-\frac{\mu}{n}\right)^{n-i} \quad [\text{as } np = \mu]$$
$$= \frac{\left(1-\frac{1}{n}\right)\left(1-\frac{2}{n}\right)\cdots\left(1-\frac{i-1}{n}\right)}{i!}\mu^i\frac{\left(1-\frac{\mu}{n}\right)^n}{\left(1-\frac{\mu}{n}\right)^i}.$$

As i, once taken, is finite,

$$\therefore \lim_{n\to\infty}\left(1-\frac{1}{n}\right)\left(1-\frac{2}{n}\right)\cdots\left(1-\frac{i-1}{n}\right) = 1\cdot1\cdot1\cdots1 = 1$$

$$\text{and } \lim_{n\to\infty}\left(1-\frac{\mu}{n}\right)^i = \lim_{n\to\infty}\left(1-\frac{\mu}{n}\right)\left(1-\frac{\mu}{n}\right)\cdots\text{to } i \text{ factors}$$
$$= 1\cdot1\cdots\text{to } i \text{ factors (as } \mu \text{ is of moderate magnitude)}$$
$$= 1.$$

To have $\lim_{n\to\infty}\left(1-\frac{\mu}{n}\right)^n$, we put $-\frac{\mu}{n} = u$. $\therefore$ as $n \to \infty, u \to 0-$ $[\because \mu > 0]$.

$$\therefore \lim_{n\to\infty}\left(1-\frac{\mu}{n}\right)^n = \lim_{u\to0-}(1+u)^{-\mu/u} = \lim_{u\to0-}\left\{(1+u)^{1/u}\right\}^{-\mu} = e^{-\mu}.$$

$\therefore$ $\lim_{n\to\infty} f_i = e^{-\mu}\frac{\mu^i}{i!}$, which is the pmf of Poisson (μ) distribution. Hence, the result.

Here we cite an example to show how much approximation takes place when $n = 30$ and $p = 0.01$.

Ex. The probability of a product produced by a machine to be defective is 0.01. If 30 products are taken at random, find the probability that exactly two will be defective.

Solution. Required probability by binomial distribution $= \binom{30}{2}(0.01)^2(1-0.01)^{30-2} = \binom{30}{2}(0.01)^2(0.99)^{28} =$ 0.03283. Here $\mu = np = 30 \times 0.01 = 0.3$.

$\therefore$ required probability of Poisson (0.3) distribution $= e^{-0.3}\frac{(0.3)^2}{2!} = 0.033337$. Error $= 1.54\%$.

A RV X follows Poisson distribution when X represents the following:

(a) number of misprint in each page of a book,

(b) number of defective articles in a packet manufactured by a company,

(c) number of any type of accident in a given interval of time,

(d) number of telephone calls in a booth in a given interval of time,

(e) number of cars passing by in a crossing of a busy road for a given interval of time,

(f) number of wars between two neighbouring countries per year,

(g) number of emission α-, β- or γ-rays from a radioactive substance per minute, etc.

4.5 Continuous distribution

When is a distribution said to be continuous?

A RV is said to have continuous probability distribution, if
(i) X is a continuous RV,
(ii) its distribution function F_X is continuous in $(-\infty,\infty)$,
(iii) the derived function F_X' is piecewise continuous everywhere.

For a continuous distribution we have the following:

1. Let $a \in \mathbb{R}$. Then, $P(X=a)=0$, i.e., no point in a continuous distribution is a step point.

Proof. $P(X=a)=F_X(a)-F_X(a-0)=0$, as F_X is continuous at a.

2. The existence of pdf.

From the properties of distribution function

$$P(a<X\leq b)=F_X(b)-F_X(a) \Rightarrow P(a<X\leq b)=\int_a^b F_X'(x)\,dx=\int_a^b f_X(x)\,dx,$$

[as F_X' is piecewise continuous in any interval, hence bdd. and integrable],

where $f_X(x)=F_X'(x)$. (1)

The function f_X is naturally called *pdf* of X as the total probability when $X\in(a,b]$ is obtained by integrating $f_X(x)$ in $(a,b]$.

3. Again, $F_X(x)=P(-\infty<X\leq x)=\displaystyle\int_{-\infty}^{x} F_X'(x)\,dx$ [as $F_X(-\infty)=0$] $=\displaystyle\int_{-\infty}^{x} f_X(x)\,dx$ [by (1)] (2)

4. As F_X is non-decreasing, $\therefore F_X'(x)\geq 0$, i.e., $f_X(x)\geq 0$ [by (1)].

As $F_X(\infty)=1$, $\therefore$ eq.(2) $\Rightarrow \displaystyle\int_{-\infty}^{\infty} f_X(x)\,dx=1.$

$\therefore$ whenever f_X is a pdf of a continuous RV X, f_X necessarily satisfies (i) $f_X(x)\geq 0,\ \forall\, x\in(-\infty,\infty)$ and (ii) $\displaystyle\int_{-\infty}^{\infty} f_X(x)\,dx=1.$

Note: These conditions are sufficient as well, i.e., whenever a function f_X satisfies the above two conditions it may be looked upon as the pdf of a RV X.

5. *Probability differential.* The differential of a distribution function F_X, i.e., dF_X is called a *probability differential* of X.

$$dF_X=F_X(x+dx)-F_X(x)=P(x<X\leq x+dx).$$

Also, $dF_X=\frac{dF_X}{dx}\,dx=F_X'\,dx=f_X\,dx$. $\therefore$ $P(x<X\leq x+dx)=f_X\,dx$.

Probability density curve. The curve of the functional relation $y=f_X(x)$ is called the *probability density curve* or simply the density curve for the distribution of X.

4.6 Some useful continuous distributions

I. Uniform or Rectangular Distribution

The probability distribution of a continuous RV X is said to be uniform in some interval (a,b) if its pdf f_X is a nonzero constant in (a,b) and vanishes outside (a,b).

So, the probability distribution of a uniform distribution is defined by the pdf

$$f_X(x)=\begin{cases}1/(b-a), & a<x<b\\ 0, & \text{otherwise.}\end{cases}$$

We see that $f_x(x) \geq 0,\ \forall\ x \in (-\infty, \infty)$ and

$$\int_{-\infty}^{\infty} f_x(x)\,dx = \int_{-\infty}^{a} 0\,dx + \int_{a}^{b} \frac{1}{b-a}\,dx + \int_{b}^{\infty} 0\,dx = 1.$$

If F_x be the distribution function, then $F_x(x) = P(-\infty < X \leq x)$. $\therefore$ $F_x(x) = 0$, when $x < a$.
When $a \leq x \leq b$,

$$F_x(x) = P(-\infty < X \leq x) = P(-\infty < X < a) + P(a \leq X \leq x) = 0 + \int_{0}^{x} \frac{1}{b-a}\,dx = \frac{x-a}{b-a}.$$

When $x > b$,

$$F_x(x) = P(-\infty < X < x) + P(a \leq X \leq b) + P(b < X \leq x) = \int_{-\infty}^{a} 0\,dx + \int_{a}^{b} \frac{1}{b-a}\,dx + \int_{b}^{x} 0\,dx = 1.$$

$$\therefore\ F_x(x) = \begin{cases} 0, & x < a \\ (x-a)/(b-a), & a \leq x \leq b \\ 1, & b < x < \infty \end{cases}$$

The figures of the density curve and the distribution curve are the following.

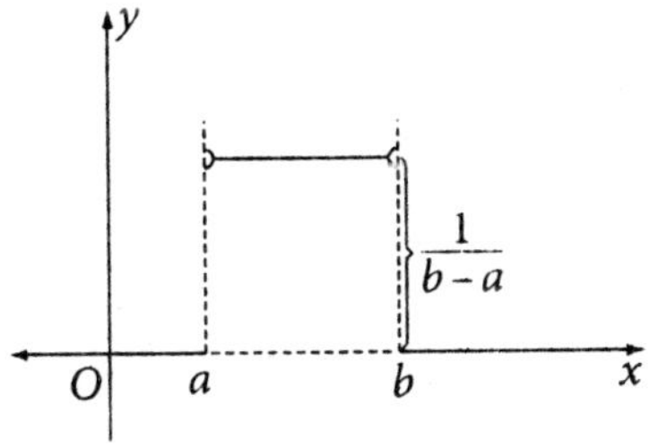

Probability density curve of uniform distribution

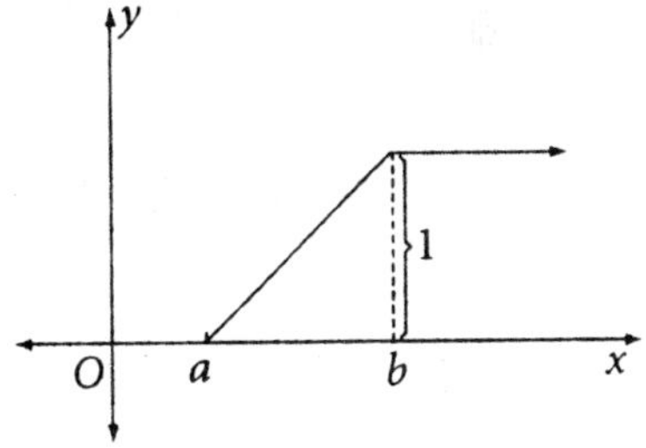

Probability distribution curve of uniform distribution

II. Normal (m, σ) distribution

This is the most important continuous distribution. Here m and $\sigma(> 0)$ are the two parameters. Later on we see that m and σ bear important significances of the distribution [m = mean and σ = standard deviation].

The normal (m, σ) distribution is defined by the pdf f_x, given by

$$f_x(x) = \frac{1}{\sqrt{2\pi}\sigma} e^{-[(x-m)^2/2\sigma^2]},\ -\infty < x < \infty.$$

Obviously, $f_x(x) > 0,\ \forall\ x \in (-\infty, \infty)$ and

$$\int_{-\infty}^{\infty} f_x(x)\,dx = \frac{1}{\sqrt{2\pi}\sigma} \int_{-\infty}^{\infty} e^{[-(x-m)^2/2\sigma^2]}\,dx = 1.$$

[For verification let us put $\frac{x-m}{\sqrt{2}\sigma} = u$. $\therefore$ $dx = \sqrt{2}\sigma\,du$.

As $\sigma > 0$, $\therefore$ $u \to -\infty$, when $x \to -\infty$ and $u \to \infty$, when $x \to \infty$.

$$\therefore\ \frac{1}{\sqrt{2\pi}\sigma} \int_{-\infty}^{\infty} e^{-[(x-m)^2/2\sigma^2]}\,dx = \frac{1}{\sqrt{2\pi}\sigma} \int_{-\infty}^{\infty} e^{-u^2} \sqrt{2}\sigma\,du$$

$$= \frac{1}{\sqrt{\pi}} \int_{-\infty}^{\infty} e^{-u^2}\,du = \frac{1}{\sqrt{\pi}} \cdot 2 \int_{0}^{\infty} e^{-u^2}\,du,$$ as integrand is an even function.

Again, substituting $u^2 = v$, we get $2u\,du = dv$, $v = 0$, when $u = 0$ and $v \to \infty$, when $u \to \infty$

$$= \frac{1}{\sqrt{\pi}} \int_0^\infty e^{-v} v^{-1/2}\,dv = \frac{1}{\sqrt{\pi}} \int_0^\infty e^{-v} v^{(1/2)-1}\,dv = \frac{\Gamma(1/2)}{\sqrt{\pi}} = 1.]$$

In particular, if we consider $m = 0$ and $\sigma = 1$, the distribution is then normal $(0, 1)$ or the standard normal distribution. Hence, the pdf for the standard normal distribution is given by

$$f_x(x) = \frac{1}{\sqrt{2\pi}} e^{-x^2/2}, \quad -\infty < x < \infty.$$

The corresponding distribution function F_x is given by

$$F_x(x) = P(-\infty < X \le x) = \frac{1}{\sqrt{2\pi}} \int_{-\infty}^{x} e^{-x^2/2}\,dx.$$

To have the figure of pdf we notice that

(i) f_x is symmetric about y-axis, as $f_x(-x) = f_x(x)$

(ii) $f_x'(x) = -\frac{x}{\sqrt{2\pi}} e^{-x^2/2} > 0$ for $-\infty < x < 0$

< 0 for $0 < x < \infty$.

$\therefore$ f_x is increasing in $(-\infty, 0)$ and f_x is decreasing in $(0, \infty)$.

$\therefore$ f_x is maximum, when $x = 0$.

(iii) $f_x(x) \to 0$ as $x \to \pm\infty$. $\therefore$ $y = 0$, i.e., x-axis is an asymptote of the density curve. Hence, the nature of pdf is given by the probability density curve as following:

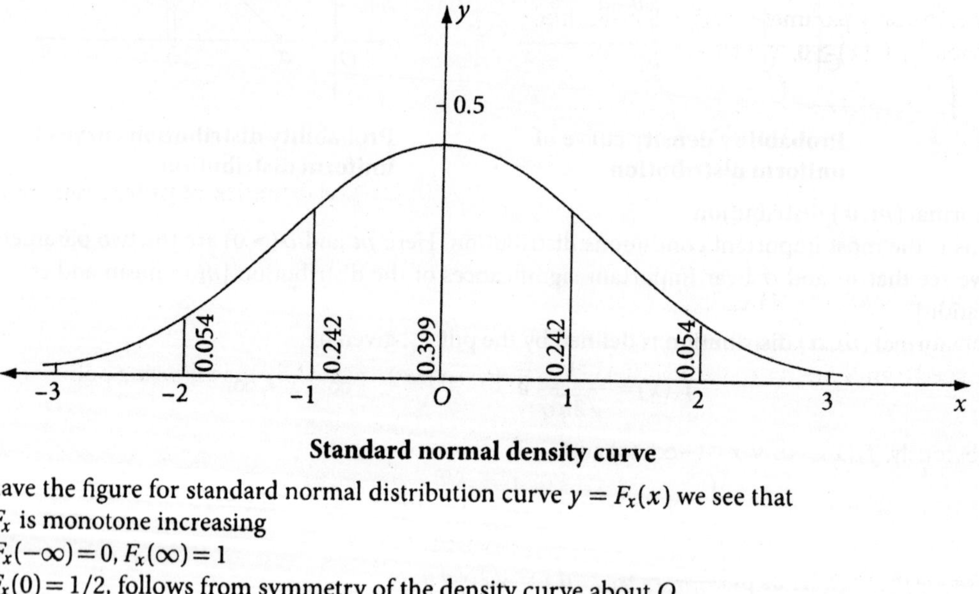

Standard normal density curve

To have the figure for standard normal distribution curve $y = F_x(x)$ we see that

(a) F_x is monotone increasing

(b) $F_x(-\infty) = 0, F_x(\infty) = 1$

(c) $F_x(0) = 1/2$, follows from symmetry of the density curve about O.

Hence, the curve is confined between $y = 0$ and $y = 1$. These lines are the asymptotes of the curve. So, the nature of the standard normal distribution curve is the following.

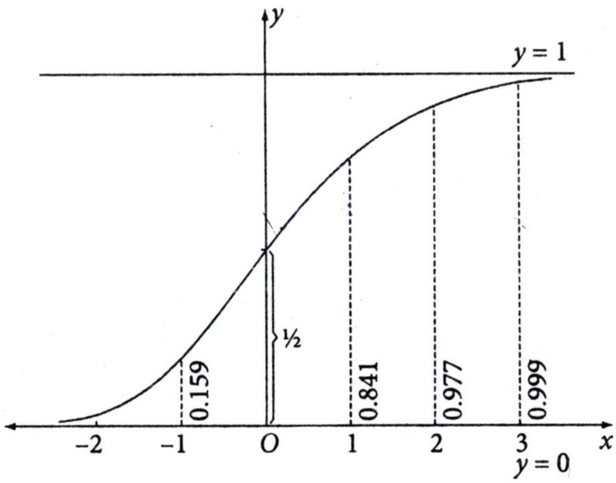

Standard normal distribution curve

III. Gamma (l) distribution

The spectrum of gamma (l) distribution is $(0,\infty)$. Gamma (l) distribution is defined by the following pdf:

$$f_x(x)=\begin{cases}\frac{e^{-x}x^{l-1}}{\Gamma(l)}, & 0<x<\infty\\ 0, & -\infty<x\leq 0.\end{cases}$$

$l(>0)$ is the only parameter of the distribution.

Obviously, $f_x(x)\geq 0,\ \forall\, x\in(-\infty,\infty)$ and

$$\int_{-\infty}^{\infty} f_x(x)\,dx=\int_{-\infty}^{0}0\,dx+\int_{0}^{\infty}\frac{e^{x}x^{l-1}}{\Gamma(l)}\,dx$$

$$=\frac{1}{\Gamma(l)}\int_{0}^{\infty}e^{-x}x^{l-1}\,dx=\frac{\Gamma(l)}{\Gamma(l)}\ \text{[by the integral definition of gamma function]}$$

$$=1.$$

The RV X is called $\gamma(l)$-variate.

IV. Beta distribution (First kind)

The spectrum is the open interval $(0,1)$. This distribution is defined by the following pdf:

$$f_x(x)=\begin{cases}\frac{x^{l-1}(1-x)^{m-1}}{B(l,m)}, & 0<x<1\\ 0, & \text{otherwise.}\end{cases}$$

$l(>0)$ and $m(>0)$ are two parameters of the distribution.

Clearly, $f_x(x)\geq 0,\ \forall\, x\in(-\infty,\infty)$ and

$$\int_{-\infty}^{\infty} f_x(x)\,dx=\int_{0}^{1}\frac{x^{l-1}(1-x)^{m-1}}{B(l,m)}\,dx=\frac{1}{B(l,m)}\int_{0}^{1}x^{l-1}(1-x)^{m-1}\,dx$$

$$=\frac{B(l,m)}{B(l,m)}\ \text{[by the integral definition of } B(l,m)\text{ function]}$$

$$=1.$$

When the distribution of X is a beta distribution of first kind with parameters l, m, we say X is $\beta_1(l,m)$-variate.

V. Beta distribution (Second kind)

The spectrum is the positive real half of the x-axis. Beta distribution of the second kind is defined by the pdf:

$$f_X(x)=\begin{cases}\frac{x^{l-1}}{B(l,m)(1+x)^{l+m}}, & 0<x<\infty\\ 0, & -\infty<x\leq 0.\end{cases}$$

Here $l(>0)$ and $m(>0)$ are the two parameters of the distribution. This is $\beta_2(l,m)$ distribution. Clearly, $f_X(x)\geq 0,\ \forall\, x\in(-\infty,\infty)$ and

$$\int_{-\infty}^{\infty} f_X(x)\,dx=\frac{1}{B(l,m)}\int_0^{\infty}\frac{x^{l-1}}{(1+x)^{l+m}}\,dx=\frac{B(l,m)}{B(l,m)}=1.$$

Such a RV X whose probability distribution is a beta function of the second kind with parameters l and m is called *$\beta_2(l,m)$ variate.*

VI. Exponential distribution

The exponential distribution of a continuous RV X is defined by the pdf:

$$f_X(x)=\begin{cases}\theta\, e^{-\theta x}, & 0<x<\infty\\ 0, & -\infty<x<0,\end{cases}$$

where $\theta(>0)$ is the only parameter of the distribution.

It is useful to study the time interval between the occurrence of two consecutive events, i.e., time to wait for the next event after the previous one. It is also useful to study the life of an electronic product.

When X is an exponential distribution with parameter θ, we say that X is an $E(\theta)$-variate.

Obviously, $f_X(x)\geq 0,\ \forall\, x\in(-\infty,\infty)$ and

$$\int_{-\infty}^{\infty} f_X(x)\,dx=\int_0^{\infty}\theta\, e^{-\theta x}\,dx=\lim_{R\to\infty}\int_0^{R}\theta\, e^{-\theta x}\,dx=\theta\lim_{R\to\infty}\left[\frac{e^{-\theta x}}{-\theta}\right]_0^R=\lim_{R\to\infty}\left[1-e^{-\theta R}\right]=1 \text{ as } \theta>0.$$

Here $f_X'(x)=-\theta^2 e^{-\theta x}<0,\ \forall\, x\in[0,\infty)$ and $f_X''(x)=\theta^3 e^{-\theta x}>0.$

$\therefore$ f_X is decreasing in $[0,\infty)$ and tends to zero as $x\to\infty$ and $f_X(x)=0$ (constant) in $(-\infty,0)$ and $y=f_X(x)$ is concave up in $(0,\infty)$.

Hence, the graph of the pdf is the following:

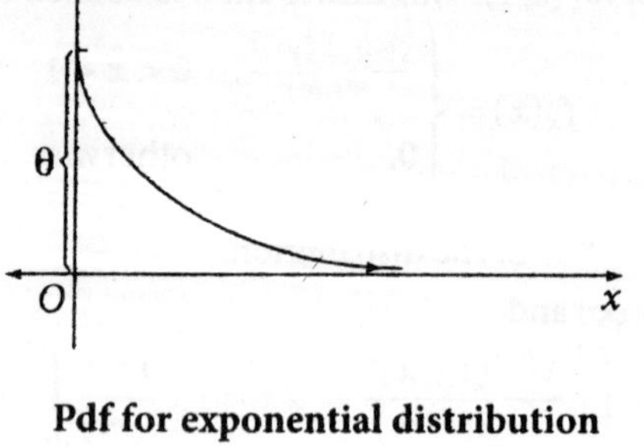

Pdf for exponential distribution

The probability distribution function is given by

$$F_X(x)=\begin{cases}0, & x<0\\ 1-e^{-\theta x}, & x\geq 0.\end{cases}$$

Thus F_x is continuous $\forall\, x \in (-\infty, \infty)$.

$F_x'(x) = \theta\, e^{-\theta x} > 0$ and $F_x''(x) = -\theta^2\, e^{-\theta x} < 0$, $\forall\, x \in (0, \infty)$. Hence, F_x is increasing and is concave down for $x > 0$.

Hence, the graph of the probability distribution curve is the following:

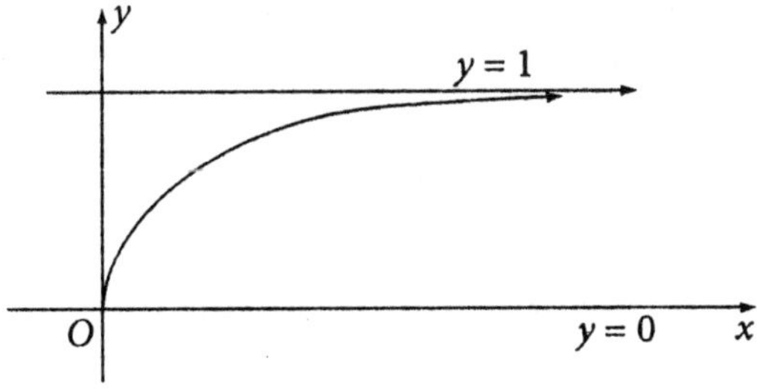

Graph: $y = F_x(x)$

VII. Cauchy distribution

The Cauchy distribution of a continuous RV X is defined by the pdf;

$$f_X(x) = \frac{1}{\pi} \frac{\lambda}{(x-\mu)^2 + \lambda^2}, \quad -\infty < x < \infty,$$

where $\lambda (> 0)$ and μ are the parameters.

The distribution function F_x is given by

$$F_x(x) = P(-\infty < X \le x) = \frac{\lambda}{\pi} \int_{-\infty}^{x} \frac{dx}{(x-\mu)^2 + \lambda^2} = \frac{1}{\pi} \lim_{L \to -\infty} \left(\tan^{-1} \frac{x-\mu}{\lambda} - \tan^{-1} \frac{L-\mu}{\lambda} \right)$$

$$= \frac{1}{\pi} \left[\tan^{-1} \frac{x-\mu}{\lambda} + \frac{\pi}{2} \right].$$

Note that $F_x'(x) > 0$ and $F_x(\infty) = 1$.

VIII. Truncated normal distribution

The truncated normal distribution of a RV X truncated to the right at $X = k$ is defined by the pdf f_x, where

$$f_x(x) = \begin{cases} \frac{1}{\phi\left(\frac{k-m}{\sigma}\right)} \cdot \frac{1}{\sqrt{2\pi}\sigma} e^{-(x-m)^2/2\sigma^2}, & x \le k \\ 0, & x > k. \end{cases}$$

ϕ is the distribution function of standard normal distribution.

Note that $f(x) \ge 0$, $\forall\, x \in (-\infty, \infty)$ and

$$\int_{-\infty}^{\infty} f_x(x)\, dx = \int_{-\infty}^{k} \frac{1}{\phi\left(\frac{k-m}{\sigma}\right)} \cdot \frac{1}{\sqrt{2\pi}\sigma} e^{-(x-m)^2/2\sigma^2}\, dx = \frac{1}{\phi\left(\frac{k-m}{\sigma}\right)} \cdot \frac{1}{\sqrt{2\pi}\sigma} \int_{-\infty}^{k} e^{-(x-m)^2/2\sigma^2}\, dx.$$

Let $\frac{x-m}{\sigma} = u$. $\therefore\ dx = \sigma\, du$, $u \to -\infty$ as $x \to -\infty$, $u = \frac{k-m}{\sigma}$, when $x = k$

$$= \frac{1}{\phi\left(\frac{k-m}{\sigma}\right)} \cdot \frac{1}{\sqrt{2\pi}\sigma} \int_{-\infty}^{\frac{k-m}{\sigma}} e^{-u^2/2} \sigma\, du = \frac{1}{\phi\left(\frac{k-m}{\sigma}\right)} \cdot \frac{1}{\sqrt{2\pi}} \int_{-\infty}^{\frac{k-m}{\sigma}} e^{-u^2/2}\, du = \frac{\phi\left(\frac{k-m}{\sigma}\right)}{\phi\left(\frac{k-m}{\sigma}\right)} = 1.$$

In particular, if $k = \infty$, $\phi\left(\frac{k-m}{\sigma}\right) = \phi(\infty) = 1$ and X becomes standard normal.

Similarly, the truncated normal distribution truncated on the left at $X = k$ is given by the following pdf:

$$f_X(x) = \begin{cases} 0, & x < k \\ \frac{1}{1-\phi\left(\frac{k-m}{\sigma}\right)} \cdot \frac{1}{\sqrt{2\pi}\sigma} e^{-(x-m)^2/2\sigma^2}, & x \geq k. \end{cases}$$

Obviously, $f_X(x) \geq 0, \ \forall\, x \in (-\infty, \infty)$ and

$$\int_{-\infty}^{\infty} f_X(x)\,dx = \int_k^{\infty} \frac{1}{1-\phi\left(\frac{k-m}{\sigma}\right)} \cdot \frac{1}{\sqrt{2\pi}\sigma} e^{-(x-m)^2/2\sigma^2}\,dx$$

$$= \frac{1}{1-\phi\left(\frac{k-m}{\sigma}\right)} \left[1 - \int_{-\infty}^{k} \frac{1}{\sqrt{2\pi}\sigma} e^{-(x-m)^2/2\sigma^2}\,dx\right]$$

$$= \frac{1}{1-\phi\left(\frac{k-m}{\sigma}\right)} \left[1 - \phi\left(\frac{k-m}{\sigma}\right)\right] = 1.$$

IX. Laplace distribution

A continuous RV X is said to have Laplace distribution if its pdf f_X is given by

$$f_X(x) = \frac{1}{2\lambda} e^{-|x-\mu|/\lambda}, \ -\infty < x < \infty,$$

$\lambda (> 0)$ and μ are the parameters of the distribution.

Obviously, $f_X(x) > 0, \ \forall\, x \in (-\infty, \infty)$ and

$$\frac{1}{2\lambda} \int_{-\infty}^{\infty} e^{-|x-\mu|/\lambda}\,dx = \frac{1}{2\lambda} \left[\int_{-\infty}^{\mu} e^{(x-\mu)/\lambda}\,dx + \int_{\mu}^{\infty} e^{-(x-\mu)/\lambda}\,dx\right]$$

$$= \frac{1}{2\lambda} \left[\lambda \int_{-\infty}^{0} e^{u}\,du + \lambda \int_{0}^{\infty} e^{-u}\,du\right], \text{ when } u = \frac{x-\mu}{\lambda}$$

$$= \frac{1}{2\lambda} [\lambda + \lambda] = 1.$$

X. Logistic distribution

The logistic distribution is defined by the cumulative distribution function F_X given by

$$F_X(x) = \frac{1}{1 + e^{-(x-\alpha)/\beta}}, \ x > 0.$$

It is symmetric about α.

4.7 Mixed distribution

So, far we have seen that there are two types of distributions, viz., discrete and continuous. They are not exhaustive. There are distributions, which are mixed in character. This is the probability distribution of RV X whose spectrum is the union is two more sets, one is a set of discrete points and the others are the sets of continuous points (intervals).

For example, consider the distribution of a RV X, which has the dual character. Let the spectrum of X consist of two open intervals, say, $(1,2)$ and $(3,4)$ and the discrete set of points, say, $\{5,6,7\}$, i.e., the spectrum is $(1,2) \cup (3,4) \cup \{5,6,7\}$. The probability distribution of X is defined by the following pdf and pmf as

$$f_X(x) = \begin{cases} x^2/6, & 1 < x < 2 \\ 1/9, & 3 < x < 4. \end{cases}$$

$f_5 = \frac{1}{14}, \ f_6 = \frac{1}{7}, \ f_7 = \frac{2}{7}.$

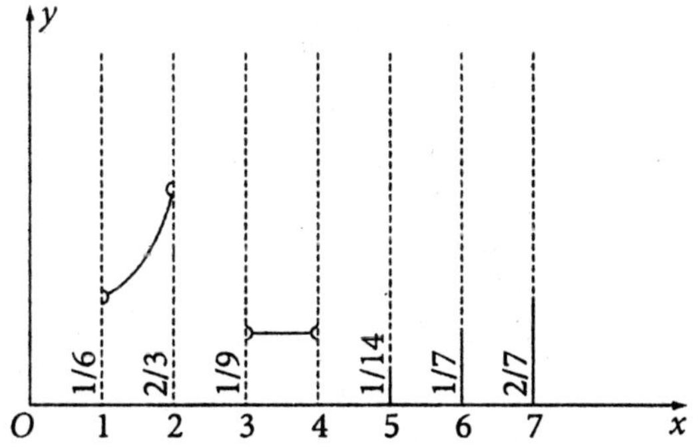

4.8 Transformation of RV

Let X be a RV on an event space S and $X(S) \subset \mathbb{R}$.

Let $g : X(S) \to \mathbb{R}$ be continuous. So, when $U \in S, X(U) \in X(S)$ and $g\{X(U)\} \in \mathbb{R}$. Therefore, for each $U \in S, g\{X(U)\}$ is a real number and so $Y = g(X)$ is also a RV on the same event space S. Suppose, we know the distribution of X. We want to find that of Y.

Continuous case. Let the function g be continuously differentiable and strictly monotonic, so that the inverse function g^{-1} exists uniquely.

When g is strictly increasing, i.e., $\frac{dy}{dx}$ is positive for $y = g(x)$, then $X \le x \Leftrightarrow g(X) \le g(x) \Leftrightarrow Y \le y$.

$\therefore P(X \le x) = P(Y \le y)$. $\therefore F_x(x) = F_y(y) \Rightarrow F_x'(x)\,dx = F_y'(y)\,dy$.

$\therefore f_x(x)\,dx = f_y(y)\,dy$. $\therefore f_y(y) = f_x(x)\frac{dx}{dy}$.

The RHS can be written in terms of y and the range of y may be determined as that of x is known.

The pdf f_y together with the range of y determines the distribution of Y, when that of X is known.

When g is strictly decreasing, then $\frac{dy}{dx}$ is negative.

$X \le x \Leftrightarrow g(X) \ge g(x) \Leftrightarrow Y \ge y$. $\therefore P(X \le x) = P(Y \ge y) = 1 - P(Y < y)$.

$\therefore F_x(x) = 1 - F_y(y)$ as probability at a point in the continuous case is zero.

$\therefore F_x'(x)\,dx = -F_y'(y)\,dy$ or, $f_y(y) = f_x(x)\left(-\frac{dx}{dy}\right)$.

This pdf together with the range of y gives us the distribution of Y. Both the formulae for increasing and decreasing cases may be put as $f_y(y) = f_x(x)\left|\frac{dx}{dy}\right|$.

Discrete case. This is very simple. The spectrum point $X = x_i$ will be $g(X) = g(x_i)$, i.e., $Y = y_i$.

$\therefore X = x_i \equiv Y = y_i$ for each possible i.

$\therefore P(X = x_i) = P(Y = y_i)$, i.e., $f_{yi} = f_{xi}$.

$\therefore$ the pmf f_{yi} together with the spectrum y_i determines the distribution of Y, when that of X is known.

WORKED-OUT EXAMPLES

Ex. 1. Find the constant k such that the function f given by

$$f_x(x) = \begin{cases} k|x|, & -2 < x < 2 \\ 0, & \text{elsewhere} \end{cases}$$

is a possible pdf and find its distribution function.

Solution. Let f be a possible pdf of a distribution. $\therefore$ by the necessary condition,

$$k\int_{-2}^{2} |x|\,dx = 1 \text{ or, } \frac{1}{k} = \int_{-2}^{0} (-x)\,dx + \int_{0}^{2} x\,dx \text{ or, } \frac{1}{k} = -\frac{1}{2}x^2\Big]_{-2}^{0} + \frac{1}{2}x^2\Big]_{0}^{2} = 2+2=4.\ \therefore\ k = \frac{1}{4}.$$

$$\therefore\ f_X(x)=\begin{cases}\frac{1}{4}|x|, & -2<x<2\\ 0, & \text{elsewhere.}\end{cases}$$

If F_X be the distribution function, then

$$F_X(x)=\int_{-\infty}^{x} f_X(x)\,dx.\ \therefore\ F_X(x)=0,\ x<-2.$$

When $-2<x<0$,

$$F_X(x)=\int_{-\infty}^{x} f_X(x)\,dx=\int_{-\infty}^{-2} 0\,dx+\int_{-2}^{x}\left(-\frac{1}{4}x\right)dx=-\frac{1}{8}\left(x^2-4\right).$$

When $0<x<2$,

$$F_X(x)=\frac{1}{4}\int_{-2}^{0}(-x)\,dx+\frac{1}{4}\int_{0}^{x} x\,dx=-\frac{1}{8}\left[x^2\right]_{-2}^{0}+\frac{1}{8}x^2=-\frac{1}{8}(0-4)+\frac{1}{8}x^2=\frac{1}{8}x^2+\frac{1}{2}.$$

When $x>2$,

$$F_X(x)=1.$$

Hence, the distribution function is

$$F_X(x)=\begin{cases}0, & x<-2\\ -\frac{1}{8}\left(x^2-4\right), & -2<x<0\\ \frac{1}{8}\left(x^2+4\right), & 0<x<2\\ 1 & x>2.\end{cases}$$

Ex. 2. Find the value of the constant k so that the function f_X given by

$$f_X(x)=\begin{cases}kx(2-x), & 0<x<2\\ 0, & \text{elsewhere}\end{cases}$$

is a pdf. Construct the distribution function and compute $P(X>1)$.

Solution. Let the given function is a possible pdf.

$$\therefore\ k\int_{0}^{2} x(2-x)\,dx=1\Rightarrow\frac{1}{k}=\int_{0}^{2}\left(2x-x^2\right)dx \text{ or, } \frac{1}{k}=\left[x^2-\frac{x^3}{3}\right]_0^2=4-\frac{8}{3}=\frac{4}{3}.$$

So, $k=3/4$.

$$\therefore\ f_X(x)=\begin{cases}\frac{3}{4}x(2-x), & 0<x<2\\ 0, & \text{elsewhere.}\end{cases}$$

If F_X be the distribution function, then $F_X(x)=0, x<0$.

When $0\le x\le 2$,

$$F_X(x)=\frac{3}{4}\int_{0}^{x}\left(2x-x^2\right)dx=\frac{3}{4}\left(x^2-\frac{x^3}{3}\right).$$

When $x>2$,

$$F_X(x)=\frac{3}{4}\int_{0}^{2}\left(2x-x^2\right)dx+\int_{2}^{x} 0\,dx=1.$$

Hence, the distribution function is given by

$$F_x(x) = \begin{cases} 0, & x < 0 \\ \frac{3}{4}\left(x^2 - \frac{x^3}{3}\right), & 0 \le x \le 2 \\ 1, & x > 2 \end{cases}$$

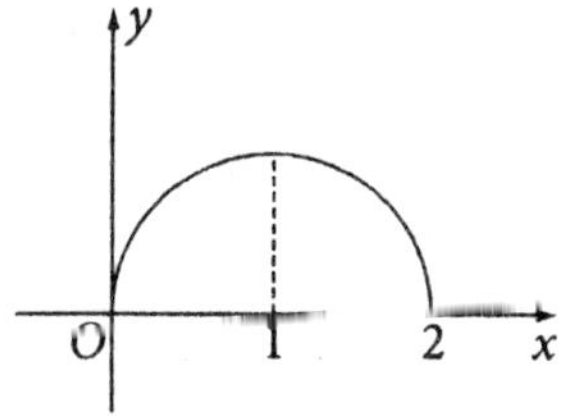

Fig $y = f_x(x)$

Now, $P(X > 1) = 1 - P(X \le 1) = 1 - F_x(1)$
$= 1 - \frac{3}{4}\left(1 - \frac{1}{3}\right) = 1 - \frac{1}{2} = \frac{1}{2}$.

Ex. 3. The probability that a product produced by a machine to be defective is 0.01. If 50 products are taken at random, find the probability that exactly 3 will be defective. Approximate it by Poisson distribution and evaluate the error of approximation.

Solution. The distribution of defective products is a binomial (50, 0.01) distribution (production of a defective bolt is taken as success).

$\therefore$ required probability $= \binom{50}{3}(0.01)^3(0.99)^{47} = 0.01222$.

Since, the probability of success p is small and n is large but $\mu = np = 0.5$, which is of moderate magnitude, therefore, the binomial (50, 0.01) distribution may be approximated by Poisson (0.5) distribution.

Hence, the probability by Poisson distribution is $e^{-0.5}\frac{(0.5)^3}{3!} = 0.01264$.

$\therefore$ error of approximation $= |0.01222 - 0.01264| = 0.00042$ (absolute error), which is very small.

Relative error $= 0.00042 \div 0.01222 \simeq 0.0344$. Percentage error $= 3.44$.

Note: If n is larger and p is smaller, the error will decrease.

For example, in the above problem, if $n = 70$ and $p = 0.007$, the absolute error become $|0.01173 - 0.01201| = 0.00028$.

Ex. 4. Let AB be a line segment whose midpoint is O'. If a point X is chosen at random on the segment AB, then find the probability that the segments AX, XB and $O'B$ may form the sides of a triangle.

Solution. Let O be chosen as origin and a, b, x are respectively the coordinates of A, B and X.

So, the coordinate of O', the midpoint of AB, is $(a+b)/2$. As the point X is chosen at random on AB, $\therefore$ x varies uniformly from a to b. So, the pdf of X is given by the following

$$f_X(x) = \begin{cases} 1/(b-a), & a < x < b \\ 0, & \text{elsewhere.} \end{cases}$$

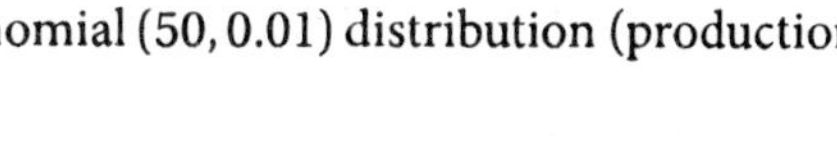

Now, the event that AX, XB and $O'B$ will be the sides of a triangle if $AX + XB > O'B$, $AX + O'B > XB$ and $XB + O'B > AX$ (the sum of two sides is greater than the third).

The first one is obvious as $AX + XB = AB = 2O'B > O'B$ is certain.

The second one $\Rightarrow x - a + b - \frac{a+b}{2} > b - x$ or, $2x > a + \frac{a+b}{2}$. $\therefore$ $x > \frac{3a+b}{4}$.

The third one $\Rightarrow b - x + b - \frac{a+b}{2} > x - a$ or, $2x < \frac{a+3b}{2}$. $\therefore$ $x < \frac{a+3b}{4}$.

$\therefore$ the required probability $= P\left(\frac{3a+b}{4} < X < \frac{a+3b}{4}\right) = \int_{(3a+b)/4}^{(a+3b)/4} \frac{1}{b-a}\,dx = \frac{1}{b-a}\left[\frac{a+3b}{4} - \frac{3a+b}{4}\right] = \frac{1}{2}$.

Ex. 5. Two numbers are chosen at random independent between 0 and 2. Find the probability that their product is less than a given number $k(0 < k < 4)$.

Solution. Let us choose Ox, Oy as rectangular Cartesian axes. Draw a square $OABC$ of side 2 such that $A = (2,0)$, $B = (2,2)$ and $C = (0,2)$.

Let x, y be any two numbers chosen independently at random between 0 and 2.

$\therefore$ (x, y) is an arbitrary point within the square $OABC$. The event $XY < K$ (where X, Y are the RV corresponding to chosen numbers) will be satisfied if (x, y) lies within the portion $OAEFC$.

So, the required probability

$$= \frac{\text{area of } OAEFC}{\text{area of } OABC} = \frac{1}{4}(4 - \text{area of } EBF)$$

$$= 1 - \frac{1}{4}\int_{k/2}^{2}\left(2 - \frac{k}{x}\right)dx = 1 - \frac{1}{4}\Big[2x - k\ln|x|\Big]_{k/2}^{2}$$

$$= 1 - \frac{1}{4}\left[4 - k - k\left(\ln 2 - \ln\frac{k}{2}\right)\right]$$

$$= \frac{k}{4}\left(1 + \ln 2\cdot\frac{2}{k}\right) = \frac{k}{4}\left(1 - \ln\frac{k}{4}\right).$$

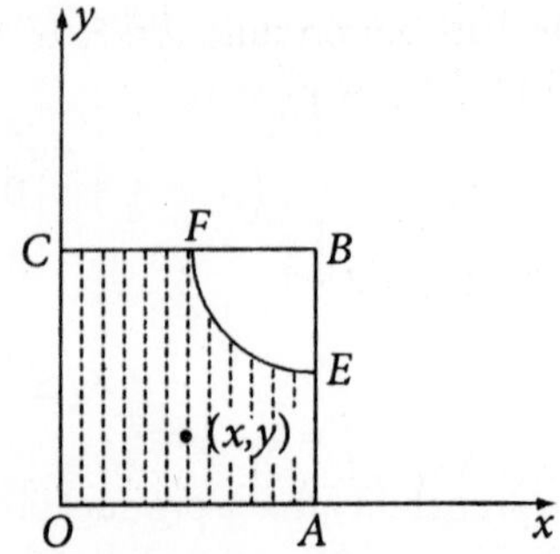

Note 1. If $k = 4$, the event $XY < k$ is certain and $P(XY < k) = 1$. Putting $k = 4$ in the required probability it becomes 1.

2. The same problem will be solved in Ch. 5 in different approach.

Ex. 6. A radioactive source emits on the average 2.5 particles per second. Find the probability that 3 or more particles will emit in an interval of 4 seconds. [C.H. 1987]

Solution. Let the unit of time be 1 second. λ = number of change per unit time on the average = 2.5. If X be the RV denoting the number of particles emitted in $(0, 4)$, then X follows Poisson distribution with the parameter $\mu = \lambda t = 2.5 \times 4 = 10$.

$\therefore$ the required probability $= P(X \geq 3) = 1 - P(X < 3) = 1 - \{P(X=0) + P(X=1) + P(X=2)\}$

$$= 1 - \left\{e^{-\mu}\frac{\mu^0}{0!} + e^{-\mu}\frac{\mu^1}{1!} + e^{-\mu}\frac{\mu^2}{2!}\right\} = 1 - 61e^{-10} \text{ as } \mu = 10.$$

Ex. 7. If X is normal (m, σ), then prove that $P(a < X < b) = \phi\left(\frac{b-m}{\sigma}\right) - \phi\left(\frac{a-m}{\sigma}\right)$ and $P(|X - m| > a\sigma) = 2\{1 - \phi(a)\}$, where ϕ is the standard normal distribution function. [C.H. 1998]

Solution. X is normal (m, σ). $\therefore$ $\frac{X-m}{\sigma}$ is normal $(0, 1)$.

$\therefore$ $P(a < X < b) = P(a < X \leq b)$ as $P(X = b) = F_X(b) - F_X(b - 0) = 0$ as F_X is continuous

$$= P\left(\frac{a-m}{\sigma} < \frac{X-m}{\sigma} \leq \frac{b-m}{\sigma}\right) = \phi\left(\frac{b-m}{\sigma}\right) - \phi\left(\frac{a-m}{\sigma}\right),$$

when ϕ is standard normal distribution function.

Now,

$$P(|X - m| > a\sigma) = 1 - P(|X - m| \leq a\sigma) = 1 - P(-a\sigma \leq X - m \leq a\sigma) = 1 - P\left(-a < \frac{X-m}{\sigma} \leq a\right)$$

$$= 1 - \{\phi(a) - \phi(-a)\} = 1 - \phi(a) + (1 - \phi(a)) = 2\{1 - \phi(a)\}.$$

Ex. 8. The RV X is distributed uniformly over the interval $(0, 2)$. Find the distribution function of the larger root of the quadratic equation $t^2 - 2t - X = 0$. [C.H. 1983, 1988, 1992]

Solution. Given that X is uniformly distributed over the interval $(0, 2)$. Hence, the pdf of X is given by

$$f_X(x) = \begin{cases} 1/2, & 0 < x < 2 \\ 0, & \text{elsewhere.} \end{cases}$$

The larger root of the equation is $-1 + \sqrt{1+X}$. Let $Y = -1 + \sqrt{1+X}$. Setting in real variables we have $y = -1 + \sqrt{1+x}$. $\therefore$ $\frac{dy}{dx} = \frac{1}{2\sqrt{1+x}} > 0$, $\forall\, x \in (0, 2)$.

$\therefore$ the pdf for Y is $f_Y(y) = f_X(x)\left|\frac{dx}{dy}\right| = \frac{1}{2}\cdot 2\sqrt{1+x} = \sqrt{1+x}$, $0 < x < 2$,

$$\text{i.e., } f_Y(y) = \begin{cases} y + 1, & 0 < y < \sqrt{3} - 1 \\ 0, & \text{elsewhere.} \end{cases}$$

The distribution function F_Y of Y is given by $F_Y(y) = 0, -\infty < y \leq 0$.
When $0 < y < \sqrt{3} - 1$,

$$F_Y(y) = \int_{-\infty}^{0} 0\,dx + \int_{0}^{y} (y+1)\,dy = \frac{y^2}{2} + y.$$

When $y \geq \sqrt{3} - 1$,

$$F_Y(y) = \int_{0}^{\sqrt{3}-1} (y+1)\,dy + \int_{\sqrt{3}-1}^{\infty} 0\,dy = 1.$$

Hence, the distribution function of the larger root is finally given by

$$F_Y(y) = \begin{cases} 0, & -\infty < y \leq 0 \\ (y^2/2) + y, & 0 < y < \sqrt{3} - 1 \\ 1, & y \geq \sqrt{3} - 1. \end{cases}$$

Ex. 9. Let a RV X is uniformly distributed over the interval $(0, 1)$. Find the distribution of the smaller root of the quadratic equation $t^2 - 2t - X^2 + 1 = 0$.

Solution. As X is uniformly distributed in $(0, 1)$, so the pdf of X is given by

$$f_X(x) = \begin{cases} 1, & 0 < x < 1 \\ 0, & \text{elsewhere.} \end{cases}$$

The smaller root of the given equation is $1 - X$.
Let $Y = 1 - X$. Setting in real variables $y = 1 - x$, we get $\frac{dy}{dx} = -1 < 0, \ \forall\, x \in (0, 1)$.
$\therefore$ the pdf of Y is given by

$$f_Y(y) = f_X(x)\left|\frac{dx}{dy}\right| = f_X(x) = 1,\ 0 < y < 1$$

and $f_Y(y) = 0$, elsewhere.

$\therefore$ the distribution of the smaller root is given by the pdf

$$f_Y(y) = \begin{cases} 1, & 0 < y < 1 \\ 0, & \text{elsewhere.} \end{cases}$$

Ex. 10. A point A is chosen at random on a semi-circle having centre at the origin and radius unity and projected on the diameter. Prove that the projection of OA on the diameter has the probability density $\frac{1}{\pi\sqrt{1-y^2}}$ for $-1 < y < 1$. [C.H. 1987, 2008]

Solution. Let A be any point on the semi-circle and B is its point of projection on the diameter.
Let $\angle AOB = X$.
X is uniformly distributed in $(0, \pi)$.
$\therefore$ the pdf for X is given by

$$f_X(x) = \begin{cases} 1/\pi, & 0 < x < \pi \\ 0, & \text{elsewhere.} \end{cases}$$

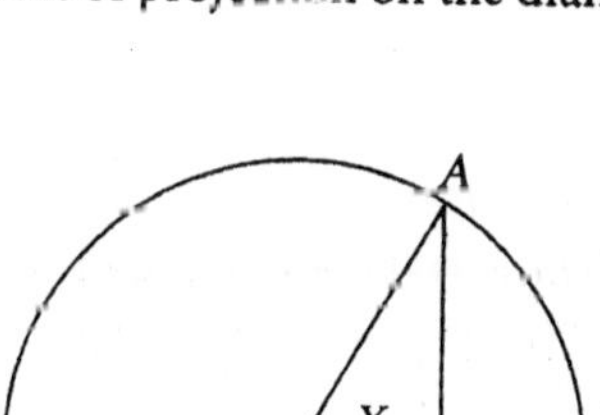

Let $Y = OB$ be the projection of OA on the diameter. $\therefore$ $Y = \cos X$. In terms of their real values

$$y = \cos x.$$

$$\therefore \frac{dy}{dx} = -\sin x < 0,\ \forall\, x \in (0, \pi).$$

$\therefore$ the probability distribution of Y is given by the following pdf: $f_Y(y) = f_X(x)\left|\frac{dx}{dy}\right| = \frac{1}{\pi}\operatorname{cosec} x,\ 0 < x < \pi$.

$$\therefore\ f_Y(y) = \begin{cases} \frac{1}{\pi\sqrt{1-y^2}}, & -1 < y < 1 \\ 0, & \text{elsewhere.} \end{cases}$$

Ex. 11. If X is a Poisson (μ) variate, then find the probability distribution of X^3.

Solution. X is a Poisson variate. $\therefore\ P(X=i) = e^{-\mu}\frac{\mu^i}{i!},\ i = 0,1,2,3,\cdots$.
Let $Y = X^3$. So, Y is also a discrete variate with spectrum $i^3 = 0, 1^3, 2^3, 3^3, \cdots$
$P(Y=i^3) = P(X^3 = i^3) = P(X=i)$, as the events are the same.
Hence, the distribution of Y is given by

$$Y = y_i = i^3, i = 0,1,2,\cdots \text{ and } f_{yi} = P(Y=y_i) = e^{-\mu}\frac{\mu^i}{i!}.$$

Ex. 12. If X is a Poisson variate with parameter μ, show that

$$P(X \le n) = \frac{1}{n!}\int_\mu^\infty e^{-x}x^n\,dx, \text{ when } n \in \mathbb{N}.$$ [C.H. 1994, 2006, 2008]

Solution.
$$\int_\mu^\infty e^{-x}x^n\,dx = \lim_{R\to\infty}\int_\mu^R e^{-x}x^n\,dx = \lim_{R\to\infty}\left[-e^{-x}x^n\right]_\mu^R + n\lim_{R\to\infty}\int_\mu^R e^{-x}x^{n-1}\,dx$$
$$= e^{-\mu}\mu^n + n\int_\mu^\infty e^{-x}x^{n-1}\,dx, \text{ as } \lim_{R\to\infty}\frac{R^n}{e^R} = 0.$$

$$\therefore\ \frac{1}{n!}\int_\mu^\infty e^{-x}x^n\,dx = e^{-\mu}\frac{\mu^n}{n!} + \frac{1}{(n-1)!}\int_\mu^\infty e^{-x}x^{n-1}\,dx = P(X=n) + \frac{1}{(n-1)!}\int_\mu^\infty e^{-x}x^{n-1}\,dx.$$

Similarly,

$$\frac{1}{(n-1)!}\int_\mu^\infty e^{x}x^{n-1}\,dx = P(X=n-1) + \frac{1}{(n-2)!}\int_\mu^\infty e^{-x}x^{n-2}\,dx$$

....

....

$$\frac{1}{1!}\int_\mu^\infty e^{x}x\,dx = P(X=1) + \frac{1}{0!}\int_\mu^\infty e^{-x}x^{1-1}\,dx = P(X=1) + \lim_{R\to\infty}\left[-e^{-x}\right]_\mu^R$$
$$= P(X=1) + e^{-\mu} = P(X=1) + P(X=0).$$

$$\therefore\ \frac{1}{n!}\int_\mu^\infty e^{-x}x^n\,dx = P(X=n) + P(X=n-1) + \cdots + P(X=1) + P(X=0) = P(X \le n), \textbf{ (proved)}.$$

Ex. 13. A point R is chosen at random on a line segment OA of length $2a$. Find the probability that the area of the rectangle with its sides OR and RA will never exceed $\frac{1}{2}a^2$.

Solution. Let X denotes the RV of the length of the segment OR. Then, X is uniformly distributed over the interval $(0, 2a)$.
$\therefore$ the pdf of X is given by

$$f_X(x) = \begin{cases} 1/2a, & 0 < x < 2a \\ 0, & \text{otherwise.} \end{cases}$$

$\therefore$ the required probability

$$= P\left\{X(2a-X) \le \frac{1}{2}a^2\right\} = 1 - P\left\{X(2a-X) > \frac{1}{2}a^2\right\} = 1 - P\left(X^2 - 2aX + \frac{1}{2}a^2 < 0\right)$$

$$= 1 - P\left\{(X-a)^2 < \left(\frac{1}{\sqrt{2}}a\right)^2\right\} = 1 - P\left\{|X-a| < \frac{1}{\sqrt{2}}a\right\} \quad [\because a > 0]$$

$$= 1 - P\left\{a - \frac{a}{\sqrt{2}} < X < a + \frac{a}{\sqrt{2}}\right\} = 1 - \int_{a-(a/\sqrt{2})}^{a+(a/\sqrt{2})} \frac{1}{2a}\,dx = 1 - \frac{1}{2a}\left(a + \frac{a}{\sqrt{2}} - a + \frac{a}{\sqrt{2}}\right)$$

$$= 1 - \frac{1}{\sqrt{2}} = \frac{1}{\sqrt{2}}\left(\sqrt{2} - 1\right).$$

Ex. 14. If X is a standard normal variate find the probability distribution of e^X.

Solution. Let $Y = e^X$. We want to have probability distribution of Y. Setting in their real variables, we get $y = e^x$.

$$\therefore \frac{dy}{dx} = e^x > 0, \ \forall\, x \in (-\infty, \infty).$$

As X is a standard normal variate, $\therefore$ pdf of X is given by

$$f_X(x) = \frac{1}{\sqrt{2\pi}} e^{-x^2/2}, \ -\infty < x < \infty.$$

If f_y be the pdf of y, then by the rule of transformation:

$$f_Y(y) = f_X(x)\left|\frac{dx}{dy}\right| = \frac{1}{\sqrt{2\pi}} e^{-x^2/2} e^{-x} = \frac{1}{\sqrt{2\pi}} \frac{e^{-\frac{1}{2}(\ln y)^2}}{y}, \ 0 < y < \infty$$

$$f_Y(y) = 0, \ y < 0$$

Ex. 15. The pdf of a continuous RV X is given by

$$f_X(x) = \begin{cases} \frac{1}{\sqrt{2\pi}} \cdot \frac{e^{-\frac{1}{2}(\ln x)^2}}{x}, & 0 < x < \infty \\ 0, & \text{elsewhere.} \end{cases}$$

Find the distribution of $\ln X$.

Solution. Let $Y = \ln X$. In real variables $y = \ln x$.

$$\therefore \frac{dy}{dx} = \frac{1}{x} > 0, \ \forall\, x > 0.$$

If f_y be the pdf of Y, then by the rule of transformation:

$$f_Y(y) = f_X(x)\left|\frac{dx}{dy}\right| = \frac{1}{\sqrt{2\pi}} \frac{e^{-\frac{1}{2}(\ln x)^2}}{x} |x| = \frac{1}{\sqrt{2\pi}} e^{-y^2/2}, \ -\infty < y < \infty \text{ as } y = \ln x \text{ and } x > 0.$$

$\therefore$ Y is a standard normal variate.

Ex. 16. Examine whether the following F_X is a distribution function:

$$F_X(x) = \begin{cases} 0, & -\infty < x < 10 \\ 1 - \frac{1}{10x}, & 10 < x < \infty. \end{cases}$$

A set of necessary conditions for a given function F_X to be a probability distribution function are the following: (i) F_X is defined in $(-\infty, \infty)$ and $F_X'(x) \ge 0, \ \forall\, x,$

(ii) $F_X(-\infty)=0$, $F_X(\infty)=1$ and
(iii) $F_X(a+0)=F_X(a)$, $\forall\, a\in(-\infty,\infty)$.
For the present problem, $F_X(10+0)=\lim_{x\to 10+}\left(1-\frac{1}{10x}\right)=1-\frac{1}{100}=\frac{99}{100}$ and $F_X(10)=0$.
$\therefore$ $F_X(10+0)\neq F_X(10)$.
$\therefore$ F_X is not a probability distribution function.

Ex. 17. If X is a gamma (l) variate find the distribution of $\sqrt{X}$. [C.H. 2006]

Solution. As X is a gamma (l) variate, $\therefore$ the pdf of X is

$$f_X(x)=\begin{cases}\frac{e^{-x}x^{l-1}}{\Gamma(l)}, & 0<x<\infty\\ 0, & \text{elsewhere,}\end{cases}$$

where $l(>0)$ is the only parameter.
Let $Y=\sqrt{X}$ be a new RV.
In real variables, $y=\sqrt{x}$. $\therefore$ $\frac{dy}{dx}=\frac{1}{2\sqrt{x}}>0$, $\forall\, x(0,\infty)$.
$\therefore$ the pdf f_y of Y is given by the rule of transformation as follows:

$$f_Y(y)=f_X(x)\left|\frac{dx}{dy}\right|=e^{-x}\frac{x^{l-1}}{\Gamma(x)}\cdot 2\sqrt{x}=e^{-y^2}\frac{y^{2l-2}}{\Gamma(l)}\cdot 2y=\frac{2}{\Gamma(l)}e^{-y^2}y^{2l-1},\ 0<y<\infty.$$

Ex. 18. If the pdf of a RV X is given by

$$f_X(x)=\begin{cases}\frac{2}{\Gamma(l)}e^{-x^2}x^{2l-1}, & 0<x<\infty\\ 0, & \text{elsewhere,}\end{cases}$$

where $l\in\mathbb{R}$ is the only parameter. Find the distribution of X^2.
Solution. Let $Y=X^2$. In terms of their real values $y=x^2$.

$$\therefore\ \frac{dy}{dx}=2x>0,\ \forall\, x\in(0,\infty).$$

$\therefore$ by the rule of transformation of RV the pdf f_y of Y is given by

$$f_Y(y)=f_X(x)\left|\frac{dx}{dy}\right|=\frac{2}{\Gamma(l)}e^{-x^2}x^{2l-1}\cdot\frac{1}{2x}\ \text{as } x>0$$
$$=\frac{e^{-x^2}x^{2l-2}}{\Gamma(l)}=\frac{e^{-y}y^{l-1}}{\Gamma(l)},\ 0<y<\infty$$

and $f_Y(y)=0$, elsewhere.
$\therefore$ Y is a gamma (l) distribution.

Ex. 19. If X is normal (m,σ), then show that $\frac{X-m}{\sigma}$ is standard normal.
Solution. Let $Y=\frac{X-m}{\sigma}$. In their real variables: $y=\frac{X-m}{\sigma}$. $\therefore$ $\frac{dy}{dx}=\frac{1}{\sigma}>0$, $\forall\, x\in\mathbb{R}$.
$\therefore$ by the rule of transformation of RV

$$f_Y(y)=f_X(x)\left|\frac{dx}{dy}\right|=\frac{1}{\sqrt{2\pi}\sigma}e^{-\frac{(x-m)^2}{2\sigma^2}}\sigma=\frac{1}{\sqrt{2\pi}}e^{-y^2/2},\ -\infty<y<\infty.$$

Hence, Y, i.e., $\frac{X-m}{\sigma}$ is standard normal.

Ex. 20. The probability distribution of a continuous RV X is defined by the pdf:

$$f_X(x)=\begin{cases}\frac{1}{4}x^3, & 0<x<2\\ 0, & \text{elsewhere.}\end{cases}$$

Find the constants c and d such that $P(X \le c) = P(X > c)$ and $P(X > d) = \frac{1}{4}$.

Solution. $P(X \le c) = P(X > c) = \dfrac{P(X \le c) + P(X > c)}{2} = \dfrac{1 - P(X > c) + P(X > c)}{2} = \dfrac{1}{2}$.

$$\therefore \int_{-\infty}^{c} f_X(x)\,dx = \frac{1}{2} \text{ or, } \int_{-\infty}^{0} 0\,dx + \int_{0}^{c} \frac{1}{4}x^3\,dx = \frac{1}{2} \text{ or, } \frac{1}{16}c^4 = \frac{1}{2} \text{ or, } c^4 = 8. \therefore c = \sqrt{2\sqrt{2}}.$$

$$\text{Again, } P(X > d) = \frac{1}{4} \Rightarrow \int_{d}^{\infty} f_X(x)\,dx = \frac{1}{4} \text{ or, } \int_{d}^{2} \frac{1}{4}x^3\,dx + \int_{2}^{\infty} 0\,dx = \frac{1}{4} \text{ or, } \frac{1}{16}\left(16 - d^4\right) = \frac{1}{4}$$

or, $16 - d^4 = 4$, or, $d^4 = 12$. $\therefore d = \sqrt{2\sqrt{3}}$.

Ex. 21. The life of an electric bulb (of a certain brand) X in months follows a continuous distribution determined by the following pdf:

$$f_X(x) = \begin{cases} (1/c)\, e^{-x/c}, & x > 0 \\ 0, & \text{elsewhere.} \end{cases}$$

(i) Find the value of c (ii) Find the probability that (a) the bulb will last for at least one year, (b) the bulb will last for at most two years and (c) the lasting period is between one to two years. (iii) Also, find the distribution function.

Solution. (i) $f_X(x) \ge 0, \ \forall x \Rightarrow c > 0$.

$$\text{Now, } \int_{-\infty}^{\infty} f_X(x)\,dx = \frac{1}{c}\int_{0}^{\infty} e^{-x/c}\,dx = \frac{1}{c}\lim_{R\to\infty}\int_{0}^{R} e^{-x/c}\,dx = -\lim_{R\to\infty}\left[e^{-x/c}\right]_0^R = -\lim_{R\to\infty}\left[e^{-R/c} - 1\right] = 1.$$

$\therefore$ c is any positive real number.

(ii) (a) The probability of lasting at least one year, i.e., 12 months

$$= P(X \ge 12) = \frac{1}{c}\int_{12}^{\infty} e^{-x/c}\,dx = \frac{1}{c}\lim_{R\to\infty}\int_{12}^{R} e^{-x/c}\,dx$$

$$= -\lim_{R\to\infty}\left[e^{-x/c}\right]_{12}^{R} = -\lim_{R\to\infty}\left[e^{-R/c} - e^{-12/c}\right] = e^{-12/c}, \ c > 0.$$

(b) The probability of lasting at most two years, i.e., 24 months

$$= P(X \le 24) = \frac{1}{c}\int_{0}^{24} e^{-x/c}\,dx = -\left[e^{x/c}\right]_0^{24} = 1 - e^{24/c}.$$

(c) The probability of lasting the bulb between one to two years

$= P(12 < X \le 24) = P(0 < X \le 24) - P(0 < X \le 12) = \left(1 - e^{-24/c}\right) - \left(1 - e^{12/c}\right) = e^{-12/c} - e^{-24/c}$.

(iii) The distribution function F_X is given by $F_X(x) = 0, \ x < 0$.

When $x \ge 0$,

$$F_X(x) = P(X \le x) = \int_{0}^{x} \frac{1}{c}e^{-x/c}\,dx = -\left[e^{-x/c}\right]_0^x = 1 - e^{-x/c}.$$

$$\therefore F_X(x) = \begin{cases} 0, & x < 0 \\ 1 - e^{-x/c}, & x \ge 0. \end{cases}$$

Ex. 22. The train services from Howrah to Bandel has the frequency within 10 minutes. A person reaches some station avail the route. If T be the RV denoting the time to wait to avail a train, find the distribution function of T. Hence, find the probability that (i) the person has to wait at least 5 minutes (ii) at most 6 minutes and (iii) the time to wait lying between 4 to 7 minutes.

Solution. As the person may reach the station at any time at random, $\therefore$ T, the time to wait, follows uniform distribution in the interval 0 to 10.
Hence, pdf of T is given by

$$f_t(t)=\begin{cases}1/10, & 0<t<10\\ 0, & \text{elsewhere.}\end{cases}$$

The probability distribution function F_t is given by $F_t(t)=0,\ t<0$.

When $0\le t\le 10$, $F_t(t)=\int_{-\infty}^{t} f_t(t)dt=\int_{-\infty}^{0} 0\,dt+\int_0^t \frac{1}{10}dt=\frac{t}{10}$.

When $t>10$, $F_t(t)=\int_{-\infty}^{0} 0\,dt+\int_0^{10}\frac{1}{10}dt+\int_{10}^{t} 0\,dt=1$,

$$\text{i.e., } F_t(t)=\begin{cases}0, & t<0\\ t/10, & 0\le t\le 10\\ 1, & t>10.\end{cases}$$

$\therefore$ the probability that the person has to wait at least 5 minutes

$=P(T\ge 5)=1-P(T<5)=1-P(T\le 5)$, [as the distribution is continuous,

$=1-F_t(5)=1-\frac{5}{10}=\frac{1}{2}$ $\therefore$ there is no step point and so $P(T=5)=0$]

The probability of waiting at most 6 minutes $=P(0<T\le 6)=F_t(6)=\frac{6}{10}=\frac{3}{5}$.
The probability that the waiting time lying between 4 minutes to 7 minutes

$$=P(4<T\le 7)=F_t(7)-F_t(4)=\tfrac{7}{10}-\tfrac{4}{10}=\tfrac{3}{10}.$$

Ex. 23. If F_1 and F_2 be two probability distribution functions of two RV, show that their convex combination $F=\lambda F_1+(1-\lambda)F_2$, $0<\lambda<1$ is also a probability distribution function of some RV.
Solution. (i) $\lim\limits_{x\to-\infty} F(x)=\lim\limits_{x\to-\infty}\{\lambda F_1(x)+(1-\lambda)F_2(x)\}$

$=\lambda\lim\limits_{x\to-\infty}F_1(x)+(1-\lambda)\lim\limits_{x\to-\infty}F_2(x)$ [as both the limits exist]

$=\lambda\times 0+(1-\lambda)\times 0=0.$

(ii) $\lim\limits_{x\to\infty}F(x)=\lambda\lim\limits_{x\to\infty}F_1(x)+(1-\lambda)\lim\limits_{x\to\infty}F_2(x)=\lambda\times 1+(1-\lambda)\times 1=1.$

(iii) For any $a\in\mathbb{R}$,

$\lim\limits_{x\to a+}F(x)=\lambda\lim\limits_{x\to a+}F_1(x)+(1-\lambda)\lim\limits_{x\to a+}F_2(x)=\lambda F_1(a+0)+(1-\lambda)F_2(a+0)$

$=\lambda F_1(a)+(1-\lambda)F_2(a)$ [as F_1 and F_2 are distribution functions]

$=F(a).$

$\therefore\ F(a+0)=F(a).$

Hence, F is continuous from right at every point.
(iv) F is defined in $(-\infty,\infty)$ as F_1 and F_2 are so and $F'(x)=\lambda F_1'(x)+(1-\lambda)F_2'(x)\ge 0,\ \forall\, x\in(-\infty,\infty)$, as $\lambda, 1-\lambda, F_1'(x)$ and $F_2'(x)$ are all ≥ 0.
Hence, F must be distribution function of some RV.
Note: Whenever F_1, F_2 are distribution functions of two continuous RV, then F_1 and F_2 are continuous everywhere. $\therefore$ F is also continuous and hence F must be the distribution function of some continuous RV.
In a similar way if F_1 and F_2 be the distribution functions of two discrete RV, then F will be the distribution function of some discrete RV.

In case, F_1 is the distribution function of a discrete RV and F_2 is the distribution function of a continuous RV, then F will determine a mixed distribution and the corresponding RV is a mixed RV.

Ex. 24. A discrete distribution is defined by $X = x_i = i$, $i = 1,2,3,4,5$, $P(X=1) = 0.2$, $P(X=2) = 0.26$, $P(X=3) = k$, $P(X=4) = 0.3$, $P(X=5) = 2k$. Find k and the distribution function. Also, draw the graph of the distribution function.

Solution. $\sum_{i=1}^{5} P(X=i) = 1 \Rightarrow 0.2 + 0.26 + k + 0.3 + 2k = 1$, or, $3k = 0.24$. $\therefore\ k = 0.08$.

If F_x be its distribution function then

$$F_x(x) = \begin{cases} 0, & \text{when } x < 1 \\ 0.2, & \text{when } 1 \le x < 2 \\ 0.46, & \text{when } 2 \le x < 3 \\ 0.54, & \text{when } 3 \le x < 4 \\ 0.84, & \text{when } 4 \le x < 5 \\ 1, & \text{when } x \ge 5 \end{cases}$$

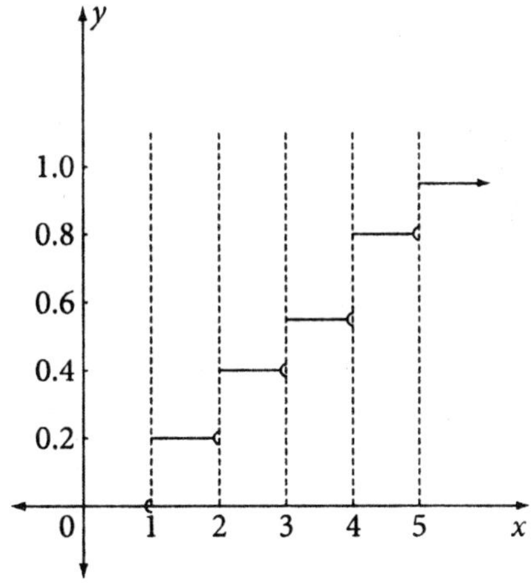

Graph of $y = F_x(x)$

Ex. 25. In a textbook of 1000 pages, total number of misprints is 1000. What is the probability that a page chosen at random will contain at most 4 misprints? Also, find the probability that a page chosen at random will contain at least 4 misprints?

Solution. The number of misprints in a page follows Poisson distribution with the parameter $\mu = 500 \times \frac{1}{500} = 1$. Let $X =$ RV denoting number of misprints in a page.

$\therefore$ probability that a page chosen at random will contain at most 4 misprints

$$= P(X=0) + P(X=1) + P(X=2) + P(X=3) + P(X=4)$$

$$= e^{-1}\frac{1^0}{0!} + e^{-1}\frac{1^1}{1!} + e^{-1}\frac{1^2}{2!} + e^{-1}\frac{1^3}{3!} + e^{-1}\frac{1^4}{4!}$$

$$= e^{-1}\left(1 + 1 + \frac{1}{2} + \frac{1}{6} + \frac{1}{24}\right) = \frac{48+12+4+1}{24e} = \frac{65}{24e}.$$

Probability of at least 4 misprints in a page

$$= P(X=4) + P(X=5) + \cdots = 1 - \{P(X=0) + P(X=1) + P(X=2) + P(X=3)\}$$

$$= 1 - e^{-1}\left(1 + 1 + \tfrac{1}{2} + \tfrac{1}{6}\right) = 1 - \tfrac{8}{3e}.$$

Ex. 26. A point A is chosen at random on a circle of radius r. Let B be a fixed point on the circle. Find the probability that the chord BA will not exceed the length of the side of an equilateral triangle inscribed within the circle.

Solution. Let X be a RV denoting $\angle B'BA$, where BB' is a diameter. The distribution of X is uniform in $(-\pi/2, \pi/2)$. Therefore, the pdf of X is given by

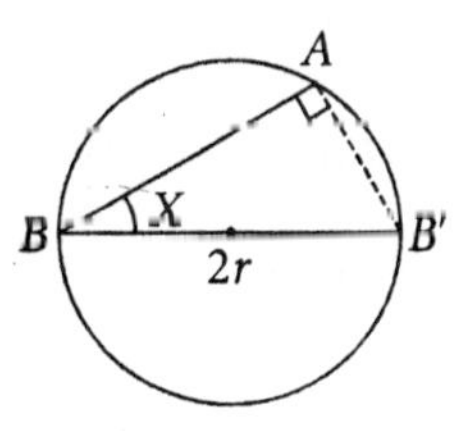

$$f_X(x) = \begin{cases} 1/\pi, & -\pi/2 < x < \pi/2 \\ 0, & \text{elsewhere.} \end{cases}$$

The side of an equilateral triangle inscribed within the circle is $r\sqrt{3}$.

$$\therefore \text{required probability} = P\left(|BA| \le r\sqrt{3}\right) = P\left(2r\cos X \le r\sqrt{3}\right) = P\left(\cos X \le \tfrac{\sqrt{3}}{2}\right)$$
$$= 1 - P\left(\cos X > \tfrac{\sqrt{3}}{2}\right) = 1 - P\left(|X| < \tfrac{\pi}{6}\right) = 1 - P\left(-\tfrac{\pi}{6} < X < \tfrac{\pi}{6}\right)$$
$$= 1 - \tfrac{1}{\pi}\int_{-\pi/6}^{\pi/6} dx = 1 - \tfrac{1}{\pi}\cdot\tfrac{\pi}{3} = \tfrac{2}{3}.$$

Ex. 27. Find the probability distribution of number of failures preceding the first success in an infinite sequence of Bernoulli's trials with probability of success p for each trial. [C.H. 2004]

Solution. Let X = the RV denoting the number of failures before the first success in an infinite sequence of Bernoulli's trials. Then, the spectrum of X is $X = i, i = 0, 1, 2, \cdots$ and

$$P(X = i) = P(F, F, \cdots, F, S) = P(F)P(F)\cdots P(F)P(S), \text{ as trials are independent}$$
$$= q^i p, \text{ when } q = 1 - p = \text{probability of failure in a trial.}$$

$\therefore$ the distribution of X is a discrete distribution and is given by

$$X = i, i = 0, 1, 2, \cdots \text{ and } f_{xi} = P(X = i) = pq^i.$$

Ex. 28. The pdf of a RV X is symmetric about the origin. Prove that X and $-X$ have the same distribution.

Solution. Let the distribution of X is defined by the pdf f_x for $x \in (-a, a), a \in \mathbb{R}$.
Let $Y = -X$. In their real values $y = -x$. $\therefore \frac{dy}{dx} = -1 < 0, \ \forall\, x \in (-a, a)$.
$\therefore$ the pdf f_y of Y is given by

$$f_Y(y) = f_X(x)\left|\frac{dx}{dy}\right| = f_X(x)|-1| = f_X(x) = f_X(-x), \text{ due to symmetry about origin}$$
$$= f_X(y),\ y \in (-a, a)$$

i.e., $f_Y(y) = f_X(y), y \in (-a, a)$. $\therefore$ X and $-X$ has the same distribution.

EXERCISE

1. Let E be a RE of rolling a die from a box till the face 6 appears for the first time. If X denotes the number of rolling required, find its probability distribution.
[Hints. Certainly X is a discrete RV with spectrum $\{1, 2, 3, \cdots\}$

$P(X = i)$ = probability that in the first $(i-1)$ rolling face 6 did not appear and 6 appeared at ith rolling

$= \frac{5}{6}\cdot\frac{5}{6}\cdots(i-1 \text{ times})\cdot\frac{1}{6}$, as the rolling are independent

$= \left(\frac{5}{6}\right)^{i-1}\cdot\frac{1}{6} = \frac{5^{i-1}}{6^i}, i = 1, 2, \cdots]$

2. Let an unbiased coin is repeatedly tossed until a tail appears for the first time. Find the distribution of the number of tosses. $\left(\textit{Ans. } P(X = i) = \frac{1}{2^i}, i = 1, 2, 3, \cdots\right)$

3. Consider the function $f_x(x) = \begin{cases} [x], & 1 < x < 2 \\ 0, & \text{elsewhere.} \end{cases}$
Is it a pdf? If so, find the distribution function.

[Hints. Here $f_X(x) = \begin{cases} 1, & 1 < x < 2 \\ 0, & \text{elsewhere.} \end{cases}$

$$\therefore f_X(x) \ge 0, \ \forall\, x \in \mathbb{R} \text{ and } \int_{-\infty}^{\infty} f_X(x)\,dx = \int_1^2 1\,dx = 1.$$

Hence, f_x is a possible pdf.
If F_x be the distribution function, then $F_x(x) = 0$, $x < 1$.
When $1 \leq x \leq 2, F_x(x) = \int_1^x dx = x$ and $F_x(x) = 1, x > 2$.]

4. Find k so that the function f_x defined by $f_x(x) = \begin{cases} k(x-[x]), & 1 < x < 2 \\ 0, & \text{elsewhere} \end{cases}$

is a pdf. Also, find the distribution function. $\left(\textit{Ans. } k=2, F_x(x) = \begin{cases} 0, & x<1 \\ (x-1)^2, & 1 \leq x \leq 2 \\ 1, & x>2 \end{cases}\right)$

5. Find k so that the function f_x defined by $f_x(x) = \begin{cases} k\left(x^{[x]}-[x]\right), & 1 < x < 2 \\ 0, & \text{elsewhere} \end{cases}$

is a pdf. Also, find the distribution function. $\left(\textit{Ans. } k=2, F_x(x) = \begin{cases} 0, & x<1 \\ (x-1)^2, & 1 \leq x \leq 2 \\ 1, & x>2 \end{cases}\right)$

6. Find k so that the function f_x defined by $f_x(x) = \begin{cases} k(x-[x]), & -1 < x < 1 \\ 0, & \text{elsewhere} \end{cases}$

is a pdf. Also, find the distribution function. $\left(\textit{Ans. } k=1, F_x(x) = \begin{cases} 0, & x<-1 \\ \frac{1}{2}(x+1)^2, & -1 \leq x < 0 \\ \frac{1}{2}(x^2+1), & 0 \leq x \leq 1 \\ 1, & x>1 \end{cases}\right)$

7. If a RV X is normal (m, σ), then show that $aX+b$ (a, b are constants) is also a normal distribution. Find its parameters. (*Ans.* $am+b, |a|\sigma$)
8. If a RV X follows $\beta_2(m,n)$ distribution, show that $\frac{1}{X}$ follows $\beta_2(n,m)$ distribution.
9. If X follows standard normal distribution, show that $\frac{1}{2}X^2$ follows gamma distribution with parameter $\frac{1}{2}$.
10. If X is $\beta_1(m,n)$ variate, show that $\frac{X}{1-X}$ is $\beta_2(m,n)$ variate.
11. If X follows Poisson (μ) distribution, find the distribution of $cX+d$ (c, d are positive constants).
[**Hints.** As X is Poisson (μ) variate, $\therefore\ X = x_i = i,\ i = 0,1,2,\cdots$ and $f_{xi} = P(X=i) = e^{-\mu}\frac{\mu^i}{i!}$.
Let $Y = cX+d$. The spectrum of Y is $Y = y_i = cx_i + d = ci + d, i = 0,1,2,\cdots$
$X = x_i \Leftrightarrow Y = y_i$. $\therefore\ f_{yi} = P(Y=y_i) = P(X=x_i) = P(X=i) = e^{-\mu}\frac{\mu^i}{i!}, i = 0,1,2,\cdots$]
12. Given that the height in cm of adult people of India is normally distributed with parameters m, σ, when $m = 165$ and $\sigma = 15$. In a gathering of 1000 Indian adults, how many of them have their heights more than 175? Given that $\frac{1}{\sqrt{2\pi}}\int_{-\infty}^{2/3} e^{-t^2/2}\,dt = 0.7475$.
[**Hints.** Let the RV X denotes the height of an Indian adult people. X is normal (165, 15). So, $\frac{X-165}{15}$ is standard normal.
$$\therefore\ P(X>175) = P\left(\frac{X-165}{15} > \frac{2}{3}\right) = 1 - P\left(\frac{X-165}{15} \leq \frac{2}{3}\right) = 1 - \int_{-\infty}^{2/3} \frac{1}{\sqrt{2\pi}} e^{-t^2} dt$$
$$= 1 - 0.7475 = 0.2525.$$
$\therefore$ required number of persons $= 0.2525 \times 1000 \simeq 252$.]
13. If the marks of Mathematics in a Higher Secondary examination is normally distributed with parameters m, σ, when $m = 55$ and $\sigma = 10$. In a gathering of 1000 such students, how many of them secured more than 65?

Given that $\frac{1}{\sqrt{2\pi}}\int_{-\infty}^{1} e^{t^2/2}\,dt = 0.8413.$ (*Ans.* 159 (approx.))

14. In the same problem (13), how many secured less than 45? (*Ans.* 159 (approx.))

15. Three concentric circles of radii $\frac{1}{2}, \frac{\sqrt{3}}{2}$ and 1 cm are drawn on the target board. If a shot falls within the inner most circle, 3 points are scored; if it falls within the next two rings, the scores are respectively 2 and 1. If the pdf of the distance of hit from the centre is $f_r(r) = \frac{2}{\pi}\cdot\frac{1}{\sqrt{1-r^2}}$, find the probability distribution of scores.
[**Hints.** If X be the RV denoting the scores, then the spectrum of X is the set $\{1,2,3\}$.

$$P(X=1) = \frac{2}{\pi}\int_{\sqrt{3}/2}^{1}\frac{dr}{\sqrt{1-r^2}} = \frac{2}{\pi}\left(\sin^{-1}1 - \sin^{-1}\frac{\sqrt{3}}{2}\right) = \frac{2}{\pi}\cdot\frac{\pi}{6} = \frac{1}{3}$$

$$P(X=2) = \frac{2}{\pi}\int_{1/2}^{\sqrt{3}/2}\frac{dr}{\sqrt{1-r^2}} = \frac{2}{\pi}\left(\sin^{-1}\frac{\sqrt{3}}{2} - \sin^{-1}\frac{1}{2}\right) = \frac{1}{3}$$

$$P(X=3) = \frac{2}{\pi}\int_{0}^{1/2}\frac{dr}{\sqrt{1-r^2}} = \frac{2}{\pi}\sin^{-1}\frac{1}{2} = \frac{1}{3}.$$

Note that with this pdf some score 1, 2 or 3 will always happen.]

16. A point A is chosen at random on a circle of radius r. Let B be a fixed point on the circle. Find the probability that the chord BA will not exceed the length of a side of an equilateral triangle inscribed within the circle. (*Ans.* 2/3)

17. A fire brigade has three cars. The number of demand of a car (which is exactly equal to the number of fire accidents) on each day follows Poisson (1.2) distribution. Find the distribution function of the number of demand of a car in a day.
[**Hints.** Let $X =$ number of demand on each day. $\therefore$ X is a discrete distribution with spectrum $X = 0,1,2,\cdots$
Given $f_{xi} = P(X=i) = e^{-1.2}\frac{(1.2)^i}{i!}, i = 0,1,2,\cdots$
If F_x be the distribution function, then $F_x(x) = P(X \le i)$, if $x_i \le x < x_{i+1}$ and $x_i = i$

$$\text{or, } F_x(x) = \frac{1}{i!}\int_{1.2}^{\infty} e^{-x}x^i\,dx, \text{ when } i \le x < i+1,\ i = 0,1,2,\cdots$$
$$= P(X=i) + P(X=i-1) + \cdots + P(X=1) + P(X=0)].$$

18. An urn contains 15 balls out of which 5 are blue and remaining 10 are red. If 6 balls are drawn from the urn, find the distribution of the number of blue balls drawn in the cases (i) with replacement and (ii) without replacement.
[**Hints.** In case (i) the number of blue ball is binomial $(6, \frac{1}{3})$ distribution. In case (ii) spectrum of X, the number of blue balls, is $\{0,1,2,3,4,5\}$ with respective probabilities:

$$P(X=0) = \frac{\binom{10}{6}}{\binom{15}{6}},\ P(X=1) = \frac{\binom{10}{5}\binom{5}{1}}{\binom{15}{6}},\ P(X=2) = \frac{\binom{10}{4}\binom{5}{2}}{\binom{15}{6}},\ \text{etc.}]$$

19. If X be a normal $(0,1)$ variate, find the density function of X^2. [C.H. 2007]
[**Hints.** Let $Y = X^2$. In their real values $y = x^2$. $\therefore$ $\frac{dy}{dx} = 2x$. As $x \in (-\infty,\infty)$, $\frac{dy}{dx}$ does not mention the same sign. Also, as x varies from $-\infty$ to ∞, y varies from 0 to ∞.

Let $x > 0$.
$\therefore$ the event $y < Y \le y + dy \Leftrightarrow x^2 < X^2 \le (x + dx)^2 \Leftrightarrow (-x - dx \le X < -x) + (x < X \le x + dx)$.
$\therefore$ $P(y < Y \le y + dy) = P(-x - dx \le X < -x) + P(x < X \le x + dx)$, as events RHS are MEE $= 2P(x < X \le x + dx)$, by the symmetry of the normal distribution.
$\therefore$ $f_Y(y)dy = 2f_X(x)dx$.
Hence, $f_Y(y) = 2\frac{1}{\sqrt{2\pi}}e^{-\frac{x^2}{2}} \cdot \frac{1}{2x}\left[\because \frac{dx}{dy} - 2x > 0\right]$
$= \frac{1}{\sqrt{2\pi}}\frac{e^{-\frac{y}{2}}}{\sqrt{y}}$, $0 < y < \infty$.
Note: Later on we see that Y is a $\chi^2(1)$ variate.

20. If X is normal $(0,1)$ variate, prove that $\frac{1}{2}X^2$ is gamma $\left(\frac{1}{2}\right)$ variate.

21. If X is gamma (l) variate, find the distribution of $\frac{1}{2}X^2$.
$\left(\textit{Ans. } f_Y(y) = \begin{cases} \frac{e^{-\sqrt{2y}}(2y)^{\frac{l}{2}-1}}{\Gamma(l)}, & y > 0 \\ 0, & \text{elsewhere} \end{cases}\right)$

22. If X is uniformly distributed in the interval $(-1, 1)$, find the probability distribution of $|X|$. [C.H. 2007]
[**Hints.** Let $Y = |X|, y = |x|, \frac{dy}{dx} = 1, x > 0, \frac{dy}{dx} = -1, \frac{dy}{dx}$ is undefined at $x = 0$. So, $\frac{dy}{dx}$ changes its sign in $(-1,0) \cup (0,1)$.
Let $x > 0$. $\therefore$ the event $y < Y \le y + dy \Leftrightarrow |x| < |X| \le |x + dx| \Leftrightarrow (x < |X| \le x + dx)$, as $x > 0$
$\Leftrightarrow (-x - dx \le X < x)(x < X \le x + dx)$.
$\therefore$ $P(y < Y \le y + dy) = 2P(x < X \le x + dx)$ since $P(-x - dx \le X < -x) = P(x < X \le x + dx)$ due to symmetry of uniform distribution.
$\therefore$ $f_Y(y) = 2f_X(x)\frac{dx}{dy} = 2 \cdot \frac{1}{2} \cdot 1 = 1$, as $f_X(x) = \begin{cases} \frac{1}{2}, & -1 < x < 1 \\ 0, & \text{elsewhere.} \end{cases}$
$\therefore$ $f_Y(y) = \begin{cases} 1, & 0 < y < 1 \\ 0, & \text{elsewhere.} \end{cases}$
$\therefore$ Y is also uniformly distributed.
In case $x < 0$, the same result follows (check).]

23. Two numbers are independently chosen between 0 and 1. Show that the probability of their product is less than a constant $k(0 < k < 1)$ is $k(1 - \log k)$. [C.H. 2006, 2007]

24. A point P is chosen at random on a circle of radius a and A is a fixed point on the circle. Find the probability that the chord AP will exceed the length of the side of an equilateral triangle inscribed in the circle. (*Ans.* 1/3)
[C.H. 2008]

25. The pdf of a RV X is symmetrical about the origin. Prove that X and $-X$ have the same distribution. [C.H. 2004]

26. The RV X has pdf $f_X(x) = \frac{1}{2}e^{-|x|}, -\infty < x < \infty$. Find the pdf of $Y = X^2$. [C.H. 2004]

27. If two numbers p and q are chosen at random from the set $\{1, 2, \cdots, 10\}$ with replacement, find the probability that the equation $x^2 + px + q = 0$ has real roots.
[**Hints.** The roots of the equation will be real iff $p^2 \ge 4q$. (1)
We now find the favourable cases in favour of (1) out of 10^2 total possible cases.
When $q = 1$, p may take any number from the set $\{2, 3, \cdots, 10\}$, to satisfy (1), which contains 9 cases.

Similarly, when $q=2, p\in\{3,4,\cdots,10\}$, which contains 8 cases,
when $q=3, p\in\{4,5,\cdots,10\}$, which contains 7 cases,
when $q=4, p\in\{4,5,\cdots,10\}$, which contains 7 cases,
when $q=5, p\in\{5,6,\cdots,10\}$, which contains 6 cases,
when $q=6, p\in\{5,6,\cdots,10\}$, which contains 6 cases,
when $q=7, p\in\{6,7,\cdots,10\}$, which contains 5 cases,
when $q=8, p\in\{6,7,\cdots,10\}$, which contains 5 cases,
when $q=9, p\in\{6,7,\cdots,10\}$, which contains 5 cases,
when $q=10, p\in\{7,8,9,10\}$, which contains 4 cases.
$\therefore$ total number of favourable cases $=9+8+2\times 7+2\times 6+3\times 5+4=62$.
$\therefore$ required probability $=\frac{62}{100}=0.62$ by classical rule.]

28. The mean and variance of a binomial distribution are μ and $k\mu$ respectively ($k>0$, a constant). Find the probability of no success, one success and at least one success. [C.H. 10]
[**Hints.** Let the distribution be binomial (n,p).]
$\therefore$ by the given conditions the mean $np=\mu$ and the variance $np(1-p)=k\mu$.
$\therefore\ \mu(1-p)=k\mu$. $\therefore\ p=1-k$ as $\mu\neq 0$ for $n\neq 0, p\neq 0$. $\therefore\ n=\frac{\mu}{p}=\frac{\mu}{1-k}$.
$\therefore$ the distribution is $B\left(\frac{\mu}{1-k},1-k\right)$.
$\therefore$ probability of no success $=(1-p)^n=k^{\frac{\mu}{1-k}}$.
Probability of at least one success $=1-k^{\frac{\mu}{1-k}}$.
Probability of one success $={}^nC_1p(1-p)^{n-1}=n(1-k)k^{\frac{\mu}{1-k}-1}$.

MISCELLANEOUS PROBLEMS

1. Two players toss an unbiased coin n times each. Show that the probability of their scoring the same number of tails is $\binom{2n}{n}2^{-2n}$. [H.S. 2001]
[**Hints.** If A_r is the event that a player will get r number of tails out of n tosses, then by binomial law $P(A_r)$ is given by $P(A_r)=\binom{n}{r}\left(\frac{1}{2}\right)^r\left(\frac{1}{2}\right)^{n-r}=\binom{n}{r}\left(\frac{1}{2}\right)^n, r=0,1,2,\cdots,n$. If B_r is the event that the other player will also get r number of tails out of n tosses is $P(B_r)=\binom{n}{r}\left(\frac{1}{2}\right)^n$.
$\therefore\ A_rB_r$ is the event that both will get r tails out of n tosses and A_r, B_r are independent.
$\therefore\ P(A_rB_r)=P(A_r)P(B_r)=\binom{n}{r}^2\frac{1}{2^{2n}}, r=0(1)n$.
$\therefore$ required probability $=\sum_{r=0}^{n}P(A_rB_r)=\left[\binom{n}{0}^2+\binom{n}{1}^2+\cdots+\binom{n}{n}^2\right]2^{-2n}=\binom{2n}{n}2^{-2n}$ [by the properties of binomial coefficients]]

2. Two players toss a biased coin n times each. Find the probability of scoring same number of tails if the probability of getting head in a single toss of the coin is 2/3.
[**Hints.** $P(A_r)=\binom{n}{r}\left(\frac{1}{3}\right)^r\left(\frac{2}{3}\right)^{n-r}$ and $P(B_r)=P(A_r)$.
$P(A_rB_r)=P(A_r)P(B_r)=\{P(A_r)\}^2=\binom{n}{r}^2\left(\frac{1}{3}\right)^{2r}\left(\frac{2}{3}\right)^{2n-2r}$
Required probability $=\sum_{r=0}^{n}\binom{n}{r}^2\left(\frac{1}{3}\right)^{2r}\left(\frac{2}{3}\right)^{2n-2r}=\frac{2^{2n}}{3^{2n}}\sum_{r=0}^{n}\binom{n}{r}^2\frac{1}{4^r}$.]

3. Two players roll an unbiased die n times each. Find the probability of their scoring the same number of sixes. $\left(Ans.\ \left(\frac{5}{6}\right)^{2n}\sum_{r=0}^{n}\binom{n}{r}^2\frac{1}{25^r}.\right)$

4. If $X\sim B(n,p)$ and $P(X=r)=P(X=n-r), r\neq n/2$, find p.
[**Hints.** $P(X=r)=P(X=n-r)\Rightarrow\binom{n}{r}p^rq^{n-r}=\binom{n}{n-r}p^{n-r}q^r$, when $q=1-p$

$$\text{or, } p^{2r-n}=q^{2r-n} \text{ or, } \left(\frac{p}{q}\right)^{2r-n}=1.$$

As $2r-n\neq 0$, $\therefore\ p=q=\frac{p+q}{2}=\frac{1}{2}$.]

5. If 3% of the bolts manufactured by a company are defective, what is the probability that in a sample of size 200 (i) 5 will be defective (ii) none is defective. [Given $e^{-6} = 0.00248$]. [C.H. 2009]
Solution. As $n = 200$ is large and p = probability of defective bolt (success) = 0.03 is small, $\therefore$ the distribution of defective bolt is approximately Poisson μ distribution, where $\mu = np = 200 \times \frac{3}{100} = 6$.
$\therefore\ P(A_5) = e^{-6} \cdot \frac{6^5}{5!} = 0.1607;\ P(A_0) = e^{-6} \cdot \frac{6^0}{0!} = e^{-6} = 0.00248.$
Alternative Solution. Probability of defective bolt is $\frac{3}{100}$.
$\therefore$ probability of defectless bolt is $\frac{97}{100}$.
$\therefore$ by binomial law,
$$P(A_5) = \binom{200}{5}\left(\frac{3}{100}\right)^5\left(\frac{97}{100}\right)^{195} = 0.1622$$
$$P(A_0) = \binom{200}{0}\left(\frac{3}{100}\right)^0\left(\frac{97}{100}\right)^{200} = \left(\frac{97}{100}\right)^{200} \simeq 0.00226,$$
where A_n = the event of n defective bolts out of 200 bolts.

6. If z is a standard normal variate and $P(0 < z < 0.8) = 0.2881$, find $P(|z| \geq 0.8)$. [C.H. 2001]
Solution.
$$\begin{aligned} P(|z| \geq 0.8) &= (z \geq 0.8 + z \leq -0.8) = (z \geq 0.8) + P(z \leq -0.8), \text{ as the events are MEE} \\ &= 2P(z \geq 0.8), \text{ as } P(z \leq -0.8) = P(z \geq 0.8) \text{ due to symmetry of standard normal distribution} \\ &= 2\{P(z \geq 0) - P(0 < z < 0.8)\} = 2(0.5 - 0.2881) = 0.4238. \end{aligned}$$

7. If $X \sim N(50, 4)$ distribution, find $P(41 \leq X \leq 56)$. [C.H. 1992]
[Hints. $X \sim N(50,4) \Rightarrow Z = \frac{X-50}{4} \sim N(0,1)$.
$$\begin{aligned} \therefore\ P(41 \leq X \leq 56) &= P\left(\frac{41-50}{4} \leq \frac{X-50}{4} \leq \frac{56-50}{4}\right) = P\left(-\frac{9}{2} \leq Z \leq \frac{6}{4}\right) \\ &= F_z\left(\frac{3}{2}\right) - F_z\left(-\frac{9}{4}\right), \ F_z \text{ is standard normal distribution function} \\ &= F_z\left(\frac{3}{2}\right) - \{1 - F_z\left(\frac{9}{4}\right) = F_z\left(\frac{3}{2}\right) + F_z\left(\frac{9}{4}\right) - 1 = 0.921.] \end{aligned}$$

8. Let there be two identical match boxes each containing N sticks. The RE is to select a box at random and a stick is withdrawn from it and use. This experiment is continued till one box is found empty. What is the probability that the other box contains exactly $r(\leq N)$ sticks.
Solution. p = probability of selecting a match box at random = $\frac{1}{2}$, as the match boxes are identical. First we find the probability that the first box is empty and the second box still contains r sticks after $2N - r$ trials. Taking the selection of first box as success, the probability is
$$\binom{2N-r}{N}\left(\frac{1}{2}\right)^N\left(\frac{1}{2}\right)^{N-r}\frac{1}{2}, \text{ i.e., } \binom{2N-r}{N}\left(\frac{1}{2}\right)^{2N-r+1}$$
because the selection of first urn when found empty has the probability $\frac{1}{2}$.
Similarly, the probability that the second urn is found empty first is also $\binom{2N-r}{N}\left(\frac{1}{2}\right)^{2N-r+1}$
$\therefore$ required probability is their sum, i.e., $\binom{2N-r}{N}\left(\frac{1}{2}\right)^{2N-r}$

9. A car-hire firm has two cars, which it hires out by the day. The number of demands for a car in each day is a Poisson distribution with parameters 1.5. Calculate the probability of no demand of car in a day. Also, calculate the probability that the demands cannot be served in a day (i.e., number of demands is more than two). [C.H. 1980]

Solution. Let X denotes the RV denoting the number of demands in a day. Then, the distribution of X is Poisson with parameter $\mu = 1.5$.

$$\therefore \; P(X=0) = e^{-1.5}\frac{(1.5)^0}{0!} = e^{-1.5} = 0.223$$

$$\text{and } P(X>2) = 1 - P(X \le 2) = 1 - \{P(X=0) + P(X=1) + P(X=2)\}$$

$$= 1 - \left(0.223 + e^{-1.5}\frac{(1.5)^1}{1!} + e^{-1.5}\frac{(1.5)^2}{2!}\right)$$

$$= 1 - 0.223(1 + 1.5 + 1.125) = 0.192.$$

10. The distribution function F_x of a continuous RV X is given by $F_X(x) = \begin{cases} 1 - e^{-x/\theta}, & x \ge 0, \theta > 0 \\ 0, & \text{elsewhere.} \end{cases}$

Find $P\{X \ge s + t \mid X \ge t\}$. [BH 02]

Solution. If f_x be the pdf, then $f_X(x) = F_X'(x) = \frac{1}{\theta} e^{-x/\theta}, x \ge 0, \theta > 0.$

$$P(X \ge s+t \mid X \ge t) = \int_{x \ge s+t \mid x \ge t} f_X(x)\,dx = 1 - \int_{x < s+t \mid x \ge t} f_X(x)\,dx = 1 - \int_{t \le x < s+t} f_X(x)\,dx$$

$$= 1 - \int_t^{s+t} \frac{1}{\theta} e^{-x/\theta}\,dx = 1 + \left[e^{-x/\theta}\right]_t^{s+t} = 1 + e^{-(s+t)/\theta} - e^{-t/\theta}$$

$$= 1 - e^{-t/\theta}\left(1 - e^{-s/\theta}\right).$$

Chapter 5

Probability Distribution of Higher Dimensions

5.1 Introduction

Probability distribution of dimension more than one is much more important from the statistical point of view. Such distributions will give us the independence, correlation coefficients, etc. of two RV X and Y, say, denoting two different characteristics of a class of items of the same kind. Suppose, we are going to study the merit of the students of W.B. in degree level in the subjects Bengali and Mathematics based on marks in those subjects in some state level common examination. If the RV X denotes the marks in Bengali and RV Y denotes the marks in Mathematics, then we can have the joint probability distribution of the RV X and Y or that of the two-dimensional RV (X, Y) and study whether such marks are correlated or not, i.e., whether a student strong in Bengali is also strong in Mathematics or whether a student strong in Bengali but weak in Mathematics or there is no such correlation. We can also study whether X, Y are independent or not, i.e., the marks in Bengali do not depend on the marks in Mathematics and vice versa, etc.

5.1.1 Let E be a RE and S be its event space. X and Y are two RV defined on S. Then, (X, Y) is called a *two-dimensional RV*. Now, we are in a position to define the distribution of the two-dimensional RV (X, Y) or the joint distribution of the RV X and Y.

We know that the probability distribution of a RV is completely determined by its distribution function. So, to have the distribution of (X, Y) we first define the distribution function of (X, Y).

The probability distribution function of a two-dimensional RV (X, Y) is denoted by $F_{x,y}(x, y)$ or sometimes by $F(x, y)$ and is defined by

$$F_{x,y}(x,y) = P(-\infty < X \le x,\ -\infty < Y \le y), \tag{1}$$

where the event $(-\infty < X \le x, -\infty < Y \le y)$ is the same as $(-\infty < X \le x)(-\infty < Y \le y)$, i.e., the event occurs only when X ranges in $(-\infty, x]$ and Y ranges in $(-\infty, y]$. The distribution of the two-dimensional RV (X, Y) is completely determined by the probability distribution function (1).

5.1.2 Properties of $F_{x,y}(x,y)$

I. $F_{x,y}(x,y)$ is non-decreasing in both the variables x and y

Let $a, b, c, d \in \mathbb{R}$ such that $a < b$ and $c < d$.

$\therefore (-\infty < X \le a, -\infty < Y \le c) + (a < X \le b, -\infty < Y \le c) = (-\infty < X \le b, -\infty < Y \le c)$.

The events on the LHS are MEE as

$$\begin{aligned}(-\infty < X \le a, -\infty < Y \le c)(a < X \le b, -\infty < Y \le c) &= (-\infty < X \le a)(-\infty < Y \le c)(a < X \le b)(-\infty < Y \le c)\\ &= (-\infty < X \le a)(a < X \le b)(-\infty < Y \le c)\\ &= 0(-\infty < Y \le c) = 0.\end{aligned}$$

$\therefore$ (2) $\Rightarrow P(-\infty < X \le a, -\infty < Y \le c) + P(a < X \le b, -\infty < Y \le c) = P(-\infty < X \le b, -\infty < Y \le c)$, [by axiom (III)]

or, $F_{x,y}(a,c) + P(a < X \le b, -\infty < Y \le c) = F_{x,y}(b,c)$

or, $F_{x,y}(b,c) - F_{x,y}(a,c) = P(a < X \le b, -\infty < Y \le c) \ge 0$, [by axiom (I)]

i.e., $F_{x,y}(b,c) - F_{x,y}(a,c) \ge 0$, when $b > a$.

$\therefore$ $F_{x,y}(x,y)$ is non-decreasing in x.

Again, $(-\infty < X \le b, -\infty < Y \le c) + (-\infty < X \le b, c < Y \le d) = (-\infty < X \le b, -\infty < Y \le d)$.

The events on LHS are MEE (like the previous).

$\therefore$ by axiom (III),

$P(-\infty < X \le b, -\infty < Y \le c) + P(-\infty < X \le b, c < Y \le d) = P(-\infty < X \le b, -\infty < Y \le d)$

or, $F_{x,y}(b,c) + P(-\infty < X \le b, c < Y \le d) = F_{x,y}(b,d)$

or, $F_{x,y}(b,d) - F_{x,y}(b,c) = P(-\infty < X \le b, c < Y \le d) \ge 0$, [by axiom (I)].

$\therefore$ $F_{x,y}(b,d) \ge F_{x,y}(b,c)$, when $d > c$.

$\therefore$ $F_{x,y}(x,y)$ is non-decreasing in y.

II. We want to find the probability that the two-dimensional RV (X, Y) lies in the close-open rectangle bounded between the lines $x = a, x = b, y = c$ and $y = d$, i.e., $P(a < X \le b, c < Y \le d)$.

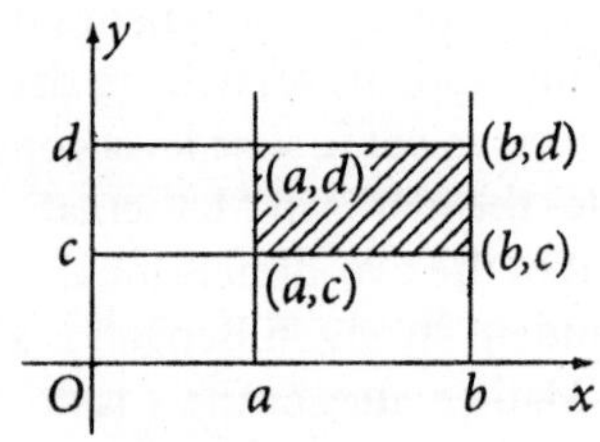

We consider $b > a$ and $d > c$.

$\therefore$ $(-\infty < X \le b, -\infty < Y \le d) = (-\infty < X \le a, -\infty < Y \le d) + (a < X \le b, -\infty < Y \le d)$.

The events on RHS are MEE, as $(-\infty < X \le a)$ and $(a < X \le b)$ are so.

$\therefore$ $P(-\infty < X \le b, -\infty < Y \le d) = P(-\infty < X \le a, -\infty < Y \le d) + P(a < X \le b, -\infty < Y \le d)$

or, $F_{x,y}(b,d) - F_{x,y}(a,d) = P(a < X \le b, -\infty < Y \le d)$. (1)

Again, $(-\infty < X \le a, -\infty < Y \le c) + (a < X \le b, -\infty < Y \le c) = (-\infty < X \le b, -\infty < Y \le c)$ and the events on LHS are MEE.

$\therefore$ considering their probabilities $F_{x,y}(b,c) - F_{x,y}(a,c) = P(a < X \le b, -\infty < Y \le c)$. (2)

Also, $(a < X \le b, -\infty < Y \le c) + (a < X \le b, c < Y \le d) = (a < X \le b, -\infty < Y \le d)$.

As the events on LHS are MEE, $\therefore$ considering their probabilities, we get

$$P(a < X \le b, -\infty < Y \le c) + P(a < X \le b, c < Y \le d) = P(a < X \le b, -\infty < Y \le d).$$

$\therefore$ $P(a < X \le b, c < Y \le d) = P(a < X \le b, -\infty < Y \le d) - P(a < X \le b, -\infty < Y \le c)$

$= F_{x,y}(b,d) - F_{x,y}(a,d) - \{F_{x,y}(b,c) - F_{x,y}(a,c)\}$ [by (1) and (2)]

or, $P(a < X \le b, c < Y \le d) = F_{x,y}(b,d) + F_{x,y}(a,c) - F_{x,y}(a,d) - F_{x,y}(b,c)$.

III. $F_{x,y}(-\infty, y) = 0, F_{x,y}(x, -\infty) = 0, F_{x,y}(-\infty, -\infty) = 0, F(\infty, \infty) = 1$.

We have the function $F_{x,y} : \mathbb{R} \times \mathbb{R} \to [0,1]$ defined by $F_{x,y}(x,y) = P(-\infty < X \le x, -\infty < Y \le y), (x,y) \in \mathbb{R} \times \mathbb{R}$.

The function is non-decreasing in both the variables x and y and is bounded on $\mathbb{R} \times \mathbb{R}$.

Taking limit as $x \to -\infty$, we get $F_{x,y}(-\infty, y) = 0$, as event on RHS is impossible.

Taking limit as $y \to -\infty$, we get $F_{x,y}(x, -\infty) = 0$, as event on RHS is impossible.

Taking limit as $x \to -\infty$ and $y \to -\infty$, we get $F_{x,y}(-\infty, -\infty) = 0$, as event on RHS is impossible.

Taking limit as $x \to \infty$ and $y \to \infty$, we get $F_{x,y}(\infty, \infty) = 1$, as the event on RHS is the certain event.

IV. $F_{x,y}(a+0, c) = F_{x,y}(a,c)$ and $F_{x,y}(a, c+0) = F_{x,y}(a,c)$. Let δa and δc are small positive increments of a and c respectively.

$\therefore$ $(-\infty < X \le a, -\infty < Y \le c) + (a < X \le a + \delta a, -\infty < Y \le c) = (-\infty < X \le a + \delta a, -\infty < Y \le c)$.

As LHS is the union of two ME events,

$\therefore P(-\infty < X \le a, -\infty < Y \le c) + P(a < X \le a + \delta a, -\infty < Y \le c) = P(-\infty < X \le a + \delta a, -\infty < Y \le c)$

or, $F_{x,y}(a, c) + P(a < X \le a + \delta a, -\infty < Y \le c) = F_{x,y}(a + \delta a, c)$

or, $F_{x,y}(a + \delta a, c) - F_{x,y}(a, c) = P(a < X \le a + \delta a, -\infty < Y \le c)$.

Taking limit as $\delta a \to 0$, $F_{x,y}(a + 0, c) - F_{x,y}(a, c) = 0$, as the event in the RHS under limit is impossible.

$\therefore F_{x,y}(a + 0, c) = F_{x,y}(a, c)$.

Again, $(-\infty < X \le a, -\infty < Y \le c) + (-\infty < X \le a, c < Y \le c + \delta c) = (-\infty < X \le a, -\infty < Y \le c + \delta c)$.

Proceeding as above, we get $F_{x,y}(a, c + \delta c) - F_{x,y}(a, c) = P(-\infty < X \le a, c < Y \le c + \delta c)$.

Taking limit as $\delta c \to 0$, $F_{x,y}(a, c + 0) - F_{x,y}(a, c) = 0$, as the event in the RHS under limit is impossible.

$\therefore F_{x,y}(a, c + 0) = F_{x,y}(a, c)$.

V. From the property (II), we have $P(a < X \le b, c < Y \le d) = F_{x,y}(b, d) + F_{x,y}(a, c) - F_{x,y}(a, d) - F_{x,y}(b, c)$.

Now, we want to find the probability when X is confined to the line $x = b$ but Y varies from c to d, i.e., $P(X = b, c < Y \le d)$.

To have it, we consider the limit as a tends to b from left (a is taken as a parameter and b is constant). Under the limit the event $(a < X \le b, c < Y \le d)$ becomes $(X = b, c < Y \le d)$.

$\therefore P(X = b, c < Y \le d) = F_{x,y}(b, d) + F_{x,y}(b - 0, c) - F_{x,y}(b - 0, d) - F_{x,y}(b, c)$.

Similarly, taking limit as c tends to d from below, we get

$$P(a < X \le b, Y = d) = F_{x,y}(b, d) + F_{x,y}(a, d - 0) - F_{x,y}(a, d) - F_{x,y}(b, d - 0).$$

Taking both the limits a tends to b from left and c tends to d from below (see Fig.) we get $P(X = b, y = d)$ as the following:

$P(X = b, Y = d) = F_{x,y}(b, d) + F_{x,y}(b - 0, d - 0) - F_{x,y}(b - 0, d) - F_{x,y}(b, d - 0)$.

Alternative proof of property IV, viz., $F_{x,y}(a, c) = F_{x,y}(a + 0, c)$ and $F_{x,y}(a, c + 0) = F_{x,y}(a, c)$, i.e., the joint distribution function $F_{x,y}$ is continuous from right and upper side at each point $(a, c) \in \mathbb{R}^2$.

Proof. Let us consider the sequence of event $\{A_n\}_n$ defined by $A_n = (a < X \le a + \frac{1}{n}, -\infty < Y \le c)$, so that

$$\lim A_n = 0 \text{ (an impossible event)}. \tag{1}$$

Now, $(-\infty < X \le a, -\infty < Y \le c) + \left(a < X \le a + \frac{1}{n}, -\infty < Y \le c\right) = \left(-\infty < X \le a + \frac{1}{n}, -\infty < Y \le c\right)$.

Note that the events on LHS are MEE.

$\therefore P(-\infty < X \le a, -\infty < Y \le c) + P\left(a < X \le a + \frac{1}{n}, -\infty < Y \le c\right) = P\left(-\infty < X \le a + \frac{1}{n}, -\infty < Y \le c\right)$

or, $F_{x,y}\left(a + \frac{1}{n}, c\right) - F_{x,y}(a, c) = P\left(a < x \le a + \frac{1}{n}, -\infty < Y \le c\right)$.

Considering limit as $n \to \infty$, we get $F_{x,y}(a + 0, c) - F_{x,y}(a, c) = 0$ [by (1)].

$\therefore F_{x,y}(a + 0, c) = F_{x,y}(a, c)$ **[Proved]**

To prove the second one we consider the sequence of events $\{B_n\}_n$ defined by $B_n = (-\infty < X \le a, c < Y \le c + \frac{1}{n})$, so that $\lim B_n = 0$. (2)

Now, $(-\infty < X \le a, -\infty < Y \le c) + \left(-\infty < X \le a, c < Y \le c + \frac{1}{n}\right) = \left(-\infty < X \le a, -\infty < Y \le c + \frac{1}{n}\right)$

Nothing that the events on LHS are MEE, we get

$P(-\infty < X \le a, -\infty < Y \le c) + P(-\infty < X \le a, c < Y \le c + \frac{1}{n}) = P(-\infty, < X \le a, -\infty < Y \le c + \frac{1}{n})$.

or, $F_{x,y}(a, c) + P\left(-\infty < X \le a, c < Y \le c + \frac{1}{n}\right) = F_{x,y}\left(a, c + \frac{1}{n}\right)$.

Taking limit as $n \to \infty$, we get $F_{x,y}(a, c + 0) - F_{x,y}(a, c) = 0$ [by (2)].

$\therefore F_{x,y}(a, c + 0) = F_{x,y}(a, c)$ **[Proved]**

5.2 Marginal distributions

From the probability distribution of a two-dimensional RV (X, Y), we may have the individual probability distributions of X and Y as follows:

We have $(-\infty < X \leq x, -\infty < Y \leq y) = (-\infty < X \leq x)(-\infty < Y \leq y)$.

Considering limit as $y \to \infty$,

$(-\infty < X \leq x, -\infty < Y < \infty) = (-\infty < X \leq x)S = (-\infty < X \leq x)$, ($S$ = certain event).

$\therefore$ $P(-\infty < X \leq x, -\infty < Y < \infty) = P(-\infty < X \leq x)$ or $F_{x,y}(x,\infty) = F_X(x)$ = distribution function of X.

Similarly, considering limit as $x \to \infty$, we get $F_{x,y}(\infty, y) = F_Y(y)$ = distribution function of y.

Thus we are getting the individual distribution functions called *marginal distribution functions* of X and Y by considering limits up to extreme margins. So, we get the distributions of X and Y separately.

Note: Distribution of X and marginal distribution of X has no difference whatever the RV Y is. The name *marginal* is probably due to the consideration of limit up to the margin.

Similar remark may be drawn for the RV Y also.

Now, consider the distribution of a three-dimensional RV (X, Y, Z) given by the probability distribution function $F_{x,y,z}$ defined by $F_{x,y,z}(x, y, z) = P(-\infty < X \leq x, -\infty < Y \leq y, -\infty < Z \leq z)$.

Considering $z \to \infty$ and keeping x, y unchanged, we get

$F_{x,y,z}(x, y, \infty) = P(-\infty < X \leq x, -\infty < Y \leq y, -\infty < z < \infty) = P(-\infty < X \leq x, -\infty, Y \leq y) = F_{x,y}(x, y)$, which is the two-dimensional marginal distribution function of the two-dimensional RV (X, Y).

Again, considering repeated limit $z \to \infty, y \to \infty$ or the simultaneous limit $(y, z) \to (\infty, \infty)$ and keeping x fixed, we get

$$F_{x,y,z}(x, \infty, \infty) = P(-\infty < X \leq x, -\infty < Y < \infty, -\infty < Z < \infty)$$

$$= F_X(x), \text{ as } -\infty < Y < \infty \text{ is certain event and } -\infty < Z < \infty \text{ is also certain event.}$$

Similarly, $F_{x,y,z}(\infty, y, z) = F_{y,z}(y, z)$, $F_{x,y,z}(\infty, y, \infty) = F_Y(y)$, etc.

Such a marginal concept can easily be extended to any finite-dimensional probability distribution.

5.3 Independence of RVs

Two RV X and Y defined on the same event space S are said to be independent if the events $-\infty < X \leq x$ and $-\infty < Y \leq y$ are independent, $\forall\, x, y \in \mathbb{R}$, i.e., when X and Y are independent, the occurrence or non-occurrence of either event does not dependent on the occurrence or non-occurrence of the other. Thus we have the following definition:

Definition. Two RV X and Y are said to be independent, iff $F_{x,y}(x, y) = F_X(x)F_Y(y)$.

Three RV X, Y, Z are said to be mutually independent, iff $F_{x,y,z}(x, y, z) = F_X(x)F_Y(y)F_Z(z)$.

In general, a set of n-variables $X_1, X_2, \cdots, X_n$ are said to be mutually independent, iff

$$F_{X_1, X_2 \cdots, X_n}(x_1, x_2 \cdots, x_n) = F_{X_1}(x_1)F_{X_2}(x_2) \cdots F_{X_n}(x_n) = \prod_{i=1}^{n} F_{X_i}(x_i),$$

where F_{X_i} are marginal distribution functions of $X_i, i = 1, 2, \cdots, n$.

When $F_{X_i, X_j}(x_i, x_j) = F_{X_i}(x_i)F_{X_j}(x_j)$, $\forall\, i, j = 1, 2, \cdots, n\ (i \neq j)$, we say that $X_1, X_2, \cdots, X_n$ are pairwise independent.

5.3.1 A two-dimensional RV (X, Y) and a RV Z are said to be independent, if $F_{x,y,z}(x, y, z) = F_{x,y}(x, y)F_Z(z)$.

Theorem 5.3.2 If the RV X, Y, Z are mutually independent, then (i) X, Y, Z are pairwise independent and (ii) X and (Y, Z) are independent.

Proof. (i) Let X, Y, Z are mutually independent. $\therefore$ $F_{x,y,z}(x, y, z) = F_X(x)F_Y(y)F_Z(z)$.

Considering limit as $z \to \infty$, $F_{x,y,z}(x, y, \infty) = F_X(x)F_Y(y)F_Z(\infty)$ or, $F_{x,y}(x, y) = F_X(x)F_Y(y) \Rightarrow X$ and Y are independent.

Similarly, proceeding to the limits as $x \to \infty$ and $y \to \infty$ one after another, we get $F_{y,z}(y, z) = F_Y(y)F_Z(z)$ and $F_{x,z}(x, z) = F_X(x)F_Z(z)$.

Hence, X, Y, Z are pairwise independent events.

(ii) Since, mutually independence implies pairwise independence.

$\therefore\ F_X(x)F_{Y,Z}(y,z) = F_X(x)F_Y(y)F_Z(z) = F_{X,Y,Z}(x,y,z)$.

Hence, X and (Y,Z) will be independent, when X, Y, Z are mutually independent.

Theorem 5.3.3 A necessary and sufficient condition (NASC) for the independence of two RV X and Y is that their joint distribution function can be expressed as the product of two functions, one is a function of x alone and the other is a function of y alone.

Proof. (Condition necessary). Let X, Y are independent. $\therefore\ F_{X,Y}(x,y) = F_X(x)F_Y(y)$, where F_X is a function of x alone and F_Y is a function of y alone. Hence, the condition is necessary.

Condition sufficient. Let $F_{X,Y}(x,y) = u(x)v(y)$, where u is a function of x alone and v is a function of y alone.

Proceeding to limit as $y \to \infty$, $F_{X,Y}(x,\infty) = u(x)v(\infty)$ or, $F_X(x) = u(x)v(\infty)$. (1)

Again, proceeding to limit as $x \to \infty$, $F_{X,Y}(\infty,y) = u(\infty)v(y)$ or, $F_Y(y) = u(\infty)v(y)$. (2)

Further proceeding to limit $y \to \infty$, $u(\infty)v(\infty) = F_Y(\infty) = 1$.

Hence, $F_{X,Y}(x,y) = u(x)v(y)u(\infty)v(\infty)$ $[\because\ u(\infty)v(\infty) = 1]$

$= \{u(x)v(\infty)\}\{v(y)u(\infty)\} = F_X(x)F_Y(y)$ [by (1) and (2)]

Hence, X, Y are independent. So, the condition is sufficient as well.

5.3.4 If X and Y are independent, then $P(a < X \le b, c < Y \le d) = P(a < X \le b)P(c < Y \le d)$.

Proof. We have $P(a < X \le b, c < Y \le d) = F_{X,Y}(b,d) + F_{X,Y}(a,c) - F_{X,Y}(a,d) - F_{X,Y}(b,c)$

$= F_X(b)F_Y(d) + F_X(a)F_Y(c) - F_X(a)F_Y(d) - F_X(b)F_Y(c)$, as X, Y are independent

$= \{F_X(b) - F_X(a)\}\left\{F_Y(d) - F_Y(c)\right\} = P(a < X \le b)P(c < Y \le d)$.

In particular, $P(X = b, Y = d) = P(X = b)P(Y = d)$, when X and Y are independent.

5.4 Discrete distributions

Let the spectrum of the two-dimensional RV (X, Y) consists of the discrete set of points $\left(x_i, y_j\right) \in \mathbb{R}^2$, so there are nonzero probability at these points and zero probability at the remaining points and the total probability is unity as usual.

Let $f_{ij} = P\left(X = x_i, Y = y_j\right)$, $\forall\ i, j = 0, \pm1, \pm2, \cdots$

Then, (a) $f_{ij} > 0$ for all possible spectrum points (x_i, y_i) and (b) $\sum_{j=-\infty}^{\infty}\sum_{i=-\infty}^{\infty} f_{ij} = 1$. [$f_{ij}$ is called the *pmf*]

To have its distribution function $F_{X,Y}$, we first consider the positions of x and y.

Let $x_\alpha < x < x_{\alpha+1}$ and $y_\beta \le y < y_{\beta+1}$.

Then, the distribution function $F_{X,Y}$ is given by $F_{X,Y}(x,y) = \sum_{j=-\infty}^{\beta}\sum_{i=-\infty}^{\alpha} f_{ij}$, $\alpha, \beta = 0, \pm1, \pm2, \cdots$, i.e., the distribution function is step function in two dimensions with steps of height $f_{\alpha\beta}$ at $\left(x_\alpha, y_\beta\right)$.

Obviously, $F_{X,Y}(\infty,\infty) = \sum_{j=-\infty}^{\infty}\sum_{i=-\infty}^{\infty} f_{ij} = 1$.

Also, $P(a < X \le b, c < Y \le d) = \sum_{c < y_j \le d}\ \sum_{a < x_i \le b} f_{ij}$

$=$ sum of the probabilities at the spectrum points $\left(x_i, y_j\right)$, which lie within the clo-open rectangular region: $a < X \le b, c < Y \le d$.

5.4.1 Marginal distributions

The marginal distribution function of X is given by

$$F_x(x) = F_{x,y}(x,\infty) = \sum_{j=-\infty}^{\infty}\sum_{i=-\infty}^{\infty} f_{ij}, \text{ when } x_\alpha \le x < x_{\alpha+1}$$

$$= \sum_{i=-\infty}^{\alpha} f_{xi}, \text{ where } f_{xi} = \sum_{j=-\infty}^{\infty} f_{ij},\ \alpha = 0, \pm 1, \pm 2, \cdots$$

This shows that F_x is a step function having step height f_{xi} at $x = x_i$. Thus the marginal distribution of X is a discrete distribution in one dimension.

The marginal distribution function of Y is given by

$$F_y(y) = F_{x,y}(\infty, y) = \sum_{j=-\infty}^{\beta}\sum_{i=-\infty}^{\infty} f_{ij} = \sum_{j=-\infty}^{\beta} f_{yj},$$

where $f_{yj} = \sum_{i=-\infty}^{\infty} f_{ij}$ and $\beta = 0, \pm 1, \pm 2, \cdots$, when $y_\beta \le y < y_{\beta+1}$.

Thus the marginal distribution of Y is also a discrete distribution in one dimension.

Note: Alternative notations for f_{xi} and f_{yj} are respectively $f_{i\cdot}$ and $f_{\cdot j}$.

Theorem 5.4.2 A NASC for the independence of two RV X and Y is $f_{ij} = f_{xi} f_{yj}$, i.e., the joint pmf is the product of individual pmfs.

Proof. Condition necessary. Let X and Y are independent.

$\therefore\ f_{ij} = P(X = x_i, Y = y_i) = P(X = x_i)P(Y = y_i) = f_{xi} f_{yj}$.

Condition sufficient. Let $f_{ij} = f_{xi} f_{yj}$ and $x_\alpha \le x < x_{\alpha+1}, y_\beta < y \le y_{\beta+1}$.

$$\therefore\ F_{x,y}(x,y) = P(-\infty < X \le x, -\infty < Y \le y) = \sum_{i=-\infty}^{\alpha}\sum_{j=-\infty}^{\beta} f_{ij} = \sum_{i=-\infty}^{\alpha}\sum_{j=-\infty}^{\beta} f_{xi} f_{yj}$$

$$= \sum_{i=-\infty}^{\alpha} f_{xi} \sum_{j=-\infty}^{\beta} f_{yj} = P(-\infty < X \le x)P(-\infty < Y \le y) = F_x(x)F_y(y).$$

Hence, the RV X and Y are independent.

5.5 Continuous distribution in higher dimensions

Let us consider a RE E and the associated event space S. Let X and Y be two continuous RV defined on the same event space S and $F_{x,y}$ be their joint probability distribution function. The probability distribution of the two-dimensional RV (X, Y) will be continuous when $F_{x,y}(x,y)$ is continuous everywhere in the entire $\mathbb{R}^2$ and all the partial derivatives of first and second orders of $F_{x,y}$ are piecewise continuous everywhere in $\mathbb{R}^2$ (which is necessary for the convergence of all integrals required to study such distributions).

5.5.1 In continuous case, there does not exist any step point, i.e., probability at any particular point vanishes, i.e., $P(X = b, Y = d) = 0$ for any $(b,d) \in \mathbb{R}^2$.

Verification. From property V of $F_{x,y}$, we have

$$P(X = b, Y = d) = F_{x,y}(b,d) + F_{x,y}(b-0, d-0) - F_{x,y}(b-0, d) - F_{x,y}(b, d-0)$$

$= 0$, as all the terms must have the same value due to continuity of $F_{x,y}$.

5.5.2 We now proceed to introduce the two-dimensional pdf of the RV (X, Y). First, try to evaluate

$$\int_{-\infty}^{x}\int_{-\infty}^{y} \frac{\partial^2 F_{x,y}(x,y)}{\partial x \partial y}\, dx\, dy.$$

$$\int_{-\infty}^{x}\int_{-\infty}^{y}\frac{\partial^2 F_{x,y}(x,y)}{\partial x\partial y}dx\,dy=\lim_{R\to-\infty}\lim_{S\to-\infty}\int_{R}^{x}\int_{S}^{y}\frac{\partial^2 F_{x,y}(x,y)}{\partial x\partial y}dx\,dy$$
$$=\lim_{R\to-\infty}\lim_{S\to-\infty}\int_{R}^{x}\left[\frac{\partial F_{x,y}(x,y)}{\partial x}\right]_{S}^{y}dx$$
$$=\lim_{R\to-\infty}\lim_{S\to-\infty}\int_{R}^{x}\left[\frac{\partial F_{x,y}(x,y)}{\partial x}-\frac{\partial F_{x,y}(x,S)}{\partial x}\right]dx$$
$$=\lim_{R\to-\infty}\int_{R}^{x}\frac{\partial F_{x,y}(x,y)}{\partial x}dx \text{ as } \lim_{S\to-\infty}\frac{\partial F_{x,y}(x,S)}{\partial x}=0$$
$$=\lim_{R\to-\infty}\left[F_{x,y}(x,y)\right]_{R}^{x}=F_{x,y}(x,y)-\lim_{R\to-\infty}F_{x,y}(R,y)$$
$$=F_{x,y}(x,y)=P(-\infty<X\le x,-\infty<Y\le y).$$

Naturally, $\frac{\partial^2 F_{x,y}(x,y)}{\partial x\partial y}$ may be taken as pdf for a two-dimensional distribution. Let us denote $\frac{\partial^2 F_{x,y}(x,y)}{\partial x\partial y}$ by $f_{x,y}(x,y)$. So, $f_{x,y}$ is the pdf of the distribution of (X, Y) and the relation between the distribution and density functions is given by

$$f_{x,y}(x,y)=\frac{\partial^2 F_{x,y}(x,y)}{\partial x\partial y} \text{ or, equivalently, } F_{x,y}(x,y)=\int_{-\infty}^{x}\int_{-\infty}^{y}f_{x,y}(x,y)dx\,dy.$$

Now, $$\int_{a}^{b}\int_{c}^{d}f_{x,y}(x,y)dx\,dy=\int_{a}^{b}\int_{c}^{d}\frac{\partial^2 F_{x,y}(x,y)}{dx\,dy}\partial x\partial y=\int_{a}^{b}\left[\frac{\partial F_{x,y}(x,y)}{\partial x}\right]_{c}^{d}dx$$
$$=\int_{a}^{b}\left[\frac{\partial F_{x,y}(x,d)}{\partial x}-\frac{\partial F_{x,y}(x,c)}{\partial x}\right]dx=\left[F_{x,y}(x,d)\right]_{a}^{b}-\left[F_{x,y}(x,c)\right]_{a}^{b}$$
$$=F_{x,y}(b,d)+F_{x,y}(a,c)-F_{x,y}(a,d)-F_{x,y}(b,c)$$
$$=P(a<X\le b,c<Y\le d), \text{ [by property II]}.$$

This result is quite natural as $f_{x,y}(x,y)$ is the pdf.

In general, $P((X, Y)\in R)=\iint_R f_{x,y}(x,y)dx\,dy,\ R\in\mathbb{R}^2.$

5.5.3 Marginal distributions

We have $F_x(x)=F_{x,y}(x,\infty)=$ marginal distribution function of X and $F_y(y)=F_{x,y}(\infty,y)=$ marginal distribution function of Y.

$$\therefore F_x(x)=F_{x,y}(x,\infty)=P(-\infty<X\le x,-\infty<Y<\infty)=\int_{-\infty}^{x}\int_{\infty}^{\infty}f_{x,y}(x,y)dy\,dx=\int_{-\infty}^{x}f_x(x)dx,$$

where $f_x(x)=\int_{-\infty}^{\infty}f_{x,y}(x,y)dy$, the marginal pdf of X and $F_x'(x)=f_x(x)$.

Similarly, $$F_y(y)=F_{x,y}(\infty,y)=P(-\infty<X<\infty,-\infty<Y\le y)=\int_{-\infty}^{\infty}\int_{\infty}^{y}f_{x,y}(x,y)dx\,dy$$
$$=\int_{-\infty}^{y}\left(\int_{-\infty}^{\infty}f_{x,y}(x,y)dx\right)dy=\int_{-\infty}^{y}f_y(y)dy,$$

where $f_y(y)=\int_{-\infty}^{\infty}f_{x,y}(x,y)dx$, the marginal pdf of Y and $F_y'(y)=f_y(y)$.

Theorem 5.5.4 Two RV X and Y will be independent, if and only if $f_{x,y}(x,y)=f_x(x)f_y(y)$.
Condition necessary. Let X and Y are independent. Hence, by definition,

$$F_{x,y}(x,y)=F_x(x)F_y(y)\Rightarrow \frac{\partial^2 F_{x,y}(x,y)}{\partial x\partial y}=F_x'(x)F_y'(y) \text{ or, } f_{x,y}(x,y)=f_x(x)f_y(y).$$

Condition sufficient

Let $f_{x,y}(x,y)=f_x(x)f_y(y)$.

$$\therefore \int_{-\infty}^{x}\int_{-\infty}^{y} f_{x,y}(x,y)\,dx\,dy=\int_{-\infty}^{x} f_x(x)dx\int_{-\infty}^{y} f_y(y)dy \Rightarrow F_{x,y}(x)=F_x(x)F_y(y).$$

$\therefore$ X and Y are independent.

5.5.5 Probability differential

Probability differential is the probability of

$$(X,Y)\in(x,x+dx;y,y+dy), \text{ i.e., } P(x<X\le x+dx, y<Y\le y+dy).$$

Now,

$$P(x<X\le x+dx, y<Y\le y+dy)=F_{x,y}(x+dx,y+dy)+F_{x,y}(x,y)-F_{x,y}(x,y+dy)-F_{x,y}(x+dx,y)$$
$$=dF_{x,y}(x,y)=\frac{\partial^2 F_{x,y}(x,y)}{\partial x\partial y}dx\,dy=f_{x,y}(x,y)dx,dy.$$

5.6 Some standard probability distributions of a continuous RV (X,Y) in two dimensions

1. Uniform distribution

Let us suppose that (X, Y) is uniformly distributed in some open rectangle $(a,b;c,d)$, i.e., pdf is constant in the rectangle and vanishes outside it.

So, $f_{x,y}(x,y)=\begin{cases} k, & a<x<b, c<y<d \\ 0, & \text{elsewhere.}\end{cases}$

The property $\int_{-\infty}^{\infty}\int_{-\infty}^{\infty} f_{x,y}(x,y)\,dx\,dy=1\Rightarrow\int_a^b\int_c^d k\,dx\,dy=1$

$$\Rightarrow\frac{1}{k}=\int_a^b dx\int_c^d dy=\left[x\right]_a^b\left[y\right]_c^d=(b-a)(d-c).$$

$\therefore$ $k=\frac{1}{(b-a)(d-c)}$.

$\therefore$ uniform distribution is thus defined by the following pdf:

$$f_{x,y}(x,y)=\begin{cases}\frac{1}{(b-a)(d-c)}, & a<X\le b, c<Y\le d\\ 0, & \text{elsewhere.}\end{cases}$$

In general, the uniform distribution of (x,y) in $R\subseteq\mathbb{R}^2$ is defined by

$$f_{x,y}(x,y)=\begin{cases}1/R, & (x,y)\in R\\ 0, & \text{elsewhere.}\end{cases}$$

Note that the necessary conditions for a pdf are satisfied, since $f_{x,y}(x,y)\ge 0$, $\forall\,(x,y)\in\mathbb{R}^2$ and

$$\iint_{\mathbb{R}^2} f_{x,y}(x,y)\,dx\,dy=\iint_R \frac{1}{R}dx\,dy+\iint_{\mathbb{R}^2-R} 0\,dx=\frac{1}{R}\iint_R dx\,dy=\frac{1}{R}\cdot R=1.$$

2. Bivariate normal distribution.

A two-dimensional continuous RV (X, Y) is said to be normally distributed, if its pdf is given by

$$f_{x,y}(x,y)=\frac{1}{2\pi\sigma_x\sigma_y\sqrt{1-\rho^2}}\exp\left[-\frac{1}{2(1-\rho^2)}\left\{\left(\frac{x-m_x}{\sigma_x}\right)^2+\left(\frac{y-m_y}{\sigma_y}\right)^2-2\rho\frac{x-m_x}{\sigma_x}\cdot\frac{y-m_y}{\sigma_y}\right\}\right],$$
$$(-\infty<x,y<\infty)$$

with five parameters $m_x, m_y, \sigma_x(>0), \sigma_y(>0)$ and $\rho\in(-1,1)$.

We now show that the marginal distributions of X and Y are respectively normal (m_x, σ_x) and normal (m_y, σ_y). Note that the parameter ρ has no room in the marginal distributions and so we may conclude that ρ is not a characteristic of X and Y alone, rather gives some jointness of X and Y.

Marginal distribution of Y is given by the pdf:

$$f_Y(y)=\int_{-\infty}^{\infty} f_{x,y}(x,y)\,dx$$
$$=\frac{1}{2\pi\sigma_x\sigma_y\sqrt{1-\rho^2}}\int_{-\infty}^{\infty}\exp\left[-\frac{1}{2(1-\rho^2)}\left\{\left(\frac{x-m_x}{\sigma_x}\right)^2+\left(\frac{y-m_y}{\sigma_y}\right)^2-2\rho\left(\frac{x-m_x}{\sigma_x}\right)\left(\frac{y-m_y}{\sigma_y}\right)\right\}\right]dx.$$

Let $\frac{x-m_x}{\sigma_x}=u$ and $\frac{y-m_y}{\sigma_y}=v$.

$\therefore\ dx=\sigma_x\,du;\ u\to\infty$ as $x\to\infty$ and $u\to-\infty$ as $x\to-\infty$ as $\sigma_x>0$.

$$\therefore\ f_Y(y)=\frac{1}{2\pi\sigma_y\sqrt{1-\rho^2}}\int_{-\infty}^{\infty}\exp\left[-\frac{1}{2(1-\rho^2)}\{u^2+v^2-2\rho uv\}\right]du$$
$$=\frac{1}{2\pi\sigma_y\sqrt{1-\rho^2}}\int_{-\infty}^{\infty}\exp\left[-\frac{1}{2(1-\rho^2)}\{(u-\rho v)^2+v^2(1-\rho^2)\}\right]du$$
$$=\frac{1}{\sqrt{2\pi}\sigma_y}e^{-v^2/2}\int_{-\infty}^{\infty}\frac{1}{\sqrt{2\pi}\sqrt{1-\rho^2}}e^{-\frac{(u-\rho v)^2}{2(1-\rho^2)}}du$$
$$=\frac{1}{\sqrt{2\pi}\sigma_y}e^{-\frac{(y-m_y)^2}{2\sigma_y^2}},\ \text{as}\ \frac{1}{\sqrt{2\pi}\sqrt{1-\rho^2}}\int_{-\infty}^{\infty}e^{-\frac{(u-\rho v)^2}{2(1-\rho^2)}}du$$

is of the form $\frac{1}{\sqrt{2\pi}\sigma}\int_{-\infty}^{\infty}e^{-\frac{(x-m)^2}{2\sigma^2}}dx$ and hence its value is 1.

$$\therefore\ f_Y(y)=\frac{1}{\sqrt{2\pi}\sigma y}e^{-\frac{(y-m_y)^2}{2\sigma_y^2}},\ -\infty<y\le\infty$$

$\therefore$ Y is $N(m_y, \sigma_y)$, i.e., Y is normally distributed with parameters $m_y, \sigma_y(>0)$.

Exactly in the same way, we get X is normally distributed with parameters $m_x, \sigma_x(>0)$.

Note: 1. When $\rho\neq 0$, $f_{x,y}(x,y)\neq f_x(x)f_y(y)$, i.e., X and Y are dependent.

2. When $\rho=0$,

$$f_{x,y}(x,y)=\frac{1}{2\pi\sigma_x\sigma_y}\exp\left[-\frac{1}{2}\left\{\left(\frac{x-m_x}{\sigma_x}\right)^2+\left(\frac{y-m_y}{\sigma_y}\right)^2\right\}\right]$$
$$=\frac{1}{\sqrt{2\pi}\sigma_x}e^{-\frac{(x-m_x)^2}{2\sigma_x^2}}\cdot\frac{1}{\sqrt{2\pi}\sigma_y}e^{-\frac{(y-m_y)^2}{2\sigma_y^2}}=f_x(x)f_y(y).$$

Hence, when $\rho=0$, X and Y are independent.

5.7 Conditional distribution

Discrete case. We know that $P(A|B)=\frac{P(AB)}{P(B)}$, provided $P(B)\neq 0$. Suppose, we have a discrete distribution of (X, Y) with spectrum points $(x_i, y_j), i, j=0,\pm1,\pm2,\cdots$

When $P(Y=y_j)\neq 0$, we confine our attention to $Y=y_j$ and find the distribution of X. Such a distribution is a one-dimensional discrete distribution of X, called the *conditional distribution* of X, when Y is confined to y_j. pmf of such a distribution is denoted by $f_{i|j}, i=0,\pm1,\pm2,\cdots$.

Therefore, $f_{i|j}=P(X=x_i\,|\,Y=y_j)=\dfrac{P\{(X=x_i)(Y=y_j)\}}{P(Y=y_j)}=\dfrac{P(X=x_i, Y=y_j)}{P(Y=y_j)}=\dfrac{f_{ij}}{f_{yj}}$.

Note that (i) $f_{i|j}>0$ since $f_{ij}>0$ and $f_{yj}>0$ and (ii) $\sum\limits_{i=-\infty}^{\infty} f_{i|j}=\sum\limits_{i=-\infty}^{\infty}\frac{f_{ij}}{f_{yj}}=\frac{\sum_{i=-\infty}^{\infty} f_{ij}}{f_{yj}}=\frac{f_{yj}}{f_{yj}}=1$, where f_{yj} is the marginal pmf for y.

Let us consider a RE of choosing a student at random from a class consisting of 100 students. Let X and Y be respectively denote their marks of Mathematics and Physics.

$\therefore$ X and Y are both discrete RV and $\{0,1,2,\cdots,100\}$ is the spectrum for both the variables X and Y.

Suppose, some of the students secured 60 in Physics, i.e., $P(Y=60)\neq 0$ $[y_j=60]$.

Then, $f_{i|j}=P(X=x_i\,|\,Y=60)$, $i=0,1,\cdots,100$ is the conditional pmf of X, given $Y=60$.

Such a distribution of X, i.e., the distribution of the marks of Mathematics for those students who secured 60 in Physics is the conditional distribution of X, given $Y=60$.

Similarly, the conditional distribution of Y, given $X=x_i$ is given by the pmf $f_{j|i}=\frac{f_{ij}}{f_{xi}}\ (f_{xi}\neq 0)$, $j=0,\pm1,\pm2,\cdots$.

Continuous case

In this case, $P(-\infty<X\leq x\,|\,Y=y)$ cannot define according to the previous process, as $P(Y=y)=0$ for a fixed y.

Similarly, $P(-\infty<Y\leq y\,|\,X=x)$ cannot defined according to the previous process as well in the continuous case there is no step point in a continuous distribution]. So, the study of conditional distribution in continuous case is as follows:

$$P(-\infty<X\leq x\,|\,y\leq Y<y+\Delta y)=\frac{P(-\infty<X\leq x, y\leq Y<y+\Delta y)}{P(y\leq Y<y+\Delta y)}=\frac{\int_{-\infty}^{x}\int_{y}^{y+\Delta y} f_{x,y}(x,y)\,dx\,dy}{\int_{y}^{y+\Delta y} f_y(y)\,dy}.$$

Assume that $f_{x,y}(x,y)$ and $f_y(y)$ are both integrable in the respective ranges of integration.

Applying mean value theorem of Integral Calculus, we get

$$\text{RHS}=\frac{\Delta y\int_{-\infty}^{x} f_{x,y}(x,\beta_1)\,dx}{\Delta y\, f_y(\beta_2)},\quad y<\beta_1,\beta_2<y+\Delta y.$$

Now, proceeding to limit $\Delta y\to 0+$, we get

$$P(-\infty<X\leq x\,|\,Y=y)=\frac{\int_{-\infty}^{x} f_{x,y}(x,y)\,dx}{f_y(y)}=\int_{-\infty}^{x}\frac{f_{x,y}(x,y)}{f_y(y)}\,dx=\int_{-\infty}^{x} f_x(x\,|\,y)\,dx,$$

where $f_x(x\,|\,y)=\frac{f_{x,y}(x,y)}{f_y(y)}$.

Naturally, $f_x(x\,|\,y)=$ pdf for the conditional distribution of X, given $Y=y$.

Note that $f_x(x|y)=\frac{f_{x,y}(x,y)}{f_y(y)}\geq 0$ as $f_{x,y}(x,y)\geq 0$ and $f_y(y)>0$ and

$$\int_{-\infty}^{\infty} f_x(x|y)\,dx=\int_{-\infty}^{\infty}\frac{f_{x,y}(x,y)}{f_y(y)}\,dx=\frac{\int_{-\infty}^{\infty} f_{x,y}(x,y)\,dx}{f_y(y)}=\frac{f_y(y)}{f_y(y)}=1,$$

$f_y(y)$ is the marginal pdf of Y.

Thus the pdf $f_x(x|y)$ gives us a one-dimensional probability distribution. Such a distribution is called *conditional distribution* of X, given $Y=y$.

$$\therefore\ P(x<X\leq b|Y=y)=\int_a^b f_x(x|y)\,dx\ \forall\ a,b\in\mathbb{R}.$$

Similarly, the conditional distribution of Y, given $X=x$, is defined by the pdf $f_y(y|x)$, when

$$f_y(y|x)=\frac{f_{x,y}(x,y)}{f_x(x)}\text{ and } P(c<Y\leq d|X=x)=\int_c^d f_y(y|x)\,dy,\ \forall\ c,d\in\mathbb{R}.$$

When $f_{x,y}(x,y)$ is the pdf of a bivariate normal distribution, then conditional probabilities of X (given $Y=y$) and of Y (given $X=x$) are given by the conditional pdf $f_x(x|y)$ and $f_y(y|x)$, where

$$f_x(x|y)=\frac{f_{x,y}(x,y)}{f_y(y)}$$

$$=\frac{\frac{1}{2\pi\sigma_x\sigma_y\sqrt{1-\rho^2}}\exp\left[-\frac{1}{2(1-\rho^2)}\left\{\left(\frac{x-m_x}{\sigma_x}\right)^2+\left(\frac{y-m_y}{\sigma_y}\right)^2-2\rho\left(\frac{x-m_x}{\sigma_x}\right)\left(\frac{y-m_y}{\sigma_y}\right)\right\}\right]}{\frac{1}{\sqrt{2\pi}\sigma_y}\exp\left[-\frac{(y-m_y)^2}{2\sigma_y^2}\right]}$$

$$=\frac{1}{\sqrt{2\pi}\sigma_x\sqrt{1-\rho^2}}\exp\left[-\frac{1}{2\sigma_x^2(1-\rho^2)}\left\{(x-m_x)^2+\left(\frac{\sigma_x}{\sigma_y}(y-m_y)\right)^2-2\rho\frac{\sigma_x}{\sigma_y}(x-m_x)(y-m_y)\right\}+\frac{(y-m_y)^2}{2\sigma_y^2}\right]$$

$$=\frac{1}{\sqrt{2\pi}\sigma_x\sqrt{1-\rho^2}}\exp\left[-\frac{1}{2\sigma_x^2(1-\rho^2)}\left\{(x-m_x)^2+\left(\frac{\sigma_x}{\sigma_y}(y-m_y)\right)^2-2\rho(x-m_x)\frac{\sigma_x}{\sigma_y}(y-m_y)-\frac{\sigma_x^2(1-\rho^2)}{\sigma_y^2}(y-m_y)^2\right\}\right]$$

$$=\frac{1}{\sqrt{2\pi}\sigma_x\sqrt{1-\rho^2}}\exp\left[-\frac{1}{2\sigma_x^2(1-\rho^2)}\left\{(x-m_x)^2+\left(\rho\frac{\sigma_x}{\sigma_y}(y-m_y)\right)^2-2(x-m_x)\rho\frac{\sigma_x}{\sigma_y}(y-m_y)\right\}\right]$$

$$=\frac{1}{\sqrt{2\pi}\sigma_x\sqrt{1-\rho^2}}\exp\left[-\frac{1}{2\sigma_x^2(1-\rho^2)}\left\{x-m_x-\rho\frac{\sigma_x}{\sigma_y}(y-m_y)^2\right\}\right],\ -\infty<x<\infty,$$

which is the pdf for normal $\left(m_x+\rho\frac{\sigma_x}{\sigma_y}(y-m_y),\ \sigma_x\sqrt{1-\rho^2}\right)$ distribution.

Thus when (X,Y) is normally distributed with parameters $m_x, m_y, \sigma_x, \sigma_y$ and ρ; the conditional distribution of X, given $Y=y$, is a one-dimensional normal distribution with parameters

$$m_x+\rho\frac{\sigma_x}{\sigma_y}(y-m_y),\ \sigma_x\sqrt{1-\rho^2}$$

Similarly, $f_y(y|x)$, the conditional pdf of Y, given $X=x$, is the pdf of a normal distribution with parameters $m_y+\rho\frac{\sigma_y}{\sigma_x}(x-m_x)$ and $\sigma_y\sqrt{1-\rho^2}$. Hence, conditional distribution of Y, given $X=x$, is a one dimensional normal distribution with parameters $m_y+\rho\frac{\sigma_y}{\sigma_x}(x-m_x)$ and $\sigma_y\sqrt{1-\rho^2}$.

5.8 Transformation of a two-dimensional RV (X, Y)

Sometimes we want to have the distribution of the RV (U, V), when U and V are continuously and functionally related to X, Y and the distribution of (X, Y) is known.

Suppose, $U = u(X, Y)$ and $V = v(X, Y)$, where u and v are continuous and continuously differentiable functions and the Jacobian of the transformation $\frac{\partial(u,v)}{\partial(x,y)}$ maintain the same sign in $\mathbb{R}^2$.

In terms of the real values of the RV the transformation is $u = u(x,y), v = v(x,y)$.

Under the above conditions the reverse transformation $x = x(u,v), y = y(u,v)$ is uniquely defined.

Under the transformation let the event $(x < X \leq x+dx, y < Y \leq y+dy)$ becomes $(u < U \leq u+du, v < V \leq v+dv)$.

$$\therefore P(u < U \leq u+du, v < V \leq v+dv) = P(x < X \leq x+dx, y < Y \leq y+dy) = f_{x,y}(x,y)\,dx\,dy$$

$$= f_{x,y}(x,y)\left|\frac{\partial(x,y)}{\partial(u,v)}\right| du\,dv.$$

$$\therefore f_{u,v}(u,v)\,du\,dv = f_{x,y}(x,y)\left|\frac{\partial(x,y)}{\partial(u,v)}\right| du\,dv \text{ or, } f_{u,v}(u,v) = f_{x,y}(x,y)\left|\frac{\partial(x,y)}{\partial(u,v)}\right|.$$

The RHS can be expressed in terms of u and v and the ranges of u and v will be obtained from the ranges of x and y from the known distribution of (X, Y).

Now, the distribution of (U, V) is completely known when that of (X, Y) is given.

Theorem 5.8.1

Let (X, Y) be a continuous two-dimensional variate whose distribution is known by the pdf $f_{x,y}$. Then, the distribution of $U = X+Y$ is given by the pdf

$$f_u(u) = \int_{-\infty}^{\infty} f_{x,y}(v, u-v)\,dv.$$

Proof. Here $U = X+Y$. Let $V = X$. In terms of their real values, $u = x+y$ and $v = x \Rightarrow x = v, y = u-v$.

$$\therefore \text{Jacobian of transformation } J = \frac{\partial(x,y)}{\partial(u,v)} = \begin{vmatrix} \frac{\partial x}{\partial u} & \frac{\partial x}{\partial v} \\ \frac{\partial y}{\partial u} & \frac{\partial y}{\partial v} \end{vmatrix} = \begin{vmatrix} 0 & 1 \\ 1 & -1 \end{vmatrix} = -1 < 0 \text{ in } \mathbb{R}^2.$$

$$\therefore \text{the pdf of } (U, V) \text{ is given by } f_{u,v}(u,v) = f_{x,y}(x,y)\left|\frac{\partial(x,y)}{\partial(u,v)}\right| = f_{x,y}(x,y) = f_{x,y}(v, u-v).$$

$\therefore f_u(u) = \int_{-\infty}^{\infty} f_{u,v}(u,v)\,dv = \int_{-\infty}^{\infty} f_{x,y}(v, u-v)\,dv$, i.e., the distribution of U, i.e., of $X+Y$ is given by the pdf f_u given above.

Note: If further X and Y are independent, then

$$f_u(u) = \int_{-\infty}^{\infty} f_{u,v}(u,v)\,dv = \int_{-\infty}^{\infty} f_{x,y}(x,y)\,dv = \int_{-\infty}^{\infty} f_x(x)f_y(y)\,dv = \int_{-\infty}^{\infty} f_x(v)f_y(u-v)\,dv.$$

Note: Sometimes it may so happen that two independent RV and X and Y having the same distribution will imply that their sum $X+Y$ possesses the same distribution. Such a property is called the *reproductive property.*

Theorem 5.8.2

If the RV X and Y are independent and two RV U, V given by $U = u(X), V = v(Y)$, where u, v are continuously differentiable and strictly monotonic functions, then U and V are also independent.

Proof. The reverse transformation is $X = x(U), Y = y(V)$ and in real values $x = x(u), y = y(v)$. The probability distributions of U and V are given the rule of transformation of RV in one dimension. If f_u, f_v, f_x, f_y are the pdf of U, V, X and Y respectively, then

$$f_u(u) = f_x(x)\left|\frac{dx}{du}\right| \text{ and } f_v(v) = f_y(y)\left|\frac{dy}{dv}\right|. \tag{1}$$

As (X, Y) transforms to $(U, V,)$ the rule of transformation of RV in two dimensions gives

$$f_{u,v}(u,v) = f_{x,y}(x,y)\left|\frac{\partial(x,y)}{\partial(u,v)}\right|.$$

$$\text{As } J = \frac{\partial(x,y)}{\partial(u,v)} = \begin{vmatrix} \frac{\partial x}{\partial u} & \frac{\partial x}{\partial v} \\ \frac{\partial y}{\partial u} & \frac{\partial y}{\partial v} \end{vmatrix} = \frac{dx}{du}\frac{dy}{dv},$$

maintains the same sign in $\mathbb{R}^2$.

$$\therefore\ f_{u,v}(u,v) = f_x(x)f_y(y)\left|\frac{dx}{du}\right|\left|\frac{dy}{dv}\right|, \text{ as } X, Y \text{ are independent}$$

$$= \left(f_x(x)\left|\frac{dx}{du}\right|\right)\left(f_y(y)\left|\frac{dy}{dv}\right|\right) = f_u(u)f_v(v) \text{ [by (1)]}.$$

$\therefore$ the RV U and V are independent.

5.9 Transformation of continuous RV in n-dimensions

Theorem Let $(X_1, X_2, \cdots, X_n)$ be a continuous RV in n dimensions whose distribution is given by the pdf $f_{x_1,x_2,\cdots,x_n}(x_1, x_2, \cdots, x_n)$.

Now, consider the bijection from $\mathbb{R}^n$ to $\mathbb{R}^n$ by $u_i = u_i(x_1, x_2, \cdots, x_n), i = 1, 2, \cdots, n$ and the inverse transformation $x_i = x_i(u_1, u_2, \cdots, u_n), i = 1, 2, \cdots, n$.

Suppose that the transformation, inverse transformation, $\frac{\partial x_i}{\partial u_j}, i, j = 1, 2, \cdots, n$ are all continuous in $\mathbb{R}^n$ and the Jacobian of transformation

$$J = \begin{vmatrix} \frac{\partial x_1}{\partial u_1} & \frac{\partial x_1}{\partial u_2} & \cdots & \frac{\partial x_1}{\partial u_n} \\ \frac{\partial x_2}{\partial u_1} & \frac{\partial x_2}{\partial u_2} & \cdots & \frac{\partial x_2}{\partial u_n} \\ \cdots & \cdots & \cdots & \cdots \\ \frac{\partial x_n}{\partial u_1} & \frac{\partial x_n}{\partial u_2} & \cdots & \frac{\partial x_n}{\partial u_n} \end{vmatrix}$$

remains the same in the corresponding domain. Then, the probability distribution of $(U_1, U_2, \cdots, U_n)$ is given by the following pdf $f_{u_1,u_2,\cdots,u_n}(u_1, u_2, \cdots, u_n) = f_{x_1,x_2,\cdots,x_n}(x_1, x_2, \cdots, x_n)|J|$, when we replace the arguments $x_1, x_2, \cdots, x_n$ on the RHS by $x_i = x_i(u_1, u_2, \cdots, u_n), i = 1, 2, \cdots, n$ and the ranges of $u_1, u_2, \cdots, u_n$ are given by those of $x_1, x_2, \cdots, x_n$.

WORKED-OUT EXAMPLES

Ex. 1. An urn contains 5 white balls bearing numbers 1, 2, 3, 4, 5; 4 black balls bearing numbers 1, 2, 3, 4 and 3 red balls bearing numbers 1, 2, 3. Consider the RE: drawing a ball from the urn at random.

Define two RV X and Y such that $X = 1,2,3$ respectively for white, black and red balls and Y denotes their numbers.
(i) Find the distribution of the RV (X,Y).
(ii) Deduce the marginal distributions of X and Y.
(iii) Check whether X and Y are independent or not.
(iv) Find the conditional distributions of X and Y.
Solution. (i) The distribution of (X,Y) is discrete. The spectrum (x_i, y_i) of (X,Y) is $(x_i, y_j) = (i,j)$, $i = 1,2,3$; $j = 1,2,3,4,5$.
The pmf is given by $f_{ij} = P(X = i, Y = j) = \frac{1}{12}$, $\forall\, i,j$ except f_{25}, f_{34}, f_{35}, which are all zero.

(ii) The marginal distribution of X is defined by the pmf $f_{xi} = f_{i\cdot} = \sum_{j=1}^{5} f_{ij}, i = 1,2,3.$

$$f_{x1} = \sum_{j=1}^{5} f_{1j} = f_{11} + f_{12} + f_{13} + f_{14} + f_{15} = \tfrac{5}{12}$$

$$f_{x2} = \sum_{j=1}^{5} f_{2j} = f_{21} + f_{22} + f_{23} + f_{24} + f_{25} = \tfrac{4}{12}$$

$$f_{x3} = \sum_{j=1}^{5} f_{3j} = f_{31} + f_{32} + f_{33} + f_{34} + f_{35} = \tfrac{3}{12}$$

Check that $f_{xi} > 0\ \forall\, i = 1,2,3$ and $\sum_{i=1}^{3} f_{xi} = 1$.

The marginal distribution of y is defined by the pmf $f_{yj} = f_{\cdot j} = \sum_{i=1}^{3} f_{ij},\ j = 1,2,3,4,5.$

$$f_{y_1} = \sum_{i=1}^{3} f_{i1} = f_{11} + f_{21} + f_{31} = \tfrac{3}{12}, \quad f_{y_2} = \sum_{i=1}^{3} f_{i2} = f_{12} + f_{22} + f_{32} = \tfrac{3}{12}$$

$$f_{y_3} = \sum_{i=1}^{3} f_{i3} = f_{13} + f_{23} + f_{33} = \tfrac{3}{12}, \quad f_{y_4} = \sum_{i=1}^{3} f_{i4} = f_{14} + f_{24} + f_{34} = \tfrac{2}{12}$$

$$f_{y_5} = \sum_{i=1}^{3} f_{i5} = f_{15} + f_{25} + f_{35} = \tfrac{1}{12}.$$

Again, check that $f_{yj} > 0$, $\forall\, j = 1,2,3,4,5$ and $\sum_{j=1}^{5} f_{yj} = 1$.
(iii) We know that NASC for independence of x, y is $f_{ij} = f_{xi} f_{yj}$, $\forall\, i,j$. Now, $f_{11} = \frac{1}{12}$ and $f_{x_1} = \frac{5}{12}$, $f_{y_1} = \frac{3}{12}$. $\therefore\ f_{11} \neq f_{x_1} f_{y_1}$. $\therefore\ X, Y$ are not independent.
(iv) $f_{i|1} = \frac{f_{i1}}{f_{y1}}$, $i = 1,2,3$ is the pmf of the conditional distribution of X, given $Y = 1$. We now check that it is a one-dimensional discrete distribution

$$f_{i|1} = \tfrac{12}{3} f_{i1} \text{ as } f_{y1} = \tfrac{3}{12}.$$

$$\therefore\ f_{1|1} = \tfrac{12}{3} f_{11} = \tfrac{12}{3} \times \tfrac{1}{12} = \tfrac{1}{3};\ f_{2|1} = \tfrac{12}{3} f_{21} = \tfrac{12}{3} \times \tfrac{1}{12} = \tfrac{1}{3};\ f_{3|1} = \tfrac{12}{3} f_{31} = \tfrac{12}{3} \times \tfrac{1}{12} = \tfrac{1}{3}.$$

$$\therefore\ \sum_{i=1}^{3} f_{i|1} = \frac{1}{3} + \frac{1}{3} + \frac{1}{3} = 1.$$

Also, $f_{i|1} > 0$ as $f_{i1} > 0$, $\forall\, i$ and $f_{y1} = \frac{3}{12} > 0$.
Similarly, $f_{i|2} = \frac{f_{i2}}{f_{y2}} = \frac{12}{3} f_{i2}, i = 1,2,3$ is the pmf of the one-dimensional conditional distribution of X, given that $Y = 2$.

... ...

$$f_{i|5} = \tfrac{f_{i5}}{f_{y5}} = 12 f_{i5},\ i = 1,2,3.$$

Here $f_{1|5}=12f_{15}=12\times\frac{1}{12}=1$; $f_{2|5}=12f_{15}=12\times 0=1$; $f_{3|5}=12f_{35}=12\times 0=0$.

$$\therefore \sum_{i=1}^{3} f_{i|5}=1 \text{ and } f_{i|5}\geq 0,\ \forall\, i.$$

Now, we proceed to find the conditional distribution of Y.

Let $X=1$ be given. The conditional probability distribution of Y, given $X=1$ is given by the pmf

$$f_{j|1}=\frac{f_{j1}}{f_{x1}}=\frac{f_{1j}}{f_{x1}},\ j=1(1)5\ [f_{j1}=P(Y=j,X=1)=P(X=1,Y=j)=f_{1j}]$$

$f_{j|1}=\frac{12}{5}f_{j1}$ as $f_{x1}=\frac{5}{12}$.

$\therefore\ f_{1|1}=\frac{12}{5}f_{11}=\frac{12}{5}\times\frac{1}{12}=\frac{1}{5}=$ probability of occurrence of no. 1 white ball, given that $X=1$, i.e., ball drawn is white.

$f_{2|1}=\frac{12}{5}f_{12}=\frac{12}{5}\times\frac{1}{12}=\frac{1}{5}=$ probability of occurrence of no. 2 white ball, given that $X=1$, i.e., ball drawn is white.

$f_{3|1}=\frac{12}{5}f_{13}=\frac{12}{5}\times\frac{1}{12}=\frac{1}{5}=$ probability of occurrence of no. 3 white ball, given that $X=1$, i.e., ball drawn is white.

$f_{4|1}=\frac{12}{5}f_{14}=\frac{12}{5}\times\frac{1}{12}=\frac{1}{5}=$ probability of occurrence of no. 4 white ball, given that $X=1$, i.e., ball drawn is white.

$f_{5|1}=\frac{12}{5}f_{15}=\frac{12}{5}\times\frac{1}{12}=\frac{1}{5}=$ probability of occurrence of no. 5 white ball, given that $X=1$, i.e., all drawn is white.

$\therefore\ f_{j|1}>0,\ \forall\, j=1(1)5$ and $\sum_{j=1}^{5} f_{j|1}=1$.

The above is the distribution of the numbers in white balls.

Similarly, the conditional distribution of Y, when $X=2$, is given by the pmf,

$$f_{j|2}=\frac{f_{j2}}{f_{x2}}=\frac{12}{4}f_{2j},\ j=1(1)5.$$

This is the distribution of the numbers in black balls. The conditional distribution of Y, when $X=3$, is given by the pmf,

$$f_{j|3}=\frac{f_{j3}}{f_{x3}}=\frac{12}{3}f_{3j},\ j=1(1)5,$$

and it is the distribution of the numbers in red balls.

Observe that the above distribution of (X, Y), the marginal distributions, are given by the following table in compact form:

X \ Y	1	2	3	4	5	$P(X=i)$
1	$\frac{1}{12}$	$\frac{1}{12}$	$\frac{1}{12}$	$\frac{1}{12}$	$\frac{1}{12}$	$\frac{5}{12}$
2	$\frac{1}{12}$	$\frac{1}{12}$	$\frac{1}{12}$	$\frac{1}{12}$	—	$\frac{4}{12}$
3	$\frac{1}{12}$	$\frac{1}{12}$	$\frac{1}{12}$	—	—	$\frac{3}{12}$
$P(Y=j)$	$\frac{3}{12}$	$\frac{3}{12}$	$\frac{3}{12}$	$\frac{2}{12}$	$\frac{1}{12}$	1

$P(Y=1)=$ probability of no. 1 balls $=\frac{3}{12}$, by classical rule, as out of 12 balls only 3 bear no. 1.

$P(X=1)=$ probability of white balls $=\frac{5}{12}$, by classical rule, as out of 12 balls only 5 are white, etc.

Ex. 2. Three fair coins are tossed simultaneously. Suppose, a RV X denotes the number of heads appeared in the coins and the RV Y denotes the absolute difference between the number of heads and tails. Find the distribution of (X, Y). Also, find the marginal distributions of X and Y. Are they independent?

Solution. The distribution of (X, Y) is discrete. The event space S is given by

$$S = \{U_1, U_2, U_3, U_4, U_5, U_6, U_7, U_8\},$$

where $U_1 = $ (H,H,H), $U_2 = $ (H,H,T), $U_3 = $ (H,T,H), $U_4 = $ (T,H,H), $U_5 = $ (T,T,H), $U_6 = $ (T,H,T), $U_7 = $ (H,T,T), $U_8 = $ (T,T,T), where first tuple corresponds to the event of first coin, second tuple for the second and third tuple for the third coin.

The spectrum of (X, Y) is the set of points $S = \{(0,3),(1,1),(2,1),(3,3)\}$ and the pmf. f_{ij} are given by $f_{03} = \frac{1}{8}, f_{11} = \frac{3}{8}, f_{21} = \frac{3}{8}, f_{33} = \frac{1}{8}$.

Note that at each step point $f_{ij} > 0$ and $\Sigma f_{ij} = 1$.

The marginal distribution of X is given by the pmf $f_{xi} = \sum_{j=1,3} f_{ij} = f_{i1} + f_{i3}, i = 0,1,2,3$, provided $(i,1),(i,3) \in S$.

$\therefore\ f_{x0} = f_{03} = \frac{1}{8}, f_{x1} = f_{11} = \frac{3}{8}, f_{x2} = f_{21} = \frac{3}{8}, f_{x3} = f_{33} = \frac{1}{8}$.

$\therefore\ f_{xi} > 0,\ \forall\ i = 0,1,2,3$ and $\sum_{i=0}^{3} f_{xi} = 1$.

The marginal distribution of Y is given by the pmf $f_{yj} = \sum_{i=0}^{3} f_{ij}, j = 1,3$, provided $(0,j)$, $(1,j)$, $(2,j)$, $(3,j) \in S$.

$f_{y_1} = f_{11} + f_{21} = \frac{3}{8} + \frac{3}{8} = \frac{6}{8}, f_{y_3} = f_{03} + f_{33} = \frac{1}{8} + \frac{1}{8} = \frac{2}{8}$. $\therefore\ f_{yj} > 0,\ \forall\ j = 1,3$ and $\sum_{j=1,3} f_{yj} = 1$.

$f_{11} = \frac{3}{8}, f_{x_1} \times f_{y_1} = \frac{3}{8} \times \frac{6}{4}$. $\therefore\ f_{11} \neq f_{xi} f_{y_1}$. $\therefore$ X and Y are dependent.

Ex. 3. Suppose, $(x_i, y_j) = (i, j), i = 1,2,3; j = 1,2,3,4$ be the spectrum of a two-dimensional RV (X, Y) with pmf $f_{ij} = P(X = i, Y = j) = K(2i + 5j)$, where k is a constant.

Determine (i) k (ii) the marginal distributions of X and Y (iii) $P(X \le 2, Y \ge 3)$ (iv) $P(X \le 2 \mid Y \ge 3)$. Also, test whether X and Y are dependent.

Solution. We have $\sum_{i=1}^{3}\sum_{j=1}^{4} f_{ij} = 1$.

$\therefore\ f_{11} + f_{12} + f_{13} + f_{14} + f_{21} + f_{22} + f_{23} + f_{24} + f_{31} + f_{32} + f_{33} + f_{34} = 1$, or, $k[7 + 12 + 17 + 22 + 9 + 14 + 19 + 24 + 11 + 16 + 21 + 26] = 1$. $\therefore\ k = \frac{1}{198}$.

[*Alternatively,* $k\sum_{i=1}^{3}\sum_{j=1}^{4}(2i + 5j) = k\sum_{i=1}^{3}\{8i + 5(1 + 2 + 3 + 4)\} = k\sum_{i=1}^{3}(8i + 50) = k\{8(1 + 2 + 3) + 150\}$

$= 198k = 1 \Rightarrow k = \frac{1}{198}$.]

(ii) Marginal distribution of X is given by the following: Spectrum is $X = x_i = i, i = 1,2,3$ and

$$f_{xi} = \sum_{j=1}^{4} f_{ij} = \frac{1}{198}\sum_{j=1}^{4}(2i+5j) = \frac{1}{198}\{8i+5(1+2+3+4)\} = \frac{1}{198}(8i+50) = \frac{1}{99}(4i+25),\ i = 1,2,3. \quad (1)$$

Note that $f_{xi} > 0,\ \forall\ i = 1,2,3$ and $\sum_{i=1}^{3} f_{xi} = \frac{1}{99}\{4(1+2+3)+75\} = 1$.

Marginal distribution of Y is given by the following: Spectrum is $Y = y_j = j, j = 1,2,3,4$ and

$$f_{yj} = \sum_{i=1}^{3} f_{ij} = \frac{1}{198}\sum_{i=1}^{3}(2i+5j) = \frac{1}{198}[2(1+2+3)+15j] = \frac{1}{198}(12+15j) = \frac{1}{66}(4+5j),\ j = 1,2,3,4. \quad (2)$$

Note that $f_{yj} > 0,\ \forall\ j = 1,2,3,4$ and $\sum_{j=1}^{4} f_{yj} = \frac{1}{66}\sum_{j=1}^{4}(4+5j) = \frac{1}{66}\{16+5(1+2+3+4)\} = 1$.

(iii) $P(X \le 2, y \ge 3) = \frac{1}{198}\sum_{i=1}^{2}\sum_{j=3}^{4}(2i+5j) = \frac{1}{198}\sum_{i=1}^{2}(4i+5(3+4)) = \frac{1}{198}\sum_{i=1}^{2}(4i+35) = \frac{1}{198}[4(1+2)+70]$
$= \frac{82}{198} = \frac{41}{99}$.

(iv) $P(X \le 2 \mid Y \ge 3) = \frac{P(X\le 2, Y\ge 3)}{P(Y\ge 3)} = \frac{41}{99P(Y\ge 3)}$.

Now, $P(Y \ge 3) = \frac{1}{198}\sum_{i=1}^{3}\sum_{j=3}^{4}(2i+5j) = \frac{1}{198}\sum_{i=1}^{3}(4i+35) = \frac{1}{198}(24+105) = \frac{129}{198}$.

$\therefore\ P(X \le 2 \mid Y \ge 3) = \frac{41}{99} \times \frac{198}{129} = \frac{82}{129}$.

Again, $f_{11} = \frac{1}{198}(2\times 1+5\times 1) = \frac{7}{198}$, $f_{x_1} = \frac{1}{99}(4\times 1+25)$ [by 1] $= \frac{29}{99}$.

$f_{y_1} = \frac{1}{66}(4\times 5\times 1) = \frac{9}{66} = \frac{3}{22}$ [by (2)].

$\therefore\ f_{11} \ne f_{x_1} f_{y_1}$. Hence, X and Y are dependent. [For independence $f_{ij} = f_{xi} f_{yi}$, $\forall\ i, j$]

Ex. 4. Buffon's needle problem. A vertical board is ruled with horizontal parallel lines at constant distance d apart. A needle of length $l(< d)$ is thrown at random on the board. Find the probability that it will intersect one of the lines. [C.H. 2003]

Solution. Let us define two RV X and Y, where $X =$ inclination of the rod with the horizontal and $Y =$ distance of the upper end of the needle from the ruling line just below it.

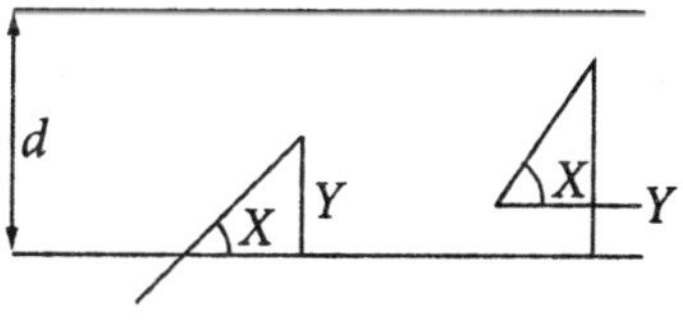

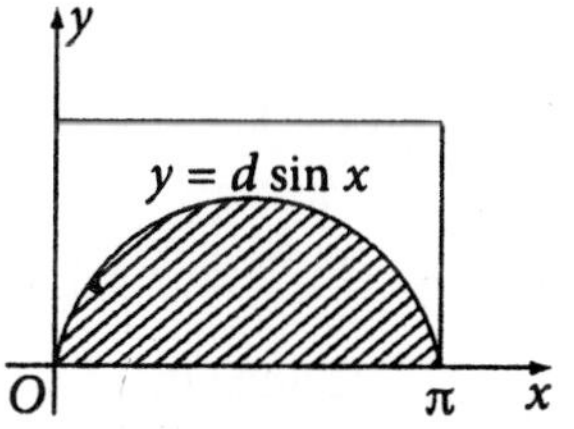

The X and Y are independent and both are uniformly distributed, the former in $(0, \pi)$ and the latter in $(0, d)$.

$$\therefore\ f_X(x) = \begin{cases} 1/\pi, & 0 < x < \pi \\ 0, & \text{elsewhere} \end{cases} \quad \text{and} \quad f_Y(y) = \begin{cases} 1/d, & 0 < y < d \\ 0, & \text{elsewhere.} \end{cases}$$

$$\therefore\ f_{X,Y}(x, y) = \begin{cases} 1/\pi d, & 0 < x < \pi, 0 < y < d \\ 0, & \text{elsewhere} \end{cases}$$

is the pdf of (X, Y).

The event that the needle will intersect one of lines is $0 \le y \le l \sin X$.

$\therefore$ required probability

$$= \int_0^{\pi}\int_0^{l\sin x} \frac{1}{\pi d}\, dx\, dy = \frac{1}{\pi d}\int_0^{\pi} \left[y\right]_0^{l\sin x} dx = \frac{1}{\pi d}\int_0^{\pi} l\sin x\, dx = -\frac{l}{\pi d}\left[\cos x\right]_0^{\pi} = \frac{2l}{\pi d}.$$

Ex. 5. Two numbers chosen at random independently between 0 and 2. Find the probability that their product is less than a given number $k (0 < k < 4)$.

Solution. Let X and Y be RV denoting the number chosen in $(0, 2)$ independently. The distributions of X and Y are both uniform in $(0, 2)$.

$$\therefore\ f_X(x) = \begin{cases} 1/2, & 0 < x < 2 \\ 0, & \text{elsewhere} \end{cases}$$

$$\text{and} \quad f_Y(y) = \begin{cases} 1/2, & 0 < y < 2 \\ 0, & \text{elsewhere.} \end{cases}$$

$\therefore$ the pdf of (X, Y) is given by

$$f_{x,y}(x,y) = f_x(x)f_y(y), \text{ as } X, Y \text{ are independent}$$

$$= \begin{cases} 1/4, & 0 < x < 2, 0 < y < 2 \\ 0, & \text{elsewhere.} \end{cases}$$

$$\therefore\ P(XY < k) = 1 - P(XY \geq k) = 1 - \int_{k/2}^{2}\int_{k/x}^{2} \frac{1}{4}\,dy\,dx$$

$$= 1 - \frac{1}{4}\int_{k/2}^{2} \Big[y\Big]_{k/x}^{2} dx = 1 - \frac{1}{4}\int_{k/2}^{2}\left(2 - \frac{k}{x}\right)dx$$

$$= 1 + \frac{1}{4}\Big[k\ln|x| - 2x\Big]_{k/2}^{2}$$

$$= 1 + \frac{1}{4}\left[k\ln 2 - 4 - k\ln\frac{k}{2} + k\right]$$

$$= \frac{k}{4}\left(1 - \ln\frac{k}{4}\right).$$

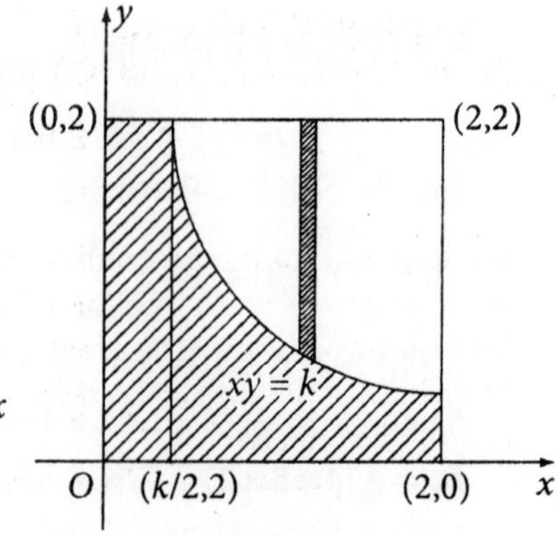

Fig 5.3

Alternatives

$$P(XY < k) = \frac{1}{4}\int_0^{k/2}\int_0^{2} dy\,dx + \frac{1}{4}\int_0^{k/x} dy\,dx = \frac{1}{4}\cdot\frac{k}{2}\cdot 2 + \frac{1}{4}\int_{k/2}^{2}\frac{k}{x}\,dx = \frac{k}{4} + \frac{k}{4}\left(\ln 2 - \ln\frac{k}{2}\right)$$

$$= \frac{k}{4} - \frac{k}{4}\ln\frac{k}{4} = \frac{k}{4}\left(1 - \ln\frac{k}{4}\right)$$

Note: The total probability that (X, Y) lying in a specific region is obtained by integrating the function $f_{x,y}$ in that region. This is why it is called *pdf*.

Ex. 6. If two numbers are chosen at random between 0 and 2, find the probability that their sum of squares is less than 2.

Solution. Following solution of Ex. 5, we get

$$f_{x,y}(x,y) = \begin{cases} 1/4, & 0 < x < 2, 0 < y < 2 \\ 0, & \text{elsewhere.} \end{cases}$$

$$\therefore\ P(X^2 + Y^2 < 2) = \int_0^{\sqrt{2}}\int_0^{\sqrt{2-x^2}} \frac{1}{4}\,dy\,dx = \frac{1}{4}\int_0^{\sqrt{2}} \sqrt{2-x^2}\,dx$$

$$= \frac{1}{4}\left[\frac{x\cdot\sqrt{2-x^2}}{2} + \frac{2}{2}\sin^{-1}\frac{x}{\sqrt{2}}\right]_0^{\sqrt{2}}$$

$$= \frac{1}{4}\cdot\frac{\pi}{2} = \frac{\pi}{8}.$$

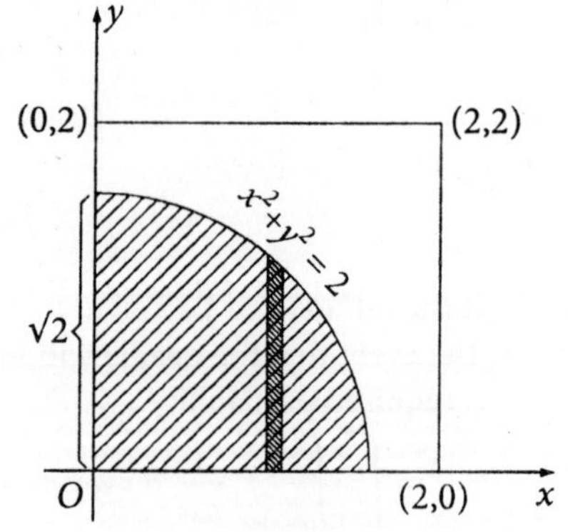

Fig 5.4

Ex. 7. Two numbers chosen at random between 0 and 2. What is the probability that their sum is less than 2, whereas sum of their squares is greater than 2.

Solution. Let X, Y be the RV denoting the two numbers chosen at random between 0 and 2.

As X is chosen at random in $(0, 2)$,

$\therefore$ the distribution of X is uniform in $(0, 2)$.

Also, the distribution of Y is uniform in $(0, 2)$ and X, Y are independent.

$\therefore$ the pdf of (X, Y) is given by

$f_{x,y}(x,y) = f_x(x)f_y(y)$, as X, Y are independent

$$= \begin{cases} 1/4, & 0 < x < 2, 0 < y < 2 \\ 0, & \text{elsewhere.} \end{cases}$$

We have to find the probability that $X + Y < 2$, whereas $X^2 + Y^2 > 2$. The shaded portion in the figure is favourable to the event. So, the required probability is obtained by integrating $f_{x,y}$ within the shaded region.

$\therefore$ required probability $= P\left(X + Y < 2 \text{ and } X^2 + Y^2 > 2\right)$

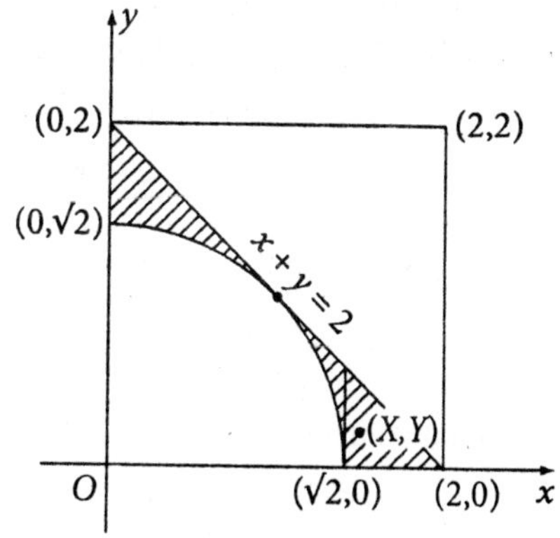

Fig 5.5

$$= \int_0^{\sqrt{2}} \int_{\sqrt{2-x^2}}^{2-x} \frac{1}{4}\, dy\, dx + \int_{\sqrt{2}}^{2} \int_0^{2-x} \frac{1}{4}\, dy\, dx$$

$$= \frac{1}{4}\int_0^{\sqrt{2}} \left(2 - x - \sqrt{2 - x^2}\right) dx + \frac{1}{4}\int_{\sqrt{2}}^{2} (2 - x)\, dx$$

$$= \frac{1}{4}\left[2x - \frac{x^2}{2} - \frac{x\sqrt{2-x^2}}{2} - \sin^{-1}\frac{x}{\sqrt{2}}\right]_0^{\sqrt{2}} + \frac{1}{4}\left[2x - \frac{x^2}{2}\right]_{\sqrt{2}}^{2}$$

$$= \frac{1}{4}\left(2\sqrt{2} - 1 - \frac{\pi}{2}\right) + \frac{1}{4}\left(4 - 2 - 2\sqrt{2} + 1\right) = \frac{1}{2} - \frac{\pi}{8}.$$

[**Note:** $P\left\{(A \cap B) \cup \left(A \cap \bar{B}\right)\right\} = \left\{A \cap \left(B \cup \bar{B}\right)\right\} = P(A)$, $\because$ $B \cup \bar{B} =$ certain event.

$\therefore$ $P(A \cap B) + P\left(A \cap \bar{B}\right) = P(A)$, as events $A \cap B$ and $A \cap \bar{B}$ are MEE.

$\therefore$ $P(A \cap B) = P(A) - P\left(A \cap \bar{B}\right)$]

First Alternative Method

$$P\left(X + Y < 2 \cap X^2 + Y^2 > 2\right) = P(X + Y < 2) - P\left(X + Y < 2 \cap X^2 + Y^2 \le 2\right)$$

$$= P(X + Y < 2) - P\left(X^2 + Y^2 \le 2\right)$$

$$= \frac{1}{4}\int_0^2 \int_0^{2-x} dy\, dx - \frac{1}{4}\int_0^{\sqrt{2}} \int_0^{\sqrt{2-x^2}} dy\, dx$$

$$= \frac{1}{4}\int_0^2 (2 - x)\, dx - \frac{1}{4}\int_0^{\sqrt{2}} \sqrt{2 - x^2}\, dx$$

$$= \frac{1}{4}\left[2x - \frac{x^2}{2}\right]_0^2 - \frac{1}{4}\left[x\sqrt{2 - x^2} + \sin^{-1}\frac{x}{\sqrt{2}}\right]_0^{\sqrt{2}}$$

$$= \frac{1}{4} \cdot 2 - \frac{1}{4} \cdot \frac{\pi}{2} = \frac{1}{2} - \frac{\pi}{8}.$$

Second Alternative Method: As (X, Y) is uniformly distributed within the square $0 < x < 2, 0 < y < 2$, so the required probability

$$= \frac{\text{area in favour of the event}}{\text{total area of the square}} = \frac{1}{4}\left(2 - \int_0^{\sqrt{2}}\int_0^{\sqrt{2-x^2}} dy\,dx\right) = \frac{1}{2} - \frac{1}{4}\int_0^{\sqrt{2}} \sqrt{2-x^2}\,dx$$

$$= \frac{1}{2} - \frac{1}{4}\left[\frac{x\sqrt{2-x^2}}{2} + \sin^{-1}\frac{x}{\sqrt{2}}\right]_0^{\sqrt{2}} = \frac{1}{2} - \frac{\pi}{8}.$$

Ex. 8. If X and Y are independent variates, both uniformly distributed over $(0,1)$, find the distributions of $X+Y$ and $X-Y$. [C.H. 2008]

Solution. Let $U = X+Y$ and $V = X-Y$. In their real values, $u = x+y, v = x-y$. $\therefore x = \frac{u+v}{2}, y = \frac{u-v}{2}$.

$$\text{Jacobian of transformation} = \frac{\partial(u,v)}{\partial(x,y)} = \begin{vmatrix} \frac{\partial u}{\partial x} & \frac{\partial u}{\partial y} \\ \frac{\partial v}{\partial x} & \frac{\partial v}{\partial y} \end{vmatrix} = \begin{vmatrix} 1 & 1 \\ 1 & -1 \end{vmatrix} = -2 < 0,\ \forall\, x, y.$$

As X and Y are independent variates both uniformly distributed over $(0,1)$. $\therefore$ the pdf of (X,Y) is given by

$$f_{x,y}(x,y) = \begin{cases} 1, & 0<x<1, 0<y<1 \\ 0, & \text{elsewhere.} \end{cases}$$

If $f_{u,v}$ be pdf of (U, V), then by the rule of transformation,

$$f_{u,v}(u\,v) = f_{x,y}(x,y)\left|\frac{\partial(x,y)}{\partial(u,v)}\right|$$

$$= \begin{cases} \frac{1}{2}, & 0 < \frac{u+v}{2} < 1, 0 < \frac{u-v}{2} < 1 \\ 0, & \text{elsewhere,} \end{cases}$$

$$\text{i.e., } f_{u,v}(u,v) = \begin{cases} \frac{1}{2}, & 0<u+v<2, 0<u-v<2 \\ 0, & \text{elsewhere.} \end{cases}$$

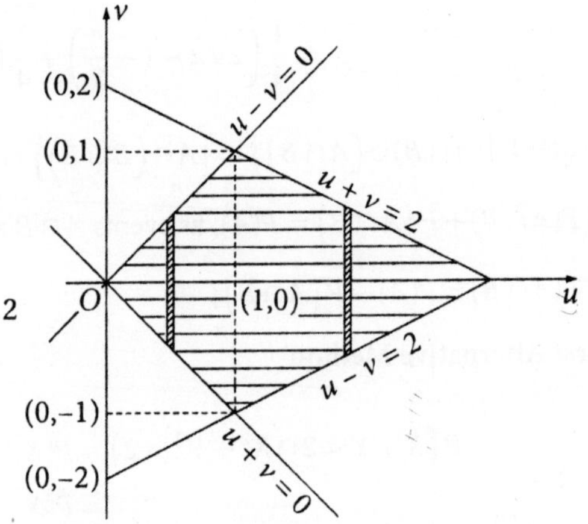

$$\therefore f_u(u) = \int_u^u \tfrac{1}{2}\,dv = u,\ 0<u<1$$

$$\text{and } f_u(u) = \int_{u-2}^{2-u} \tfrac{1}{2}\,dv = 2-u,\ 1<u<2$$

and $f_u(u) = 0$, elsewhere.

$$f_v(v) = \int_{-v}^{v+2} \tfrac{1}{2}\,du = v+1,\ -1<v<0$$

$$f_v(v) = \int_v^{2-v} \tfrac{1}{2}\,du = 1-v,\ 0<v<1$$

$$f_v(v) = \begin{cases} 1-|v|, & -1<v<1 \\ 0, & \text{otherwise.} \end{cases}$$

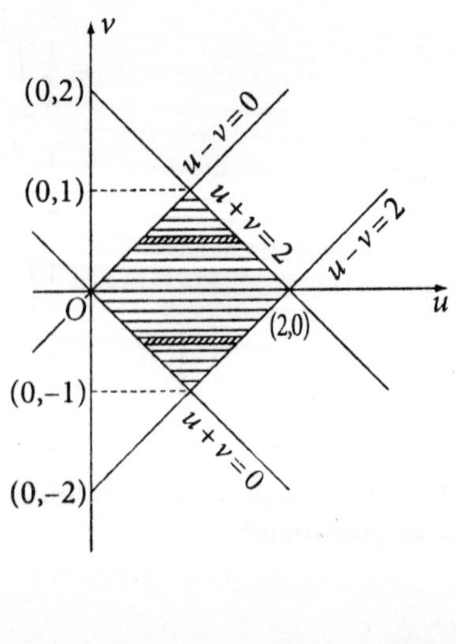

Ex. 9. If p, q are independent variates each uniformly distributed over the interval $(-1, 1)$, find the probability that the equation $x^2 + 2px + q = 0$ has real roots. [C.H. 2007]

Solution. The pdf of p and q are respectively given by

$$f_p(p) = \begin{cases} 1/2, & -1 < p < 1 \\ 0, & \text{elsewhere} \end{cases} \quad \text{and} \quad f_q(q) = \begin{cases} 1/2, & -1 < q < 1 \\ 0, & \text{elsewhere.} \end{cases}$$

As p, q are independent, $\therefore$ the pdf of (p, q) is given by

$$f_{p,q}(p, q) = \begin{cases} 1/4, & -1 < p < 1, -1 < q < 1 \\ 0, & \text{elsewhere.} \end{cases}$$

The condition that the roots of $x^2 + 2px + q = 0$ are real is $4p^2 - 4q \geq 0$, i.e., $q \leq p^2$.

$\therefore$ the required probability

$$= P\left(q \leq p^2\right) = \int_{-1}^{1} \int_{-1}^{p^2} \frac{1}{4}\, dq\, dp = \frac{1}{4} \int_{-1}^{1} \left(p^2 + 1\right) dp = \frac{1}{4}\left[\frac{p^3}{3} + p\right]_{-1}^{1} = \frac{1}{4}\left(\frac{4}{3} + \frac{4}{3}\right) = \frac{2}{3}.$$

Ex. 10. Show that if (X, Y) is a two-dimensional RV with probability density

$$f_{x,y}(x, y) = \begin{cases} 4xy, & 0 < y < x < 1 \\ 0, & \text{elsewhere,} \end{cases}$$

then X and Y are not independent. [C.H. 2006]

Solution. If f_x and f_y be the marginal pdf for X and Y respectively, then

$$f_x(x) = \int_0^x 4xy\, dy = 4x\left[\frac{y^2}{2}\right]_0^x = 2x^3,\ 0 < x < 1$$

and $f_x(x) = 0$, elsewhere.

$$f_y(y) = \int_y^1 4xy\, dx = 4y\left[\frac{x^2}{2}\right]_y^1 = 2y(1 - y^2),\ 0 < y < 1$$

$$= 0, \text{ elsewhere.}$$

$\therefore\ f_{x,y}(x, y) \neq f_x(x) f_y(y),\ \forall\, x, y.$

Hence, X and Y are not independent.

Ex. 11. If the joint pdf of the RV X, Y is

$$f_{x,y}(x, y) = \begin{cases} k(3x + y), & 1 \leq x \leq 3, 0 \leq y \leq 2 \\ 0, & \text{elsewhere.} \end{cases}$$

Find (i) $P(X + Y < 2)$ (ii) the marginal distributions of X and Y. Investigate whether X and Y are independent. [C.H. 2002, 1994]

Solution. First we find k. We have $\int_{-\infty}^{\infty} \int_{-\infty}^{\infty} f_{x,y}(x, y)\, dx\, dy = 1$. $\therefore\ k \int_1^3 \int_0^2 (3x + y)\, dy\, dx = 1$

or, $\frac{1}{k} = \int_1^3 \left[3xy + \frac{y^2}{2}\right]_0^2 dx = \int_1^3 (6x + 2)\, dx = \left[3x^2 + 2x\right]_1^3 = 27 + 6 - 3 - 2 = 28$. $\therefore\ k = \frac{1}{28}$.

$\therefore\ f_{x,y}(x, y) = \frac{1}{28}(3x + y), 1 \leq x \leq 3, 0 \leq y \leq 2$

$= 0$, elsewhere.

$$\text{(i) } P(X+Y<2)=\int_1^2\int_0^{2-x}\frac{1}{28}(3x+y)\,dy\,dx$$
$$=\frac{1}{28}\int_1^2\left[3xy+\frac{y^2}{2}\right]_0^{2-x}dx$$
$$=\frac{1}{28}\int_1^2\left(-\frac{5}{2}x^2+4x+2\right)dx$$
$$=\frac{1}{28}\left[-\frac{5}{6}x^3+2x^2+2x\right]_1^2=\frac{13}{168}.$$

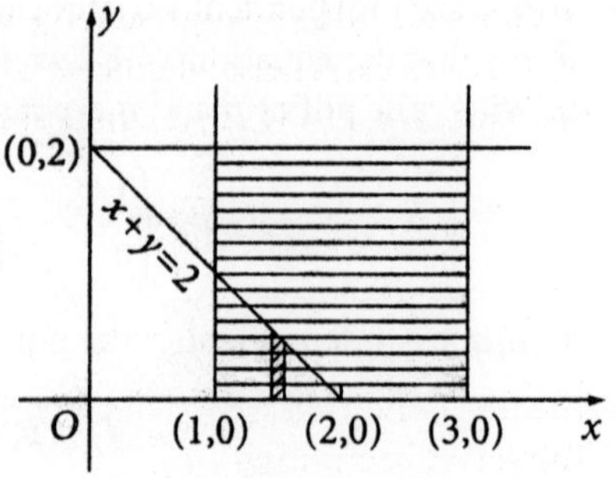

Marginal distributions are given by the pdf,

$$f_X(x)=\frac{1}{28}\int_0^2(3x+y)\,dy=\frac{1}{28}\left[3xy+\frac{y^2}{2}\right]_0^2=\frac{1}{14}(3x+1)\text{ or, } f_X(x)=\begin{cases}\frac{1}{14}(3x+1), & 1<x<3\\ 0, & \text{elsewhere.}\end{cases}$$

$$\text{and } f_Y(y)=\frac{1}{28}\int_1^3(3x+y)\,dx=\frac{1}{28}\left[\frac{3x^2}{2}+xy\right]_1^3=\frac{1}{14}(y+6)$$

$$\therefore\ f_Y(y)=\frac{1}{14}(y+6),\ 0<y<2$$
$$=0,\text{ elsewhere.}$$

As $f_{X,Y}(x,y)\neq f_X(x)f_Y(y)$. $\therefore$ X and Y are not independent.

Ex. 12. Two points P, Q are independently chosen at random on a circle and A is a fixed point also on the circle. Find the probability that the three points A, P, Q will lie on the same semi-circle.

Solution. Let X, Y respectively denote the angles $\angle AOP$ and $\angle AOQ$.

As the points P, Q are chosen independently at random on the circle, X and Y are distributed uniformly in $(0, 2\pi)$.

$\therefore$ the pdf of (X, Y) is given by

$$f_{X,Y}(x,y)=f_X(x)f_Y(y)=\tfrac{1}{2\pi}\cdot\tfrac{1}{2\pi}$$
$$=\tfrac{1}{4\pi^2},\ 0<x<2\pi, 0<y<2\pi$$

and $f_{X,Y}(x,y)=0$, elsewhere.

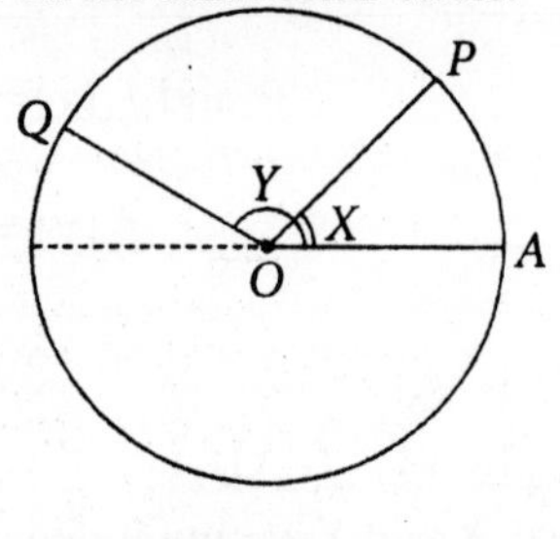

Fig 5.9

The points A, P, Q will lie on the same semi-circle if either $(0<X<\pi, 0<Y<\pi)$ or $(\pi<X<2\pi, \pi<Y<2\pi)$. The events are MEE.

$\therefore$ required probability

$$=P\{(0<X<\pi,0<Y<\pi)+(\pi<X<2\pi,\pi<Y<2\pi)\}$$
$$=P(0<X<\pi,0<Y<\pi)+P(\pi<X<2\pi,\pi<Y<2\pi)$$
$$=\int_0^\pi\int_0^\pi\frac{1}{4\pi^2}\,dx\,dy+\int_\pi^{2\pi}\int_\pi^{2\pi}\frac{1}{4\pi^2}\,dx\,dy=\frac{1}{4\pi^2}\cdot\pi^2+\frac{1}{4\pi^2}(2\pi-\pi)(2\pi-\pi)$$
$$=\frac{1}{4}+\frac{1}{4\pi^2}\cdot\pi^2=\frac{1}{4}+\frac{1}{4}=\frac{1}{2}.$$

Ex. 13. Show that the function $f_{X,Y}$ defined by

$$f_{X,Y}(x,y)=\begin{cases}k\sin x\sin y, & 0<x<\pi/2, 0<y<\pi/2\\ 0, & \text{elsewhere}\end{cases}$$

is a possible pdf for some suitable value of the constant k. Find (i) k (ii) the marginal distributions of X, Y and (iii) check whether X and Y are independent or not.

Solution. The NASC that a function $f_{x,y}$ to be a possible pdf are $f_{x,y}(x,y) \geq 0$ and

$$\int_{-\infty}^{\infty}\int_{-\infty}^{\infty} f_{x,y}(x,y)\,dx\,dy = 1.$$

The first condition is satisfied, if $k > 0$.
The second condition

$$\Rightarrow k\int_0^{\pi/2}\int_0^{\pi/2} \sin x \sin y \, dx\, dy = 1$$

$$\text{or, } \frac{1}{k} = \left[-\cos x\right]_0^{\pi/2}\left[-\cos y\right]_0^{\pi/2} = (-0+1)(-0+1) = 1 \Rightarrow k = 1 > 0.$$

Hence, $k = 1$,

$$\therefore f_{x,y}(x,y) = \begin{cases} \sin x \sin y, & \text{for } 0 < x < \pi/2, 0 < y < \pi/2 \\ 0, & \text{elsewhere.} \end{cases}$$

(ii) The marginal distribution of X is given by the following pdf,

$$f_x(x) = \int_0^{\pi/2} \sin x \sin y \, dy = \sin x\left[-\cos y\right]_0^{\pi/2} = \sin x, 0 < x < \pi/2$$

and $f_x(x) = 0$, elsewhere.

Similarly, $f_y(y) = \begin{cases} \sin y, & 0 < y < \pi/2 \\ 0, & \text{elsewhere.} \end{cases}$

(iii) As $f_{x,y}(x,y) = f_x(x)f_y(y)$, $\forall\, x,y$. $\therefore$ X and Y are independent RV.

Ex. 14. If two numbers are independently chosen in $(0,2)$, show that the probability that the distance between them is less than a fixed constant $k(0 < k < 2)$ is $\frac{1}{4}k(4-k)$.

Solution. Let X, Y be the RV representing the chosen numbers. Both X and Y are uniformly and independently distributed over $(0,2)$.

$\therefore$ the pdf for (X, Y) is $f_{x,y}(x,y) = f_x(x)f_y(y) = \frac{1}{2}\cdot\frac{1}{2} = \frac{1}{4}$, $0 < x < 2, 0 < y < 2$, i.e.,

$$f_{x,y}(x,y) = \begin{cases} \frac{1}{4}; & 0 < x < 2, 0 < y < 2 \\ 0; & \text{elsewhere.} \end{cases}$$

The event that the distance between the chosen numbers less than k is $|X - Y| < k$, i.e., $X - Y < k$ or, $Y - X < k$.

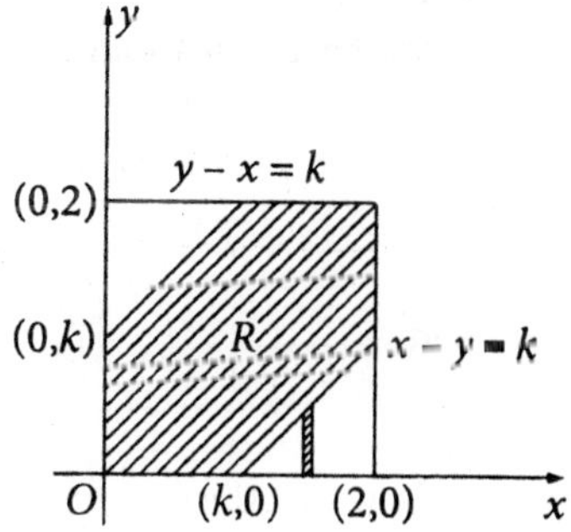

Fig 5.10

As the pdf is constant,

$$\therefore \text{ required probability} = \frac{\text{area in favour of the event}}{\text{total area}} = \frac{4 - 2\displaystyle\int_k^2 \int_0^{k-x} dy\, dx}{4}$$

$$= 1 - \frac{1}{2}\left\{\int_k^2 (k-x)\,dx\right\} = 1 - \frac{1}{2}\left[kx - \frac{x^2}{2}\right]_k^2$$

$$= 1 - \frac{1}{2}\left(2k - 2 - k^2 + \frac{k^2}{2}\right) = 1 - \frac{1}{4}\left(4 - 4k + k^2\right)$$

$$= 1 - \frac{1}{4}(2-k)^2 = \frac{1}{4}k(4-k).$$

Ex. 15. Suppose, the RV X and Y have join probability density function

$$f_{x,y}(x,y) = \begin{cases} e^{-y}, & \text{if } 0 < x < y < \infty \\ 0, & \text{elsewhere.} \end{cases}$$

Find the conditional density function of X, given that $Y = y$. [C.H. 1997]

Solution. We are to find $f_x(x\,|\,y)$. We have

$$f_x(x\,|\,y) = \frac{f_{x,y}(x,y)}{f_y(y)} = \frac{e^{-y}}{\displaystyle\int_0^y e^{-y}\,dx} = \frac{e^{-y}}{y\,e^{-y}} = \frac{1}{y},\ 0 < x < y$$

$$\therefore\ f_x(x\,|\,y) = \begin{cases} 1/y, & 0 < x < y \\ 0, & \text{elsewhere.} \end{cases}$$

Ex. 16. RV X and Y have joint density function

$$f_{x,y}(x,y) = \begin{cases} cx, & \text{if } 0 < y < x < 1 \\ 0, & \text{otherwise,} \end{cases}$$

for some constant c. Find c and the marginal density functions of X and Y. Are X and Y independent? [C.H. 1996]

Solution. We have $f_{x,y}(x,y) \geq 0,\ \forall\, x, y$ and $\displaystyle\int_{-\infty}^{\infty}\int_{-\infty}^{\infty} f_{x,y}(x,y)\,dx\,dy = 1.$

$\therefore\ c > 0$ and $\displaystyle\int_0^1\int_0^x cx\,dy\,dx = 1$ or, $\displaystyle\int_0^1 cx\left[y\right]_0^x dx = 1$ or, $c\displaystyle\int_0^1 x^2\,dx = 1$ or, $\dfrac{c}{3} = 1.\ \therefore\ c = 3.$

$$\therefore\ f_{x,y}(x,y) = \begin{cases} 3x, & 0 < y < x < 1 \\ 0, & \text{otherwise.} \end{cases}$$

Marginal density function f_x and f_y of X and Y respectively are given by

$$f_x(x) = \begin{cases} \displaystyle\int_0^x 3x\,dy = 3x^2, & 0 < x < 1 \\ 0, & \text{otherwise} \end{cases}$$

and

$$f_y(y) = \begin{cases} \displaystyle\int_y^1 3x\,dx = \frac{3}{2}\left[x^2\right]_y^1 = \frac{3}{2}\left(1 - y^2\right), & 0 < y < 1 \\ 0, & \text{otherwise.} \end{cases}$$

As $f_{x,y}(x,y) \neq f_x(x) f_y(y),\ \forall\, x, y.\ \therefore\ X$ and Y are not independent.

Ex. 17. Let RV X and Y have joint density function

$$f_{x,y}(x,y)=\begin{cases} e^{-x-y}, & \text{if } x,y>0 \\ 0, & \text{otherwise,} \end{cases}$$

and let $U=X+Y$ and $V=\frac{X}{X+Y}$. Find the joint density function of U and V. [C.H. 1996]

Solution. Here $U=X+Y$ and $V=\frac{X}{X+Y}$. In their real values $u=x+y, v=\frac{x}{x+y}$.
$\therefore\ v=\frac{x}{u}$. So, $x=uv$ and $y=u-uv$. Now, Jacobian of transformation

$$J=\frac{\partial(u,v)}{\partial(x,y)}=\begin{vmatrix} \frac{\partial u}{\partial x} & \frac{\partial u}{\partial y} \\ \frac{\partial v}{\partial x} & \frac{\partial v}{\partial y} \end{vmatrix}=\begin{vmatrix} 1 & 1 \\ \frac{y}{(x+y)^2} & -\frac{x}{(x-y)^2} \end{vmatrix}=-\frac{1}{x+y}<0,\ \forall\, x,y>0.\ \therefore\ J=-\frac{1}{u}.$$

$\therefore$ the joint density function $f_{u,v}$ of U and V is given by

$$f_{u,v}(u,v)=f_{x,y}(x,y)\left|\frac{\partial(x,y)}{\partial(u,v)}\right|=ue^{-u},\ u>0,\ 0<v<1$$

$$\therefore\ f_{u,v}(u,v)=\begin{cases} ue^{-u}, & u>0,\ 0<v<1 \\ 0, & \text{elsewhere.} \end{cases}$$

Ex. 18. Let (X,Y) has the pdf

$$f_{x,y}(x,y)=\begin{cases} e^{-x-y}; & x>0,\ y>0 \\ 0; & \text{elsewhere.} \end{cases}$$

Let $U=X+Y$ and $V=\frac{X}{X+Y}$. Find the distribution of (U,V) and the marginal distributions of U and V.

Solution. Proceeding as in Ex 17, we get the distribution of (U,V) by the pdf

$$f_{u,v}(u,v)=\begin{cases} ue^{-u}, & \text{for } u>0, 0<v<1 \\ 0, & \text{otherwise.} \end{cases}$$

$\therefore$ the marginal distributions of U and V are given by their pdf f_u and f_v as follows:

$$f_u(u)=\int_{-\infty}^{\infty} f_{u,v}(u,v)\,dv=\int_0^1 ue^{-u}\,dv=ue^{-u},\ u>0$$

$$=0,\ \text{elsewhere.}$$

$$f_v(v)=\int_0^{\infty} ue^{-u}\,du=\lim_{R\to\infty}\int_0^R ue^{-u}\,du=\lim_{R\to\infty}\left[-ue^{-u}\Big|_0^R+\int_0^R e^{-u}\,du\right]$$

$$=\lim_{R\to\infty}\left[-e^{-u}\right]_0^R=\lim_{R\to\infty}\left(-e^{-R}+1\right)=1,\ 0<v<1.$$

$$\therefore\ f_v(v)=\begin{cases} 1, & 0<v<1 \\ 0, & \text{elsewhere.} \end{cases}$$

$\therefore$ V is uniformly distributed over $(0,1)$.

Ex. 19. Let the distribution of the two-dimensional RV (X,Y) is given by

$$f_{x,y}(x,y)=\begin{cases} e^{-x-y}, & x\geq 0, y\geq 0 \\ 0, & \text{elsewhere.} \end{cases}$$

Find (a) the joint distribution function, (b) the marginal distributions of X and Y, (c) $P(X = Y)$, (d) $P(X+Y \leq 5)$, (e) $P(X \geq 2)$, (f) $P(X \leq Y)$, (g) $P(c < X+Y < d)$, where $0 < c < d$. Show that X and Y are independent.

Solution. (a) The joint distribution function $F_{x,y}$ is given by

$$F_{x,y}(x,y) = P(-\infty < X \leq x, -\infty < Y \leq y) = \int_{-\infty}^{x}\int_{-\infty}^{y} f_{x,y}(x,y)\,dx\,dy$$

$$= \int_{-\infty}^{x} e^{-x}\,dx \int_{-\infty}^{y} e^{-y}\,dy = \int_{0}^{x} e^{-x}\,dx \int_{0}^{y} e^{-y}\,dy$$

$$= (1-e^{-x})(1-e^{-y}),\ x \geq 0,\ y \geq 0.$$

$$\therefore\ F_{x,y}(x,y) = \begin{cases} (1-e^{-x})(1-e^{-y}), & \text{for } x \geq 0,\ y \geq 0 \\ 0, & \text{elsewhere.} \end{cases}$$

(b) The marginal distributions of X and Y are given by their pdf f_x and f_y as follows:

$$f_x(x) = \int_{-\infty}^{\infty} f_{x,y}(x,y)\,dy = \int_{0}^{\infty} e^{-x-y}\,dy = \lim_{R\to\infty}\int_{0}^{R} e^{-x-y}\,dy = \lim_{R\to\infty} e^{-x}\left[-e^{-y}\right]_0^R = e^{-x},\ x \geq 0.$$

$$\therefore\ f_x(x) = \begin{cases} e^{-x}, & x \geq 0 \\ 0, & \text{elsewhere.} \end{cases}$$

$$f_y(y) = \int_{-\infty}^{\infty} f_{x,y}(x,y)\,dx = \int_{0}^{\infty} e^{-x-y}\,dx = e^{-y},\ y \geq 0$$

$$\therefore\ f_y(y) = \begin{cases} e^{-y}, & y \geq 0 \\ 0, & \text{elsewhere.} \end{cases}$$

(c) $P(X = Y) = P(X \leq Y) - P(X < Y) = 0.$

(d) $$P(X+Y \leq 5) = P(Y \leq 5-X) = \int_0^5 \left\{\int_0^{5-x} e^{-x}\cdot e^{-y}\,dy\right\} dx = \int_0^5 e^{-x}\left(1-e^{x-5}\right) dx$$

$$= \int_0^5 e^{-x}\,dx - \int_0^5 e^{-5}\,dx = 1 - e^{-5} - 5e^{-5} = 1 - 6e^{-5}.$$

(e) $$P(X \geq 2) = \int_2^{\infty} f_x(x)\,dx = \int_2^{\infty} e^{-x}\,dx = e^{-2}.$$

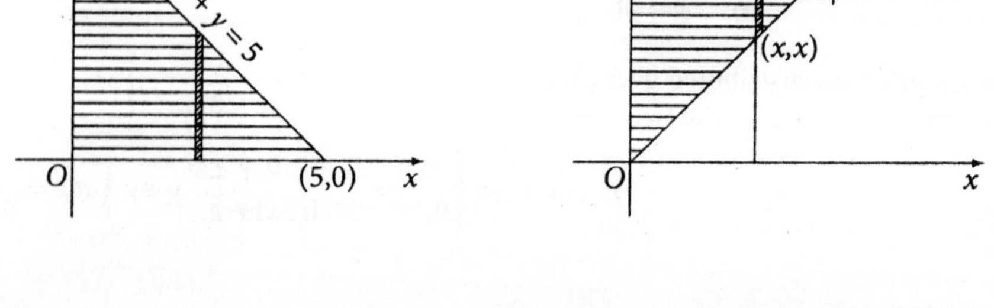

(f) $P(X \le Y) = \int_0^{\infty} e^{-x}\left(\int_x^{\infty} e^{-y}\,dy\right)dx = \int_0^{\infty} e^{-2x}\,dx = \lim_{R\to\infty}\left[\frac{e^{-2x}}{-2}\right]_0^R = \lim_{R\to\infty}\left[\frac{e^{-2R}}{-2}+\frac{1}{2}\right] = \frac{1}{2}.$

(g) $P(c \le X+Y \le d) = P(c-X \le Y \le d-X)$

$$= \int_0^{c} e^{-x}\left(\int_{c-x}^{d-x} e^{-y}\,dx\right)dx + \int_c^{d} e^{-x}\left(\int_0^{d-x} e^{-y}\,dy\right)dx$$

$$= \int_0^{c} e^{-x}\left(e^{x-c}-e^{x-d}\right)dx + \int_c^{d} e^{-x}\left(1-e^{x-d}\right)dx$$

$$= \int_0^{c}\left(e^{-c}-e^{-d}\right)dx + \int_c^{d}\left(e^{-x}-e^{-d}\right)dx$$

$$= c\left(e^{-c}-e^{-d}\right)+e^{-c}-e^{-d}-(d-c)e^{-d} = (1+c)e^{-c}-(1+d)e^{-d}.$$

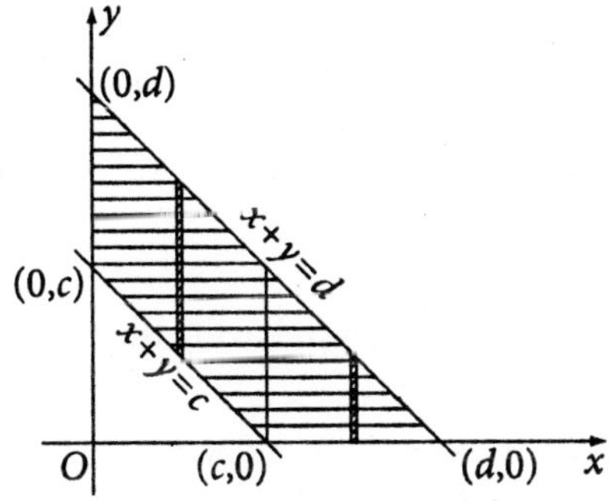

Since, $f_{x,y}(x,y) = e^{-x-y} = e^{-x}\cdot e^{-y} = f_x(x)f_y(y)$, $\forall\, x,y$, the RV X and Y are independent.

Ex. 20. The joint pdf of the RV X and Y is given by

$$f_{x,y}(x,y) = \begin{cases} cxy, & \text{for } 0<x<4, 1<y<5 \\ 0, & \text{elsewhere.} \end{cases}$$

Find (a) the value of the constant c, (b) $P(1<X<2, 2<Y<3)$, (c) $P(X \ge 3, Y \le 2)$, (d) the marginal distribution functions of X and Y.

Solution. (a) $\int_{-\infty}^{\infty}\int_{-\infty}^{\infty}(x,y)\,dx\,dy = 1$ and $f_{x,y}(x,y) \ge 0$, $\forall\, x,y \Rightarrow c>0$ and $\int_0^4\int_1^5 cxy\,dx\,dy = 1$,

i.e., $\frac{c}{4}\left[x^2\right]_0^4\left[y^2\right]_1^5 = 1$ or, $c = \frac{4}{16\times 24} = \frac{1}{96}$ (which is positive).

(b) $P(1<X<2, 2<Y<3) = \frac{1}{96}\int_1^2\int_2^3 xy\,dx\,dy = \frac{5}{128}.$

(c) $P(X \ge 3, Y \le 2) = \frac{1}{96}\int_3^4\int_1^2 xy\,dx\,dy = \frac{7}{128}.$

(d) Marginal distribution function F_X of X:

When $x<0$, $F_X(x) = 0$

When $0 \le x < 4$, $F_X(x) = P(X \le x) = \int_{-\infty}^{x} f_X(x)\,dx = \int_{-\infty}^{x}\left(\int_{-\infty}^{\infty}\frac{xy}{96}\,dy\right)dx$

$$= \frac{1}{96}\int_0^x x\left(\int_1^5 y\,dy\right)dx = \frac{12}{96\times 2}x^2 = \frac{x^2}{16}.$$

When $x \geq 4, F_x(x) = \frac{1}{96}\int_0^4\int_1^5 xy\,dx\,dy = 1.$

Thus

$$F_x(x) = \begin{cases} 0, & x < 0 \\ x^2/16, & 0 \leq x < 4 \\ 1, & x \geq 4. \end{cases}$$

Marginal distribution function F_y of Y:

When $y < 1, F_y(y) = 0$

When $1 \leq y < 5, F_y(y) = P(Y \leq y) = \int_{-\infty}^{y}\left(\int_{-\infty}^{\infty}\frac{xy}{96}dx\right)dy = \frac{1}{96}\int_1^y y\left(\int_0^4 x\,dx\right)dy$

$$= \frac{8}{96 \times 2}(y^2 - 1) = \frac{1}{24}(y^2 - 1).$$

When $y \geq 5, F_y(y) = \int_1^5\int_0^4 \frac{xy}{96}dx\,dy = 1.$

Thus $F_y(y) = \begin{cases} 0, & y < 1 \\ \frac{1}{24}(y^2 - 1), & 1 \leq y < 5 \\ 1, & y \geq 5 \end{cases}$

Ex. 21. X_1 and X_2 are two independent RV each having the same pdf $2x\exp(-x^2)$; $0 < x < \infty$. Find the pdf of $\sqrt{X_1^2 + X_2^2}$. [C.H. 1995]

Solution. Let $U = \sqrt{X_1^2 + X_2^2}$ and $V = X_1$. In their real values $u = \sqrt{x_1^2 + x_2^2}, v = x_1$.
As x_1, x_2 ranges from 0 to ∞, u, v also ranges from 0 to ∞, and $v < u$.
$\therefore\ x_1^2 = v^2$. Again, $u^2 = x_1^2 + x_2^2$. $\therefore\ x_2^2 = u^2 - x_1^2 = u^2 - v^2$.
$\therefore\ x_1 = v$ and $x_2 = \sqrt{u^2 - v^2}$ as all of them ≥ 0.

$$\therefore\ \frac{\partial(x_1, x_2)}{\partial(u, v)} = \begin{vmatrix} \frac{\partial x_1}{\partial u} & \frac{\partial x_1}{\partial v} \\ \frac{\partial x_2}{\partial u} & \frac{\partial x_2}{\partial v} \end{vmatrix} = \begin{vmatrix} 0 & 1 \\ \frac{u}{\sqrt{u^2 - v^2}} & -\frac{v}{\sqrt{u^2 - v^2}} \end{vmatrix} = -\frac{u}{\sqrt{u^2 - v^2}} < 0,\ \forall\, u, v.$$

$\therefore$ if $f_{u,v}$ be the pdf of (U, V), then

$$f_{u,v}(u, v) = f_{x_1,x_2}(x_1, x_2)\left|-\frac{u}{\sqrt{u^2 - v^2}}\right| = 2x_1 e^{-x_1^2} \cdot 2x_2 e^{-x_2^2}\frac{u}{\sqrt{u^2 - v^2}}.$$

$\because$ X_1, X_2 are independent, $\therefore\ f_{x_1,x_2}(x_1, x_2) = f_{x_1}(x_1) f_{x_2}(x_2)$

$$= 4v\sqrt{u^2 - v^2}\, e^{-u^2}\frac{u}{\sqrt{u^2 - v^2}} = 4uve^{-u^2},\ 0 < v < u < \infty.$$

$$\therefore\ f_u(u) = 4\int_0^u uve^{-u^2}dv = 4ue^{-u^2}\left[\frac{v^2}{2}\right]_0^u = 2u^3 e^{-u^2},\ 0 < u < \infty.$$

Hence, the distribution of $U = \sqrt{X_1^2 + X_2^2}$ is given by the pdf:

$$f_u(u) = \begin{cases} 2u^3 e^{-u^2}, & 0 < u < \infty \\ 0, & \text{elsewhere.} \end{cases}$$

Ex. 22. The distribution of (X, Y) is given by the following pdf:

$$f_{x,y}(x,y)=\begin{cases}4xy\exp\{-(x^2+y^2)\}, & x>0,\ y>0\\ 0, & \text{elsewhere.}\end{cases}$$

Test whether X and Y are independent. Find the conditional pdf of X, given $Y=y$.

Solution. The marginal pdfs f_x and f_y of X and Y respectively are given by

$$f_x(x)=\int_0^\infty 4xy\,e^{-x^2-y^2}\,dy=\lim_{R\to\infty}2xe^{-x^2}\int_0^R 2ye^{-y^2}\,dy=2xe^{-x^2}\lim_{R\to\infty}\left[1-e^{-R^2}\right]\ (\text{Putting } y^2=t)$$

$$\therefore\ f_x(x)=\begin{cases}2xe^{-x^2}, & x>0\\ 0, & \text{elsewhere.}\end{cases}$$

$$f_y(y)=\int_0^\infty 4xye^{-x^2-y^2}\,dx=2ye^{-y^2}\lim_{R\to\infty}\int_0^R 2xe^{-x^2}\,dx.$$

Let $x^2=u$, $\therefore\ 2x\,dx=du$.

$$=2ye^{-y^2}\lim_{R\to\infty}\int_0^{R^2}e^{-u}\,du=2ye^{-y^2}\lim_{R\to\infty}\left[-e^{-u}\right]_0^{R^2}$$

$$=2ye^{-y^2}\lim_{R\to\infty}\left(1-e^{-R^2}\right)=2ye^{-y^2},\ y>0$$

$$\therefore\ f_y(y)=\begin{cases}2ye^{-y^2}, & y>0\\ 0, & \text{elsewhere.}\end{cases}$$

$\therefore\ f_{x,y}(x,y)=f_x(x)f_y(y),\ \forall\, x,y\in\mathbb{R}$.
$\therefore$ the RV X and Y are independent.
Now, conditional density of X, given $Y=y$, is given by

$$f_x(x\,|\,y)=\frac{f_{x,y}(x,y)}{f_y(y)}=\frac{4xye^{-x^2-y^2}}{2ye^{-y^2}}=2xe^{-x^2},\ x>0,$$

i.e., $f_x(x\,|\,y)=\begin{cases}2xe^{-x^2}, & x>0\\ 0, & \text{elsewhere.}\end{cases}$

Note: (1) $f_x(x\,|\,y)$ is the same as $f_x(x)$ as X and Y are independent. (2) In the same way the conditional density of Y, given $X=x$, is given by $f_y(y\,|\,x)=\begin{cases}2ye^{-y^2}, & y>0\\ 0, & \text{elsewhere.}\end{cases}$

Ex. 23. Let the pdf of the two-dimensional RV (X, Y) is a constant c, say, when $x^2+y^2<a^2$, otherwise vanishes. Find (a) the value of c, (b) the marginal distributions of X and Y. Check whether X and Y are independent.

Solution. Given the pdf of (X, Y): $f_{x,y}(x,y)=\begin{cases}c, & x^2+y^2<a^2\\ 0, & \text{elsewhere.}\end{cases}$

We have $f_{x,y}(x,y)\geq 0,\ \forall\, x,y\in\mathbb{R}$. $\therefore\ c>0$ and

$$\int_{-\infty}^{\infty}\int_{-\infty}^{\infty}f_{x,y}(x,y)\,dx\,dy=1,\ \text{i.e., } c\iint_{x^2+y^2<a^2}dx\,dy=1 \text{ or, } c\pi a^2=1.\ \therefore\ c=\frac{1}{\pi a^2}.$$

$$\therefore\ f_{x,y}(x,y)=\begin{cases}1/\pi a^2, & x^2+y^2<a^2\\ 0, & \text{elsewhere.}\end{cases}$$

The marginal distribution of X is given by the pdf f_X as

$$f_X(x)=\int_{-\infty}^{\infty} f_{X,Y}(x,y)\,dy=\frac{1}{\pi a^2}\int_{-\sqrt{a^2-x^2}}^{\sqrt{a^2-x^2}} dy=\frac{2}{\pi a^2}\sqrt{a^2-x^2},\ -a<x<a.$$

$$\therefore\ f_X(x)=\begin{cases}\frac{2}{\pi a}\sqrt{1-(x/a)^2}, & -a<x<a\\ 0, & \text{elsewhere.}\end{cases}$$

The marginal distribution of Y is given by the pdf f_Y as

$$f_Y(y)=\int_{-\infty}^{\infty} f_{X,Y}(x,y)\,dx=\frac{1}{\pi a^2}\int_{-\sqrt{a^2-y^2}}^{\sqrt{a^2-y^2}} dx=\frac{2}{\pi a^2}\sqrt{a^2-y^2},\ -a<y<a$$

$$\therefore\ f_Y(y)=\begin{cases}\frac{2}{\pi a}\sqrt{1-\left(\frac{y}{a}\right)^2}, & -a<y<a\\ 0, & \text{elsewhere.}\end{cases}$$

$\therefore\ f_{X,Y}(x,y)\neq f_X(x)f_Y(y),\ \forall\ x,y\in\mathbb{R}.$
$\therefore$ the RV X ad Y are not independent.

Ex. 24. If the RV X is gamma (l) variate, RV Y is gamma (m) variate and X, Y are independent, show that (a) $X+Y$ is gamma $(l+m)$ variate and (b) $\frac{X}{Y}$ is a $\beta_2(l,m)$ variate. [C.H. 1987]

Solution. Let $U=X+Y$ and $V=\frac{X}{Y}$. $\therefore$ in their real values $u=x+y, v=\frac{x}{y}$.
$\therefore\ x=\frac{uv}{1+v},\ y=\frac{u}{1+v}$.
$\therefore$ Jacobian of transformation $=\frac{\partial(u,v)}{\partial(x,y)}=-\frac{x+y}{y^2}<0$, as $x,y>0$.
$\therefore$ the pdf $f_{U,V}$ of (U,V) is given by

$$f_{U,V}(u,v)=f_{X,Y}(x,y)\left|\frac{\partial(x,y)}{\partial(u,v)}\right|=f_X(x)f_Y(y)\left|\frac{\partial(x,y)}{\partial(u,v)}\right| \text{ as } X, Y \text{ are independent}$$

$$=\frac{e^{-x}x^{l-1}}{\Gamma(l)}\frac{e^{-y}y^{m-1}}{\Gamma(m)}\frac{y^2}{x+y}=\frac{e^{-u}}{\Gamma(l)\Gamma(m)}\left(\frac{uv}{1+v}\right)^{l-1}\left(\frac{u}{1+v}\right)^{m+1}\frac{1}{u}$$

$$=\frac{e^{-u}u^{l+m-1}}{\Gamma(l+m)}\cdot\frac{v^{l-1}}{B(l,m)(1+v)^{l+m}},\ u>0, v>0\ \left[\because\ B(l,m)=\frac{\Gamma(l)\Gamma(m)}{\Gamma(l+m)}\right].$$

$$\therefore\ f_U(u)=\frac{e^{-u}u^{l+m-1}}{\Gamma(l+m)}\int_0^{\infty}\frac{v^{l-1}}{B(l,m)(1+v)^{l+m}}dv=\frac{e^{-u}u^{l+m-1}}{\Gamma(l+m)},\ u>0$$

$$\because\ \int_0^{\infty}\frac{v^{l-1}}{B(l,m)(1+v)^{l+m}}du=1$$

$=0$, elsewhere.

$\therefore\ U$, i.e., $X+Y$ is gamma $(l+m)$ variate.
Again

$$f_V(v)=\frac{v^{l-1}}{B(l,m)(1+v)^{l+m}}\int_0^{\infty}\frac{e^{-u}u^{l+m-1}}{\Gamma(l+m)}du=\frac{v^{l-1}}{B(l,m)(1+v)^{l+m}},\ v>0.$$

$$\because\ \int_0^{\infty}\frac{e^{-u}u^{l+m-1}}{\Gamma(l+m)}du=1$$

$=0$, elsewhere.

$\therefore$ V, i.e., $\frac{X}{Y}$ is a $\beta_2(l,m)$ variate.

Note: 1. $X+Y$ is gamma $(l+m)$ variate, also follows from the reproductive property of gamma distribution.

2. Regarding Jacobian of transformation.

If $x_i=\phi_i(u_1,u_2,\cdots,u_n), i=1,2,\cdots,n$ are differentiable functions of $u_1,u_2,\cdots,u_n$, then the Jacobian of the functions $x_1,x_2,\cdots,x_n$ w.r.t. the variables $u_1,u_2,\cdots,u_n$ is denoted by J or by

$$\frac{\partial(x_1,x_2,\cdots,x_n)}{\partial(u_1,u_2,\cdots,u_n)}$$

and is defined by

$$J=\frac{\partial(x_1,x_2,\cdots,x_n)}{\partial(u_1,u_2,\cdots,u_n)}=\begin{vmatrix}\frac{\partial x_1}{\partial u_1} & \frac{\partial x_1}{\partial u_2} & \cdots & \frac{\partial x_1}{\partial u_n}\\ \frac{\partial x_2}{\partial u_1} & \frac{\partial x_2}{\partial u_2} & \cdots & \frac{\partial x_2}{\partial u_n}\\ \cdots & \cdots & & \cdots\\ \frac{\partial x_n}{\partial u_1} & \frac{\partial x_n}{\partial u_2} & \cdots & \frac{\partial x_n}{\partial u_n}\end{vmatrix}$$

Jacobian of transformation follows the property:

$$\frac{\partial(x_1,x_2,\cdots,x_n)}{\partial(u_1,u_2,\cdots,u_n)}\cdot\frac{\partial(u_1,u_2,\cdots,u_n)}{\partial(x_1,x_2,\cdots,x_n)}=1.$$

Ex. 25. The numbers $X_1,X_2,\cdots,X_n$ are independently chosen at random in the interval (a,b). Prove that the pdf of the RV $X=\min\{X_1,X_2,\cdots,X_n\}$ is given by $f_x(x)=\frac{n(b-x)^{n-1}}{(b-a)^n}$, $(a<x<b)$.

Solution. Let x be any real values of X, then $a < x < b$.

The event $X>x \Leftrightarrow (X_1>x,X_2>x,\cdots,X_n>x)$.

$\therefore$ $P(X>x)=P(X_1>x,X_2>x,\cdots,X_n>x)=P(X_1>x)P(X_2>x)\cdots P(X_n>x)$ as $X_1,X_2,\cdots,X_n$ are independent. (1)

As $X_1,X_2,\cdots,X_n$ are independently chosen in (a,b), so they are independent and each is uniformly distributed over (a,b), i.e., pdf for X_i is

$$f_{x_i}(x)=\begin{cases}1/(b-a), & a<x<b\\ 0, & \text{elsewhere},\end{cases} \qquad i=1,2,\cdots,n.$$

$\therefore$ from (1),

$$\begin{aligned}P(X>x)&=\left(\int_x^b\frac{1}{b-a}dx_1\right)\left(\int_x^b\frac{1}{b-a}dx_2\right)\cdots\left(\int_x^b\frac{1}{b-a}dx_n\right)\\ &=\frac{b-x}{b-a}\cdot\frac{b-x}{b-a}\cdots\text{to } n \text{ factors}\\ &=\left(\frac{b-x}{b-a}\right)^n\end{aligned}$$

$\therefore$ the probability distribution function F_x of X is given by

$$F_x(x)=P(X\le x)=1-P(X>x)=1-\left(\frac{b-x}{b-a}\right)^2, \quad a<x<b$$

$$\text{and } F_x(x)=\begin{cases}0, & \text{when } x\le a\\ 1, & \text{when } x\ge b.\end{cases}$$

Therefore, the pdf f_X of X is given by

$$f_X(x) = F_X'(x) = \begin{cases} \frac{n(b-x)^{n-1}}{(b-a)^n}, & a < x < b \\ 0, & x \le a \text{ or } x \ge b. \end{cases}$$

Note: $f_X(a)$ is not defined as $LF_X'(a) \neq RF_X'(a)$.

Ex. 26. The joint pdf of the RV X and Y is

$$f_{X,Y}(x,y) = \begin{cases} k(1-x-y), & \text{for } x \ge 0, y \ge 0 \text{ and } x+y \le 1 \\ 0, & \text{elsewhere,} \end{cases}$$

where k is a constant.

Find (a) the constant k, (b) the marginal pdf of X, (c) the pdf of Y, when $X = 1/2$. [C.H. 1992]

Solution. (a) We have $f_{X,Y}(x,y) \ge 0$, $\forall\, x, y$ and $\int_{-\infty}^{\infty}\int_{-\infty}^{\infty} f_{X,Y}(x,y)\,dx\,dy = 1$.

$$\therefore k > 0 \text{ and } k\int_0^1\int_0^{1-x}(1-x-y)\,dy\,dx = 1$$

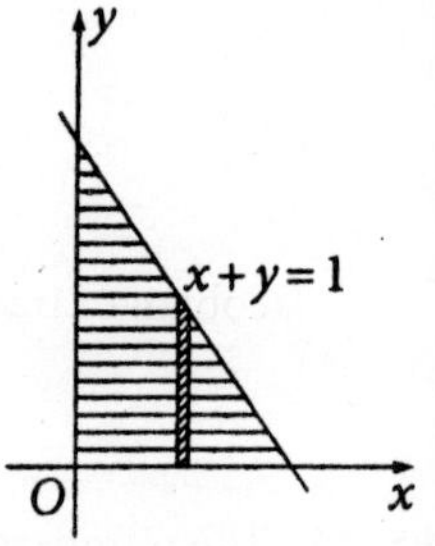

$$\text{or, } \frac{1}{k} = \int_0^1 \left[y - xy - \tfrac{1}{2}y^2\right]_0^{1-x} dx = \int_0^1 \left\{(1-x)^2 - \tfrac{1}{2}(1-x)^2\right\} dx$$

$$= \tfrac{1}{2}\int_0^1 \left(1 - 2x + x^2\right) dx = \tfrac{1}{2}\left[x - x^2 + \frac{1}{3}x^3\right]_0^1 = \tfrac{1}{6}.$$

$\therefore k = 6.$

(b) If f_X be the marginal pdf of X, then

$$f_X(x) = 6\int_0^{1-x}(1-x-y)\,dy = 6\left[(1-x)y - \tfrac{1}{2}y^2\right]_0^{1-x}$$

$$\therefore f_X(x) = \begin{cases} 3(1-x)^2, & 0 < x < 1 \\ 0, & \text{elsewhere.} \end{cases}$$

(c) $f_Y\left(y \mid \tfrac{1}{2}\right)$ = conditional density function of Y, given $X = \tfrac{1}{2}$.

$$f_Y(y \mid x) = \frac{f_{X,Y}(x,y)}{f_X(x)} = \frac{6(1-x-y)}{3\,(1-x^2)} = \frac{2(1-x-y)}{(1-x)^2},\ 0 < y \le 1-x,$$

where x is any fixed number such that $0 < x < 1$.

When $x = \tfrac{1}{2}$, $f_Y\left(y \mid \tfrac{1}{2}\right) = \begin{cases} 4(1-2y), & 0 < y < \tfrac{1}{2} \\ 0, & \text{elsewhere.} \end{cases}$

Ex. 27. Two people agree to meet at a definite place between 12 and 1 o'clock with the understanding that each will wait 20 minutes for the other. What is the probability that they will meet. [C.H. 1989]

Solution. Time interval between 12 and 1 o'clock is 1 hour = 60 minutes.

Taking 12 noon as origin, let X and Y be the RV denoting the time of arrival of the two people between 12 and 1 o'clock. As their times of reaching the place are random in nature (any time is equally likely) they are uniformly distributed over the interval $(0, 60)$.

$\therefore$ their pdfs are

$$f_X(x) = \begin{cases} 1/60, & 0 < x < 60 \\ 0, & \text{elsewhere} \end{cases};\quad f_Y(y) = \begin{cases} 1/60, & 0 < y < 60 \\ 0, & \text{elsewhere.} \end{cases}$$

As X, Y are independent, $\therefore$ the pdf of (X, Y) is

$$f_{x,y}(x,y) = f_x(x)f_y(y) = \begin{cases} 1/3600, & 0 < x < 60,\ 0 < y < 60 \\ 0, & \text{elsewhere.} \end{cases}$$

The event that they will meet is $|X - Y| < 20$.

$\therefore\ P(|X - Y| < 20) = P(X - Y < 20 \cup Y - X < 20)$

$= \dfrac{\text{Area in favour of the event}}{\text{Total area}}$, as it is a geometric probability for constant pdf

$= \frac{1}{3600}\left[3600 - 2 \cdot \frac{1}{2} \cdot 40 \cdot 40\right] = \frac{2000}{3600} = \frac{5}{9}.$

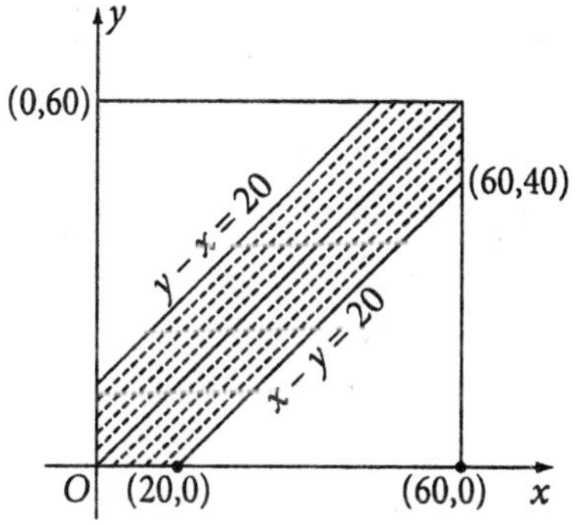

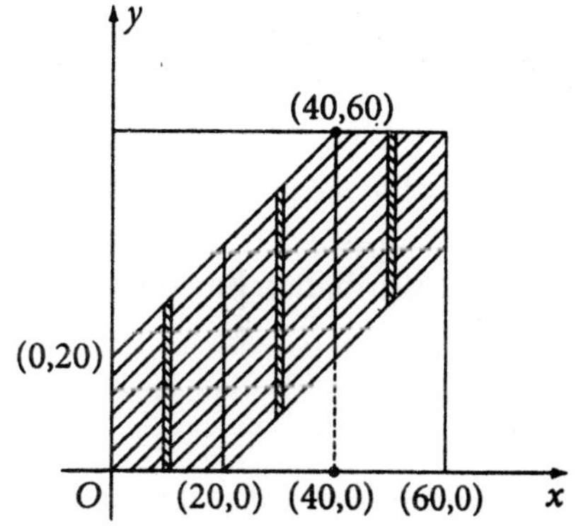

Alternative Method

$$P(|X - Y| < 20) = P(X - Y < 20 \cup Y - X < 20)$$

$$= \int_0^{20}\int_0^{x+20} \frac{1}{3600}\,dy\,dx + \int_{20}^{40}\int_{x-20}^{x+20} \frac{1}{3600}\,dy\,dx + \int_{40}^{60}\int_{x-20}^{60} \frac{1}{3600}\,dy\,dx$$

$$= \frac{1}{3600}\left[\left(\frac{x^2}{2} + 20x\right)\Bigg|_0^{20} + 40x\Bigg|_{20}^{40} + \left(80x - \frac{x^2}{2}\right)\Bigg|_{40}^{60}\right]$$

$$= \frac{1}{3600}[200 + 400 + 800 + 1600 - 1000] = \frac{2000}{3600} = \frac{5}{9}.$$

EXERCISE

1. Let X and Y have the joint distribution given by the pdf

$$f_{x,y}(x,y) = \begin{cases} \frac{1}{\pi c^2}, & \text{if } x^2 + y^2 < c^2 \\ 0, & \text{otherwise.} \end{cases}$$

Find the marginal distribution of X and the conditional distribution of Y, given $X = x, (|x| < c)$. Hence, examine if X and Y are mutually independent. [C.H. 1990]

2. If X and Y are two independent continuous variates with pdfs f_x and f_y respectively, then show that the pdf h_u of $U = X + Y$ is given by $h_u(u) = \int_{-\infty}^{\infty} f_x(v)f_y(u - v)\,dv$.

Hence, show that if the independent RV X_1 and X_2 are distributed normally with parameters (m_1, σ_1) and (m_2, σ_2) respectively, then the RV $X = X_1 + X_2$ is also normally distributed with parameters (m, σ), where $m = m_1 + m_2$ and $\sigma = \sqrt{\sigma_1^2 + \sigma_2^2}$. [C.H. 1989]

3. If X and Y are two independent $\gamma(l)$ and $\gamma(m)$ variates, then prove that $U=X+Y$ and $V=\frac{X}{X+Y}$ are independent $\gamma(l+m)$ and $\beta_1(l,m)$ variates respectively. [C.H. 1989]
4. Two gentlemen agree to meet at Howrah railway station between 5 p.m. and 6 p.m. with the understanding that each will wait 10 minutes for the other. What is the probability that they will meat. (*Ans.* 11/36)
5. Two friends agree to meet at Alipore zoo between 10 a.m. to 12 noon with the understanding that each will wait up to 30 minutes for the other. What is the probability that they will meet. (*Ans.* 7/16)
6. The RV X and Y are independent and their pdfs are respectively given by $f_X(x)=\frac{1}{\pi}\frac{1}{\sqrt{1-x^2}}, |x|<1$ and $f_Y(y)=ye^{-y^2/2}, y>0$.
   Find the joint pdfs of Z and W, where $Z=XY$ and $W=X$. Deduce the pdf of Z. [C.H. 1985]

   [*Ans.* When $z>0$, $f_{Z,W}(z,w)=\begin{cases}\frac{z}{\pi w^2\sqrt{1-w^2}}\exp\left(-\frac{z^2}{2w^2}\right), & 0<w<1\\ 0, & \text{elsewhere}\end{cases}$

   and $f_Z(z)=\frac{1}{\sqrt{2\pi}}e^{-z^2/2}$. When $z<0$, $f_Z(z)$ is also has the same expression.
   $\therefore f_Z(z)=\frac{1}{\sqrt{2\pi}}e^{-z^2/2}, -\infty<z<\infty$. Hence, Z is normal $(0,1)$.]
7. Let the distribution function $F_{X,Y}$ of (X,Y) is given by

$$F_{X,Y}(x,y)=\begin{cases}1+e^{-x-y}-e^{-x}-e^{-y}, & \text{if } x,y>0\\ 0, & \text{otherwise.}\end{cases}$$

   Find the marginal distribution functions F_X and F_Y of X and Y respectively. Hence, show that X and Y are independent.

   [Hints. $F_X(x)=\lim_{y\to\infty}F_{X,Y}(x,y)=\begin{cases}1-e^{-x}, & x>0\\ 0, & \text{otherwise.}\end{cases}$
   Thus the distribution of X is exponential with the value of parameter 1, etc.]
8. Find the constant k such that the function $f_{X,Y}$ given by $f_{X,Y}(x,y)=\begin{cases}k(x+y)e^{-x-y}, & x,y>0\\ 0, & \text{elsewhere,}\end{cases}$

   is a pdf of the RV (X,Y).
   Find the distribution of $X+Y$. $\left(\textit{Ans. } k=1/2, f_U(u)=\begin{cases}\frac{1}{2}u^2e^{-u}, & u>0\\ 0, & \text{elsewhere}\end{cases} \text{ where } u=x+y\right)$

   [Hints. Let $U=X+Y$ and $V=Y$.]
9. If $f_{X,Y}(x,y)=\begin{cases}\frac{1}{4}\exp\left[-\frac{1}{2}(x+y)\right], & \text{if } x>0, y>0\\ 0, & \text{elsewhere,}\end{cases}$

   be the pdf of the two-dimensional RV (X,Y), then show that the joint distribution of U and V where $U=\frac{1}{2}(X-Y)$ and $V=Y$ is

$$f_{U,V}(u,v)=\begin{cases}\frac{1}{2}\exp(u-v), & \text{if } 2u+v>0, v>0\\ 0, & \text{elsewhere.}\end{cases}$$

   Hence, deduce that U follows the bilateral exponential distribution with pdf:
   $f_U(u)=\frac{1}{2}e^{-|u|}, -\infty<u<\infty$.

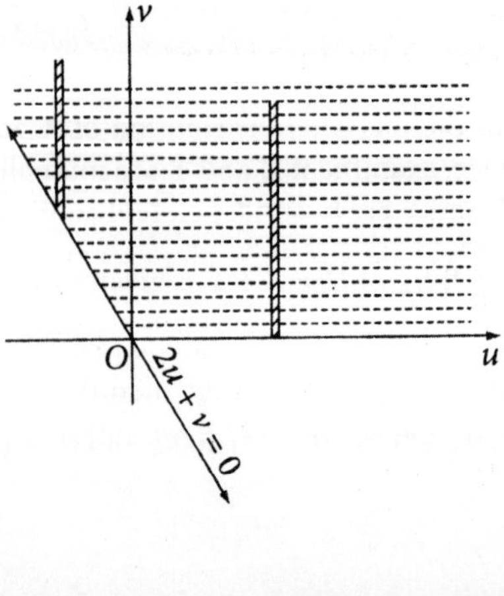

[**Hints.** When $u<0$,

$$f_u(u)=\frac{1}{2}\int_{-2u}^{\infty} e^{-u}\,e^{-v}\,dv=\frac{1}{2}e^{u}.$$

When $u>0$,

$$f_u(u)=\frac{1}{2}\int_{0}^{\infty} e^{-u}\,e^{-v}\,dv=\frac{1}{2}e^{-u}.$$

$$\therefore\ f_u(u)=\frac{1}{2}e^{-|u|},-\infty<u<\infty.]$$

10. The distribution of (X, Y) is given by the following pdf:

$$f_{x,y}(x,y)=\begin{cases}2e^{-x-y}, & \text{if } 0<x<y<\infty\\ 0, & \text{elsewhere.}\end{cases}$$

Find the conditional pdf of Y, given $X=x$, and the conditional pdf of X, given $Y=y$.

$\left(\textit{Ans. } f_{Y|X}(y\,|x)=e^{x-y} \text{ for } y>x;\ f_{X|Y}(x\,|y)=\frac{e^{-x}}{1-e^{-y}} \text{ for } 0<x<y\right)$

11. Let each of the independent RV X and Y follow the exponential distribution with parameter θ. Find the joint distribution of X and $X+Y$ and deduce that the conditional density function of X, given $X+Y=a$, is uniform over the interval $(0,a)$ for each $a>0$.

$\left(\textit{Ans. } f_{u,v}(u,v)=\theta^2 e^{-\theta u}, u>0, v>0; \text{ when } U=X+Y \text{ and } V=X\right)$

12. Let X, Y be independent normal variates and each be distributed with common mean zero and unit variance. Find the pdf of $U=\sqrt{X^2+Y^2}$. [C.H. 2003]

[**Hints.** X and Y are standard normal variates. So, $\frac{1}{2}X^2$ and $\frac{1}{2}Y^2$ are each gamma $(\frac{1}{2})$ variate and they are independent. $\therefore$ by reproductive property of gamma distribution $\frac{1}{2}X^2+\frac{1}{2}Y^2$, i.e., $\frac{1}{2}(X^2+Y^2)$ is a gamma (1) distribution.

Thus $U=\sqrt{X^2+Y^2}=\sqrt{2V}$, where $V=\frac{1}{2}(X^2+V^2)$ is Gamma (1) variate. Set in terms of their real values $u=\sqrt{2v}$. $\therefore\ \frac{du}{dv}=\frac{1}{\sqrt{2v}}>0,\ \forall\ v>0$.

$\therefore$ by the transformation of RV $f_u(u)=f_v(v)\left|\frac{dv}{du}\right|=e^{-v}\frac{v^{1-1}}{\Gamma(1)}\sqrt{2v}=u\,e^{-u^2/2},\ 0<u<\infty$ and $f_u(u)=0$, otherwise.]

13. The joint pdf of the RV X and Y is given by

$$f_{x,y}(x,y)=\begin{cases}x+y, & 0<x<1, 0<y<1\\ 0, & \text{otherwise.}\end{cases}$$

Find the distribution of XY. [C.H. 2000]

$$\left(\textit{Ans. } f_u(u)=\begin{cases}2(1-u), & 0<u<1\\ 0, & \text{elsewhere,}\end{cases} \text{ where } U=XY.\right)$$

14. If X and Y be two independent RV both uniformly distributed over $(0,1)$, find the distribution of $X+Y$ and $X-Y$. [C.H. 2008]

$$\left(\textit{Ans. } f_u(u)=\begin{cases}2-u, & 1<u<2\\ 0, & \text{elsewhere}\end{cases} \text{ and } f_v(v)=\begin{cases}1-v, & 0<v<1\\ 0, & \text{elsewhere}\end{cases} \text{ when } U=X+Y \text{ and } V=X-Y.\right)$$

15. The bivariate RV (X, Y) follows the joint pdf $f_{x,y}(x,y)=\begin{cases}kx^2(8-y), & x<y<2x, 0\le x\le 2\\ 0, & \text{elsewhere.}\end{cases}$

Find (i) k, (ii) the marginal pdfs of X and Y, (iii) the conditional pdfs $f_{X|Y}(x\,|y)$ and $f_{Y|X}(y\,|x)$. [B.H. 2006]

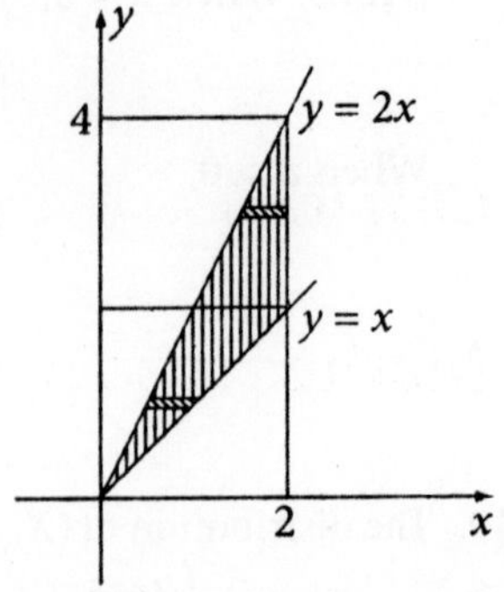

[**Hints.** (i) $\displaystyle\int_{-\infty}^{\infty}\int_{-\infty}^{\infty} f_{x,y}(x,y)\,dx\,dy = 1 \Rightarrow k = \frac{5}{112}$.

(ii) $\displaystyle f_x(x) = \int_{-\infty}^{\infty} f_{x,y}(x,y)\,dy = \tfrac{5}{112}x^3\left(8 - \tfrac{3}{2}x\right),\ 0 \le x \le 2$

$= 0$, elsewhere.

$$f_y(y) = \frac{5}{112}\int_{y/2}^{y} x^2(8-y)\,dx = \frac{5}{112}(8-y)\cdot\frac{1}{3}\left(y^3 - \frac{y^3}{8}\right)$$

$$= \frac{35}{2688}y^3(8-y), 0 \le y \le 2$$

$$f_y(y) = \frac{5}{112}\int_{y/2}^{2} x^2(8-y)\,dx = \frac{5}{112}(8-y)\left(\frac{8}{3} - \frac{y^3}{24}\right) = \frac{5}{2688}(8-y)\left(64-y^3\right),\ 2 < y \le 4.$$

$f_y(y) = 0$, elsewhere.

(iii) $\displaystyle f_{x|y}(x\,|\,y) = \frac{f_{x,y}(x,y)}{f_y(y)}$ and $\displaystyle f_{y|x}(y\,|\,x) = \frac{f_{x,y}(x,y)}{f_x(x)}$, etc.]

Chapter 6

Mathematical Expectation I

6.1 Introduction

Where there is a chance there is expectation, i.e., expectation comes into play, whenever there is any chance of occurrence. Where there is no chance the expectation must be zero. Expectation increases with the increase of probability. So, most expected value occurs with maximum probability. As mean value is most likely to occur (as most of the values of a RV lie near the mean), therefore, we define the expected value of a RV $g(X)$ to be the weighted mean of $g(X)$, where the weights are the respective probabilities of the events related to $g(X)$. The idea was first introduced by Huygens (1629–1695). It is very much relevant in the games of chance.

Thus the expected or mean value of a RV $g(X)$ is denoted by $E\{g(X)\}$ and is defined as

$$E\{g(X)\} = \sum_{i=-\infty}^{\infty} g(x_i) f_{xi}, \text{ when } g(X) \text{ is a discrete RV}$$

$$\text{and } E\{g(X)\} = \int_{-\infty}^{\infty} g(x) f_x(x)\, dx, \text{ when } g(X) \text{ is a continuous RV,}$$

provided the infinite series or the improper integral converges absolutely, i.e., the mere convergence does not guarantee the existence of expectation.

Follow the example:

Ex. 1 Suppose, the spectrum of a discrete RV X are $x_i = (-1)^i \frac{2^i}{i}, i = 1, 2, 3, \cdots$ and $f_{xi} = P(X = x_i) = \frac{1}{2^i}$.

$$\therefore\ E(X) = \sum_{i=1}^{\infty} x_i f_{xi} = \sum_{i=1}^{\infty} (-1)^i \frac{2^i}{i} \cdot \frac{1}{2^i} = -\sum_{i=1}^{\infty} \frac{(-1)^{i-1}}{i} = -\log 2,$$

which is not true as the series

$$\sum_{i=1}^{\infty} |x_i|\, f_{xi} = \sum_{i=1}^{\infty} \frac{2^i}{i} \cdot \frac{1}{2^i} = \sum_{i=1}^{\infty} \frac{1}{i} = 1 + \frac{1}{2} + \frac{1}{3} + \cdots \text{ (harmonic series), diverges.}$$

To make the idea and importance of expectation clear we first cite two examples.

Ex. 2 Suppose, a game is played with a single die assumed fair. In the game a player wins ₹20 if face 2 turns up, ₹40 if face 4 turns up and loses ₹30 if face 6 turns up; while he neither wins nor loses if any other face turns up. Find the expected sum of his money won.

Solution. Let X be a RV giving the amount of money won by the player on any toss.

$$\therefore\ E(X) = \sum_{i=1}^{6} x_i f_{xi} = 0 \cdot \tfrac{1}{6} + 20 \cdot \tfrac{1}{6} + 0 \cdot \tfrac{1}{6} + 40 \cdot \tfrac{1}{6} + 0 \cdot \tfrac{1}{6} + (-30) \cdot \tfrac{1}{6} = 5,$$

which is actually the mean value of X.

It follows that the player can expect to win ₹5 on any toss. That is, the game is in his favour. On any toss 5 will not be the value of the game. ∴ the player can expect to win ₹5 on the average at every toss. That is, he plays a series of 100 tosses he will win more or less a total of ₹500.

x. 3 A pair of fair dice is tossed. What is the expectation of the sum of points on the two dice?

olution. If the RV X denotes the sum of points, then the discrete distribution is given as follows:

x_i	2	3	4	5	6	7	8	9	10	11	12	Total
f_{xi}	1/36	2/36	3/36	4/36	5/36	6/36	5/36	4/36	2/36	2/36	1/36	1

$$\therefore\ E(X)=\sum_{i=2}^{12} x_i f_{xi}=2\cdot\tfrac{1}{36}+3\cdot\tfrac{2}{36}+4\cdot\tfrac{4}{36}+5\cdot\tfrac{4}{36}+6\cdot\tfrac{5}{36}+7\cdot\tfrac{6}{36}+8\cdot\tfrac{5}{36}+9\cdot\tfrac{4}{36}+10\cdot\tfrac{3}{36}+11\cdot\tfrac{2}{36}+12\cdot\tfrac{1}{36}$$
$$=\tfrac{1}{36}[2+6+12+20+30+42+40+36+30+22+12]=7.$$

Note that $P(X=7)=\frac{6}{36}$, which is the maximum.

Alternative Method: Let X and Y be the RV denoting points on the dice.

$$\therefore\ E(X+Y)=E(X)+E(Y),\ \text{as expectation is a linear operator}$$
$$=2E(X),\ \text{by symmetry } E(Y)=E(X)$$
$$=2\sum_{i=1}^{6} x_i f_{xi}=2\left(1\cdot\tfrac{1}{6}+2\cdot\tfrac{1}{6}+3\cdot\tfrac{1}{6}+4\cdot\tfrac{1}{6}+5\cdot\tfrac{1}{6}+6\cdot\tfrac{1}{6}\right)=2\cdot\tfrac{7}{2}=7.$$

.2 Properties of expectations

. $E(k)=k$, when k is a constant.

. $E\{c_1g_1(X)+c_2g_2(X)+\cdots+c_ng_n(X)\}=c_1E(g_1(X))+c_2E(g_2(X))+\cdots+c_nE(g_n(X))$, provided $E(g_i(X))$ exists for all $i=1,2,\cdots,n$.

. $|E(g(X))|\le E\{|g(X)|\}$.

. If $g(x)\ge 0$, $\forall\, x\in\mathbb{R}$, then $E\{g(X)\}\ge 0$.

roof. 1. $E(k)=\displaystyle\int_{-\infty}^{\infty} k f_x(x)\,dx=k\int_{-\infty}^{\infty} f_x(x)\,dx=k\cdot 1=k$, for continuous case.

$E(k)=\displaystyle\sum_{i=-\infty}^{\infty} k f_{xi}=k\sum_{i=-\infty}^{\infty} f_{xi}=k\cdot 1=k$, for discrete case.

2. $$E\{c_1g_1(X)+c_2g_2(X)+\cdots+c_ng_n(X)\}=\int_{-\infty}^{\infty}\{c_1g_1(x)+c_2g_2(x)+\cdots+c_ng_n(x)\}f_x(x)\,dx$$
$$=c_1\int_{-\infty}^{\infty} g_1(x)f_x(x)\,dx+c_2\int_{-\infty}^{\infty} g_2(x)f_x(x)\,dx+\cdots+c_n\int_{-\infty}^{\infty} g_n(x)f_x(x)\,dx,$$
where $f_x=$ pdf of X
$$=c_1E(g_1(X))+c_2E(g_2(X))+\cdots c_nE(g_n(X)),\ \text{if } E\{g_i(X)\} \text{ exists } \forall\, i=1(1)n,$$

e., $E\left[\displaystyle\sum_{i=1}^{n} c_ig_i(X)\right]=\displaystyle\sum_{i=1}^{n} c_iE(g_i(X))$, for continuous case.

Again,
$$E[c_1g_1(X)+c_2g_2(X)+\cdots+c_ng_n(X)]$$
$$=\sum_{i=-\infty}^{\infty}\{c_1g_1(x_i)+c_2g_2(x_i)+\cdots+c_ng_n(x_i)\}f_{xi},\ f_{xi}=P(X=x_i)$$
$$=c_1\sum_{i=-\infty}^{\infty} g_1(x_i)f_{xi}+c_2\sum_{i=-\infty}^{\infty} g_2(x_i)f_{xi}+\cdots+c_n\sum_{i=-\infty}^{\infty} g_n(x_i)f_{xi}$$
$$=c_1E\{g_1(X)\}+c_2E\{g_2(X)\}+\cdots+c_nE\{g_n(X)\},\ \text{for discrete case [if } E\{g_i(X)\} \text{ exists } \forall i]$$

3. $|E\{g(X)\}| = \left|\int_{-\infty}^{\infty} g(x)f_x(x)\,dx\right| \le \int_{-\infty}^{\infty} \left|g(x)f_x(x)\right|\,dx = \int_{-\infty}^{\infty} |g(x)|f_x(x)\,dx$ as $f_x(x) \ge 0$

$= E(|g(X)|)$, provided both $E\{g(X)\}$ and $E(|g(X)|)$ both exist.

$\therefore\ |E(g(X))| \le E(|g(X)|)$, for continuous case and

$$|E\{g(X)\}| = \left|\sum_{i=-\infty}^{\infty} g(x_i)f_{xi}\right| \le \sum_{i=-\infty}^{\infty} \left|g(x_i)f_{xi}\right| = \sum_{i=-\infty}^{\infty} \left|g(x_i)\right| f_{xi},\ \because\ f_{xi} > 0,\ \forall\ i$$

$$= E(|g(X)|).$$

$\therefore\ |E(g(X))| \le E(|g(X)|)$, for discrete case.

4. $E(g(X)) = \int_{-\infty}^{\infty} g(x)f_x(x)\,dx \ge 0$, as the integrand $g(x)f_x(x) \ge 0$ for $g(x) \ge 0$ (given) and $f_x(x) \ge 0$

and $E(g(X)) = \sum_{i=-\infty}^{\infty} g(x_i)f_{xi} \ge 0$ as each term is ≥ 0,

provided the improper integral or the infinite sum converges absolutely.

6.3 Mean

The mean of a RV X is the mean of the distribution of X. The mean of X is defined by its expected value $E(X)$ and is denoted by m_x, i.e., $m_x = E(X)$. It represents the centre of mass of the probability mass distribution of X (the total mass being unity).

When we know the position of m_x in the probability mass distribution line we must have a rough idea of location of the distribution and as such it is an important measure of location of the distribution.

Now, we find means of some standard probability distribution.

1. Binomial (n, p) distribution

$$m_x = E(X) = \sum_{i=0}^{n} i\binom{n}{i}p^i q^{n-i},\ p+q=1$$

$$= \sum_{i=1}^{n} i\frac{n!}{i!\,(n-i)!}p^i q^{n-i},\ \text{as the first term is 0.}$$

$$= np\sum_{i=1}^{n}\frac{(n-1)!}{(i-1)!\,(n-1)!}\,p^{i-1}\,q^{n-i}$$

$$= np\sum_{j=0}^{n-1}\frac{(n-1)!}{j!\,(n-j-1)!}\,p^j\,q^{n-j-1},\ \text{where } i-1=j$$

$$= np\sum_{j=0}^{n-1}\binom{n-1}{j}p^j\,q^{n-1-j} = np(p+q)^{n-1} = np,\ \because\ p+q=1.$$

2. Poisson (μ) distribution

$$m_x = E(X) = \sum_{i=0}^{\infty} i\frac{e^{-\mu}\mu^i}{i!} = \sum_{i=1}^{\infty}\frac{e^{-\mu}\mu^i}{(i-1)!},\ \text{as the first term is 0.}$$

$$= e^{-\mu}\cdot\mu\sum_{i=1}^{\infty}\frac{\mu^{i-1}}{(i-1)!} = e^{-\mu}\cdot\mu\left(1+\frac{\mu}{1!}+\frac{\mu^2}{2!}+\cdots\right) = e^{-\mu}\cdot\mu e^{\mu} = \mu e^0 = \mu.$$

3. Discrete uniform distribution

$$m_x = E(X) = \sum_{i=1}^{n} x_i f_{xi} = \sum_{i=1}^{n} \frac{x_i}{n} = \frac{x_1 + x_2 \cdots + x_n}{n} = \bar{x}, \text{ say.}$$

4. Gamma (l) distribution

$$m_x = E(X) = \int_0^\infty x \frac{e^{-x} x^{l-1}}{\Gamma(l)} dx = \frac{1}{\Gamma(l)} \int_0^\infty e^{-x} x^{(l+1)-1} dx = \frac{\Gamma(l+1)}{\Gamma(l)} = \frac{l\Gamma(l)}{\Gamma(l)} = l.$$

5. Normal (m, σ) distribution

$$m_x = \int_{-\infty}^{\infty} x \cdot \frac{1}{\sqrt{2\pi}\sigma} e^{-\frac{(x-m)^2}{2\sigma^2}} dx, \text{ provided it converges absolutely.}$$

Let $x - m = u$. $\therefore dx = du$ and $x \to \pm\infty, u \to \pm\infty$.

$$\therefore m_x = \int_{-\infty}^{\infty} (u+m) \cdot \frac{1}{\sqrt{2\pi}\sigma} e^{-u^2/2\sigma^2} du = \frac{1}{\sqrt{2\pi}\sigma} \int_{-\infty}^{\infty} u e^{-u^2/2\sigma^2} du + m \cdot \frac{1}{\sqrt{2\pi}\sigma} \int_{-\infty}^{\infty} e^{-u^2/2\sigma^2} du$$

$$= 0 + m \cdot \frac{1}{\sqrt{2\pi}\sigma} \int_{-\infty}^{\infty} e^{-\frac{(x-m)^2}{2\sigma^2}} dx, \text{ as the integrand in the first integral is an odd function.}$$

$$= m, \text{ as } \frac{1}{\sqrt{2\pi}\sigma} \int_{-\infty}^{\infty} e^{-\frac{(x-m)^2}{2\sigma^2}} dx = 1.$$

So, the parameter m in the normal distribution bears the natural meaning the *mean* of the distribution.

6. Geometric distribution

$$\text{Mean} = E(X) = \sum_{i=0}^{\infty} i f_i = \sum_{i=0}^{\infty} i p q^i, \; q = 1 - p \text{ and } 0 < p < 1$$

$$= p\left(q + 2q^2 + 3q^3 + \cdots\right).$$

Let

$$\begin{aligned} s &= q + 2q^2 + 3q^3 + 4q^4 + \cdots \\ sq &= \phantom{q+{}} q^2 + 2q^3 + 3q^4 + \cdots \\ \hline (1-q)s &= q + q^2 + q^3 + \cdots = \frac{q}{1-q}, \text{ as } 0 < q < 1 \end{aligned}$$

$$\therefore s = \frac{q}{(1-q)^2} = \frac{1-p}{p^2}. \therefore \text{mean} = p \cdot \frac{1-p}{p^2} = \frac{1-p}{p}.$$

7. Uniform distribution over (a, b)

$$\text{Mean} = E(X) = \int_{-\infty}^{\infty} x f_x(x) dx = \int_a^b x \cdot \frac{1}{b-a} dx = \frac{1}{2(b-a)} \left[x^2\right]_a^b = \frac{1}{2(b-a)} \left(b^2 - a^2\right) = \frac{a+b}{2}.$$

8. Beta distribution of first kind

$$\text{Mean} = E(X) = \int_0^1 x \cdot \frac{x^{l-1}(1-x)^{m-1}}{B(l,m)} dx = \frac{1}{B(l,m)} \int_0^1 x^{l+1-1}(1-x)^{m-1} dx = \frac{B(l+1,m)}{B(l,m)}$$

$$= \frac{\Gamma(l+1)\Gamma(m)}{\Gamma(l+m+1)} \cdot \frac{\Gamma(l+m)}{\Gamma(l)\Gamma(m)} = \frac{l\Gamma(l)\Gamma(l+m)}{(l+m)\Gamma(l+m)\Gamma(l)} = \frac{l}{l+m}.$$

9. Beta distribution of second kind

$$\text{Mean} = E(X) = \int_0^\infty x \frac{x^{l-1}}{B(l,m)(1+x)^{l+m}}\,dx = \frac{1}{B(l,m)} \int_0^\infty \frac{x^{l+1-1}}{(1+x)^{l+m}}\,dx$$

$$= \frac{1}{B(l,m)} \int_0^\infty \frac{x^{l+1-1}}{(1+x)^{l+1+m-1}}\,dx = \frac{B(l+1,m-1)}{B(l,m)} = \frac{\Gamma(l+1)\Gamma(m-1)}{\Gamma(l+m)} \cdot \frac{\Gamma(l+m)}{\Gamma(l)\Gamma(m)}$$

$$= \frac{l\Gamma(l)\Gamma(m-1)}{\Gamma(l)(m-1)\Gamma(m-1)} = \frac{l}{m-1}, \text{ if } m > 1.$$

10. Exponential distribution

$$\text{Mean} = E(X) = \theta \int_0^\infty x e^{-\theta x} = \theta \lim_{R\to\infty} \int_0^R x e^{-\theta x}\,dx$$

$$= \theta \lim_{R\to\infty} \left\{ R\frac{e^{-\theta R}}{-\theta} + \frac{1}{\theta}\int_0^R e^{-\theta x}\,dx \right\} = \lim_{R\to\infty}\left(\frac{1}{\theta} - \frac{e^{-\theta R}}{\theta}\right) = \frac{1}{\theta}.$$

11. Cauchy distribution

Mean of this distribution does not exist as the integral

$$\frac{\lambda}{\mu}\int_{-\infty}^\infty \frac{x\,dx}{(x-\mu)^2+\lambda^2} = \frac{\lambda}{\mu}\int_{-\infty}^\infty \frac{(x-\mu)+\mu}{(x-\mu)^2+\lambda^2}\,dx = \lambda\int_{-\infty}^\infty \frac{dx}{(x-\mu)^2+\lambda^2} + \frac{\lambda}{\mu}\int_{-\infty}^\infty \frac{x-\mu}{(x-\mu)^2+\lambda^2}\,dx$$

does not converge absolutely, as integral of the form $\int_{-\infty}^\infty \frac{u}{u^2+a^2}\,du$ does not converge absolutely.

12. Laplace distribution

$$\text{Mean} = E(X) = \frac{1}{2\lambda}\int_{-\infty}^\infty x e^{-\frac{|x-\mu|}{\lambda}}\,dx. \text{ Let } y = \frac{x-\mu}{\lambda}. \therefore \lambda dy = dx$$

$$= \frac{1}{2}\int_{-\infty}^0 (\lambda y+\mu)e^y\,dy + \frac{1}{2}\int_0^\infty (\lambda y+\mu)e^{-y}\,dy$$

$$= \frac{1}{2}\left[(\lambda y+\mu)e^y\Big|_{-\infty}^0 - \lambda\int_{-\infty}^0 e^y\,dy\right] + \frac{1}{2}\left[-(\lambda y+\mu)e^{-y}\Big|_0^\infty + \lambda\int_0^\infty e^{-y}\,dy\right]$$

$$= \tfrac{1}{2}(\mu-\lambda) + \tfrac{1}{2}(\mu+\lambda) = \mu.$$

6.4 Moment

$E\{(X-a)^n\}$ is defined to be the moment of the distribution of X of order n about the point $x = a$, when n is zero or a positive integer.

Physically, it represents the nth order moment of inertia (MI) of the distribution about a line passing through the point $x = a$ and perpendicular to the line of distribution.

Explanation. Let $L'L$ be the line of probability distribution. Let O be some origin.

$\therefore P(x < X \le x+dx) = f_x(x)\,dx$ is the probability mass in $(x, x+dx)$.

Its nth order MI about the line $M'M$, passing through $x = a$ and perpendicular to $L'L$,

is $(x-a)^n f_x(x)\,dx$.

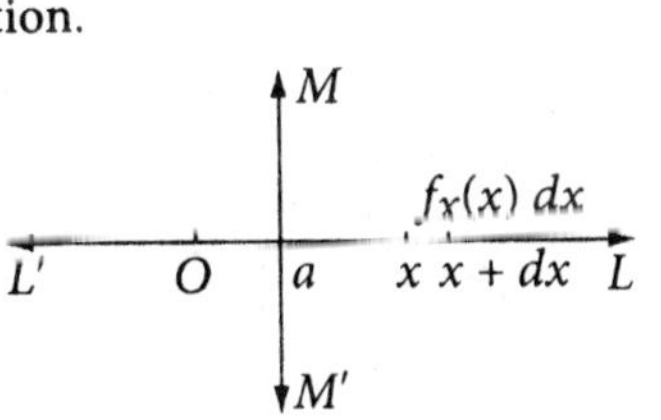

Fig 6.1

$\therefore$ MI of order n of the whole probability mass distribution about $M'M$

$$= \int_{-\infty}^{\infty} (x-a)^n f_x(x)\,dx = E\{(X-a)^n\}.$$

When $n=2$, MI is of order 2, which is usually called the MI. Similar explanation for discrete case is usually left for exercise.

We are now interested to have moment about the origin and about the mean of the distribution. Let $\alpha_n(X)$ and $\mu_n(X)$ respectively denote moment of order n about origin and moment of order n about the mean.

$\therefore\ \alpha_n(X) = E(X^n)$.

$\alpha_n(X)$ is sometimes called ordinary moment of X of order n and $\mu_n(X) = E(X-m_x)^n$, called central moment of X of order n.

An important property

If $E(X^n)$ exists, then $E(X^r)$ exists for all r satisfying $1 \le r < n$. That is, if an ordinary moment exists for certain order, then the ordinary moments of lesser orders also exist.

Proof. $$\int_{-\infty}^{\infty} |x|^r f_x(x)\,dx = \int_{|x|\le 1} |x|^r f_x(x)\,dx + \int_{|x|>1} |x|^r f_x(x)\,dx = \int_{-1}^{1} |x|^r f_x(x)\,dx + \int_{|x|>1} |x|^r f_x(x)\,dx$$

$$\le \int_{-1}^{1} f_x(x)\,dx + \int_{|x|>1} |x|^n f_x(x)\,dx, \quad \because \text{ for } |x|\le 1, |x|^r \le 1 \text{ and } |x|>1, |x|^r < |x|^n \text{ for } r<n$$

$$\le 1 + E(|X|^n), \text{ which is finite as } E(X^n) \text{ exists.}$$

$\therefore\ E(X^r)$ exists for r lying in $1 \le r < n$.

Note 1. For $r=0$, $E(X^0) = E(1) = 1$ for all RV X.

2. This is true for moment about any point a.
That is, if $E(X-a)^n$ exists, then $E(X-a)^r$ exists $\forall\ r$ in $1 \le r < n$.

[Hints. $$\int_{-\infty}^{\infty} |x-a|^r f_x(x)\,dx = \int_{|x-a|\le 1} |x-a|^r f_x(x)\,dx + \int_{|x-a|>1} |x-a|^r f_x(x)\,dx$$

$$\le \int_{|x-a|\le 1} f_x(x)\,dx + \int_{|x-a|>1} |x-a|^n f_x(x)\,dx$$

$\because$ for $|x-a| \le 1, |x-a|^r \le 1$ and
for $|x-a| > 1, |x-a|^r \le |x-a|^n$ for $r<n$

$$\le 1 + \int_{-\infty}^{\infty} |x-a|^n f_x(x)\,dx \le 1 + E(|X-a|^n),$$

which is finite, as $E(X-a)^n$ exists.

$\therefore\ E(X-a)^r$ exists $\forall\ r$ in $1 \le r < n$.]

3. In particular, when $a = m_x$, then mean of X, μ_r exists $\forall\ r$ in $1 \le r < n$, whenever μ_n exists, i.e., the existence of the central moment of a RV X of order n implies the existence of all central moments of order $< n$.
4. Similar proof when X is a discrete RV.

Central moment

$\mu_n(X) = E\{(X-m_x)^n\}$. $\mu_n(x)$ is sometimes called the central moment of order n. Obviously, $\mu_1(x) = E(X-m_x) = E(X) - m_x = 0$, for any RV X.

Now,

$$\mu_n(X) = E\{(X-m_x)^n\} = E\left\{\sum_{i=0}^{n}\binom{n}{i}X^{n-i}(-m_x)^i\right\} = \sum_{i=0}^{n}(-1)^i\binom{n}{i}E\left(X^{n-i}\right)m_x^i, \text{ as } E \text{ is linear.}$$

$$= \sum_{i=0}^{n}(-1)^i\binom{n}{i}\alpha_{n-i}(X)m_x^i.$$

Hence, central moment of order n can be expressed as a linear combination of ordinary moments of order $\leq n$.

$$\therefore\ \mu_2(X) = \sum_{i=0}^{2}(-1)^i\binom{2}{i}\alpha_{2-i}(X)m_x^i = \alpha_2(X) - 2\alpha_1(X)m_x + \alpha_0(X)m_x^2$$

$$= \alpha_2(X) - m_x^2 \text{ as } \alpha_1(X) = E\left(X^1\right) = m_x \text{ and } \alpha_0(X) = E\left(X^0\right) = 1.$$

$$\mu_3(X) = \sum_{i=0}^{3}(-1)^i\binom{3}{i}\alpha_{3-i}(X)m_x^i = \alpha_3(X) - 3\alpha_2(X)m_x + 3\alpha_1(X)m_x^2 - \alpha_0(X)m_x^3$$

$$= \alpha_3(X) - 3\alpha_2(X)m_x + 2m_x^3.$$

Similarly, $\mu_4(X) = \alpha_4(X) - 4\alpha_3(X)m_x + 6\alpha_2(X)m_x^2 - 3m_x^4$, etc.

Now, $\mu_n(aX+b) = E\{aX+b-(am_x+b)\}^n = a^n E(X-m_x)^n = a^n\mu_n(X)$, i.e., central moment is independent of the choice of origin but depends upon the scale of measurement.

Variance

The second order central moment is known as *variance*. It plays an important role for understanding the dispersion of the distribution, i.e., how widely the values of the RV X are deviated from the mean. Variance increases with the increase of deviation. Variance of X is denoted by $\text{var}(X)$,

i.e., $\text{var}(X) = \mu_2(X) = E\left\{(X-m_x)^2\right\}$, provided $\displaystyle\int_{-\infty}^{\infty}(x-m_x)^2 f_x(x)\,dx$ or, $\displaystyle\sum_{i=-\infty}^{\infty}(x_i-m_x)^2 f_{xi}$

converges absolutely. Physically, it represents the MI (second order) of the probability distribution about a line passing through the mean and perpendicular to the line of distribution.

Standard deviation (SD)

The positive square root of variance of X is known as standard deviation of X and is denoted by σ_x, i.e., $\sigma_x = \sqrt{\text{var}(X)}$.

Standard deviation also measures the dispersion of the distribution. Its dimension is the same as that of the dimension of the RV.

Whenever $\text{var}(X)$ is small σ_x is also small and the values of X are concentrated near the mean m_x and we say that the values of X are consistent.

Suppose, the RV X and Y represent runs in consecutive 10 tests of two players A and B respectively. Let X and Y are given by the following table:

X	42	58	43	45	62	48	49	50	52	51
Y	10	90	15	45	90	100	0	30	42	78

$m_x = \frac{500}{10} = 50,\ m_y = \frac{500}{10} = 50;\ \sigma_x \simeq 6, \sigma_y \simeq 31$ [Here $f_{xi} = \frac{1}{10}, f_{yi} = \frac{1}{10}, \forall\ i = 1,2,\cdots,10$]

Although mean run for both the players are the same, viz., 50, the standard derivations are significantly different. Here the player A is much more consistent than B.

Note that mean and SD are two independent characters of the distribution. A player with low mean may be highly consistent and a player with high mean may be highly inconsistent. In our example above, the players A and B have the same means but one is consistent and the other is inconsistent.

6.5 Some related results

1. Second order moment about any point is minimum, when taken about the mean.

Proof. Let $a \neq m_x$.

$$\therefore\ E\{(X-a)^2\} = E\{(X-m_x)+(m_x-a)\}^2 = E(X-m_x)^2 + 2(m_x-a)E(X-m_x) + (m_x-a)^2$$
$$= E(X-m_x)^2 + (m_x-a)^2, \text{ as } E(X-m_x) = E(X) - m_x = 0. \quad (1)$$
$$\therefore\ E(X-m_x)^2 < E(X-a)^2, \text{ when } a \neq m_x.$$

[(1) gives the rule of parallel axes, i.e., the MI of the whole probability mass about a line $x = a$ is equal to MI of the whole mass about a line $x = m_x$, through the mean m_x and parallel to $x = a$ plus MI of the whole mass (= 1) concentrated at the point m_x about the line $x = a$.]

2. Two useful formulae for variance (i) $\mu_2(X) = \alpha_2(X) - m_x^2$ (ii) $\mu_2(X) = E(X(X-1)) - m_x(m_x-1)$.

Proof. (i) $\mu_2(X) = E(X-m_x)^2 = E(X^2) - 2m_x E(X) + m_x^2 = \alpha_2(X) - m_x^2$ as $E(X) = m_x$.

(ii) $E(X(X-1)) = E(X^2 - X) = \alpha_2(X) - m_x$.

$$\therefore\ E(X(X-1)) - m_x(m_x-1) = \alpha_2(X) - m_x - m_x^2 + m_x = \alpha_2(X) - m_x^2 = \mu_2(X).$$

3. $\sigma_{ax+b} = |a|\sigma_x$.

$$\sigma_{ax+b} = \sqrt{\text{var}(aX+b)} = \sqrt{E(aX+b-am_x-b)^2} = \sqrt{E(aX-am_x)^2} = \sqrt{a^2E(X-m_x)^2}$$
$$= |a|\sqrt{\text{var}(X)} = |a|\sigma_x.$$

Normalised/Standardised RV

The normalised RV of X is denoted by X^* and $X^* = \frac{X-m_x}{\sigma_x}$.

Advantages of X^*

(a) X^* is dimensionless as $X - m_x$ and σ_x are of same dimension.

(b) $m_{x^*} = 0$.

$$m_{x^*} = E(X^*) = E\left(\frac{X-m_x}{\sigma_x}\right) = \frac{1}{\sigma_x}E(X-m_x) = \frac{1}{\sigma_x}[E(X) - m_x] = 0.$$

(c) $\sigma_{x^*} = 1$.

$$\sigma_{x^*} = \sqrt{\text{var}(X^*)} = \sqrt{E(X^*)^2} \text{ as } m_{x^*} = 0$$
$$= \sqrt{E\left(\frac{X-m_x}{\sigma_x}\right)^2} = \sqrt{\frac{1}{\sigma_x^2}E(X-m_x)^2} = \sqrt{\frac{\sigma_x^2}{\sigma_x^2}} = 1.$$

(d) As the mean of X^* is zero. $\therefore$ all the central moments coincide with the corresponding ordinary moments.

Variances and standard deviations of some well-known standard distributions

1. Binomial (n, p) distribution

For a discrete probability distribution the formula

$$\text{var}(X) = E(X(X-1)) - m_x(m_x-1) \text{ is helpful.}$$

Now,

$$E(X(X-1))=\sum_{i=0}^{n} i(i-1)\binom{n}{i}p^i q^{n-i},\ p+q=1$$

$$=\sum_{i=2}^{n}\frac{n!}{(i-2)!\,(n-i)!}p^i\,q^{n-i},\ \text{as the first two terms vanish}$$

$$=\sum_{i=2}^{n}\frac{n(n-1)(n-2)}{(i-2)!(n-1)!}p^2p^{i-2}q^{n-i}$$

$$=n(n-1)p^2\sum_{j=0}^{n-2}\frac{(n-2)!}{j!(n-2-j)!}p^iq^{n-2-j},\ \text{where } j=i-2$$

$$=n(n-1)p^2(p+q)^{n-2}=n(n-1)p^2,\ \because\ p+q=1.$$

$\because$ $\text{var}(X)=n(n-1)p^2-np(np-1)$, as $m_x=np$

$=n^2p^2-np^2-n^2p^2+np=np(1-p)=npq.$

$\because$ $\sigma_x=\sqrt{npq}$.

Note that $\text{var}(X)=npq<np=m_x$ and $\max\{\text{var}(X)\}=n/4$, when $p=q=1/2$.

[**Hints.** $\text{var}(X)=np(1-p)$. $\therefore$ $\frac{d}{dp}(\text{var}(X))=n(1-2p)$ and $\frac{d^2}{dp^2}(\text{var}(X))=-2n<0$.]

2. Poisson (μ) distribution

$$E(X(X-1))=\sum_{i=0}^{\infty} i(i-1)e^{-\mu}\frac{\mu^i}{i!}=e^{-\mu}\sum_{i=2}^{\infty}\frac{\mu^i}{(i-2)!},\ \text{as the first two terms vanish}$$

$$=\mu^2e^{-\mu}\sum_{j=0}^{\infty}\frac{\mu^j}{j!},\ \text{when } j=i-2$$

$$=\mu^2e^{-\mu}e^{\mu}=\mu^2.$$

$\therefore$ $\text{var}(X)=E(X(X-1))-m_x(m_x-1)=\mu^2-\mu(\mu-1)$ [$\because$ $m_x=\mu$ for Poisson (μ) distribution] $=\mu$.

$\therefore$ $\sigma_x=\sqrt{\text{var}(X)}=\sqrt{\mu}$.

Note that mean and variance are the same for a Poisson distribution.

3. Gamma (l) distribution

For a continuous distribution the formula $\mu_2(X)=\alpha_2(X)-m_x^2$ is helpful.

$$\alpha_2(X)=E\left(X^2\right)=\int_0^{\infty}x^2\frac{e^{-x}x^{l-1}}{\Gamma(l)}dx=\int_0^{\infty}\frac{e^{-x}x^{(l+2)-1}}{\Gamma(l)}=\frac{\Gamma(l+2)}{\Gamma(l)}=\frac{(l+1)l\Gamma(l)}{\Gamma(l)}=l(l+1).$$

$\therefore$ $\mu_2(X)=l(l+1)-l^2$ [$\because$ $m_x=l$ for a gamma (l) distribution] $=l$. $\therefore$ $\sigma_x=\sqrt{l}$.

Here also mean and variance are the same.

Note: In statistics, suppose, we want to find the probability distribution of a population by studying a sample drawn from the population. If it is seen that the sample mean and sample variance are nearly equal we may test the hypothesis: (1) the distribution of the population is Poisson distribution or, (2) the distribution of the population in gamma distribution with some unknown parameter (to be estimated) and apply the test of goodness of fit and test whether the hypothesis is accepted or rejected with some level of confidence.

4. Normal (m,σ) distribution

$\text{var}(X) = E(X-m)^2$, as $m_x = m$ for normal (m,σ) distribution.

$$= \int_{-\infty}^{\infty} (x-m)^2 \cdot \frac{1}{\sqrt{2\pi}\sigma} e^{-\frac{(x-m)^2}{2\sigma^2}} dx$$

$$= \frac{1}{\sqrt{2\pi}\sigma} \int_{-\infty}^{\infty} u^2 e^{-u^2/2\sigma^2} du, \text{ where } u = x-m. \therefore du = dx \text{ and } u \to \pm\infty \text{ as } x \to \pm\infty$$

$$= \frac{2}{\sqrt{2\pi}\sigma} \int_0^{\infty} u^2 e^{-u^2/2\sigma^2} du, \text{ as the integrand is an even function of } u.$$

Let $\frac{u^2}{2\sigma^2} = v$. $\therefore u\,du = \sigma^2 dv$ and $v = 0$, when $u = 0$, $v \to \infty$ as $u \to \infty$.

$$\therefore \text{var}(X) = \frac{2}{\sqrt{2\pi}\sigma} \int_0^{\infty} \sqrt{2}\sigma\sqrt{v} e^{-v} \sigma^2 dv = \frac{2\sigma^2}{\sqrt{\pi}} \int_0^{\infty} e^{-v} v^{\frac{3}{2}-1} dv = \frac{2\sigma^2}{\sqrt{\pi}} \Gamma\left(\frac{3}{2}\right)$$

$$= \frac{2\sigma^2}{\sqrt{\pi}} \frac{1}{2} \Gamma\left(\frac{1}{2}\right) = \frac{\sigma^2}{\sqrt{\pi}} \cdot \sqrt{\pi} \quad \left[\because \Gamma\left(\frac{1}{2}\right) = \sqrt{\pi}\right]$$

$$= \sigma^2.$$

$\therefore \sigma_x = \sigma$ (standard deviation).

Note that like m, the mean, the other parameter σ also bears the natural meaning of SD of the normal (m,σ) distribution.

We now find $\mu_k(X), k \in \mathbb{N}$ for normal (m,σ) distribution. Also, find $\alpha_k(X)$ for gamma (l) distribution.

5. $\mu_k(X)$ for normal (m,σ) distribution

$\mu_k(X) = E(X-m)^k$, as $E(X) = m$ for normal (m,σ) distribution.

$$= \frac{1}{\sqrt{2\pi}\sigma} \int_{-\infty}^{\infty} (x-m)^k e^{-\frac{(x-m)^2}{2\sigma^2}} dx, \text{ provided the integral converges absolutely } \forall\, k \in \mathbb{N}$$

$$= \frac{1}{\sqrt{2\pi}\sigma} \lim_{\substack{M \to -\infty \\ N \to \infty}} \int_M^N (x-m)^{k-1} \left\{(x-m) e^{-\frac{(x-m)^2}{2\sigma^2}}\right\} dx$$

$$= \frac{1}{\sqrt{2\pi}\sigma} \lim_{\substack{M \to -\infty \\ N \to \infty}} \left[-\sigma^2 (x-m)^{k-1} e^{-\frac{(x-m)^2}{2\sigma^2}} \Big|_M^N + \int_M^N (k-1)\sigma^2 (x-m)^{k-2} e^{-\frac{(x-m)^2}{2\sigma^2}} dx \right]$$

$$= (k-1)\sigma^2 \cdot \frac{1}{\sqrt{2\pi}\sigma} \int_{-\infty}^{\infty} (x-m)^{k-2} e^{-\frac{(x-m)^2}{2\sigma^2}} dx \text{ [as the first term vanishes by the limit]}$$

$$= (k-1)\sigma^2 \mu_{k-2}(X).$$

Case (i) Let k be odd, i.e., $k = 2n-1, n \in \mathbb{N}$.

$$\therefore \mu_{2n-1}(X) = (2n-2)\sigma^2 \mu_{2n-3}(X) = (2n-2)(2n-4)\sigma^4 \mu_{2n-5}(X)$$

$$\cdots\cdots\cdots\cdots\cdots\cdots\cdots\cdots\cdots\cdots$$

$$= (2n-2)(2n-4)\cdots 4 \cdot 2\sigma^{2n-2} \mu_1(X)$$

$$= 0, \text{ as } \mu_1(X) = E(X-m_x)^1 = 0 \text{ for all distributions.}$$

Case (ii) Let k be even, i.e., $k=2n, n\in\mathbb{N}$.

$$\therefore\ \mu_{2n}(X)=(2n-1)\sigma^2\mu_{2n-2}(X)=(2n-1)(2n-3)\sigma^2\mu_{2n-4}(X)$$

$$\cdots\cdots\cdots\cdots\cdots\cdots\cdots\cdots\cdots\cdots\cdots\cdots$$

$$=(2n-1)(2n-3)\cdots5\cdot3\cdot1\sigma^{2n}\mu_0(X)$$

$$=1\cdot3\cdot5\cdots(2n-3)(2n-1)\sigma^{2n},\text{ as } \mu_0(X)=E(X-m_x)^0=1,\text{ for all distributions.}$$

$\therefore\ \mu_2(X)=\sigma^2.\ \therefore\ \sigma(X)=\sigma.$

6. $\boldsymbol{\alpha_k(X)}$ for gamma ($\boldsymbol{l}$) distribution

$$\alpha_k(X)=E\left(X^k\right)=\int_0^\infty x^k\frac{e^{-x}x^{l-1}}{\Gamma(l)}dx=\frac{1}{\Gamma(l)}\int_0^\infty e^{-x}x^{(k+l-1)}dx=\frac{\Gamma(k+l)}{\Gamma(l)}$$

$$=\frac{(k+l-1)(k+l-2)\cdots l\Gamma(l)}{\Gamma(l)}=(k+l-1)(k+l-2)\cdots(l+1)l.$$

7. Geometric distribution

$$E\{X(X-1)\}=\sum_{i=0}^{\infty}i(i-1)pq^i, q=1-p$$

$$=\sum_{i=2}^{\infty}i(i-1)pq^i=p\left(1\cdot2q^2+2\cdot3q^3+3\cdot4q^4+\cdots\infty\right)$$

$$=pq^2\left(1\cdot2+2\cdot3q+3.4q^2+\cdots\right).$$

Let $s=1.2+2.3q+3.4q^2+\cdots$

$sq=\qquad 1.2q+2.3q^2+\cdots$

$(1-q)s=1.2+2.2q+3.2q^2+\cdots$

$\quad=2\left(1+2q+3q^2+\cdots\right)$

Let $k=1+2q+3q^2+4q^3+\cdots$

$kq=\qquad q+2q^2+3q^3+\cdots$

$(1-q)k=1+q+q^2+q^3+\cdots$

$\quad=\frac{1}{1-q}.\ \therefore\ k=\frac{1}{p^2}\ \because\ p+q=1.$

$$\therefore\ s=\frac{2}{p^3},\ \therefore\ E(X(X-1))=pq^2\cdot\frac{2}{p^3}=\frac{2q^2}{p^2}.$$

$$\therefore\ \operatorname{var}X=E\{X(X-1)\}-E(X)\{E(X)-1\}$$

$$=\frac{2q^2}{p^2}-\frac{q}{p}\left(\frac{q}{p}-1\right)=\frac{q^2}{p^2}+\frac{q}{p}=\frac{q^2+pq}{p^2}=\frac{q}{p^2}\ [p+q=1]$$

$$=\frac{1-p}{p^2}$$

$\therefore\ \sigma(X)=\frac{\sqrt{1-p}}{p}.$

8. Exponential distribution

$$\alpha_2(X)=\theta\int_0^\infty x^2e^{-\theta x}dx=\lim_{R\to\infty}\left[\theta\cdot\left.\frac{x^2e^{-\theta x}}{-\theta}\right|_0^R+\frac{\theta}{\theta}\int_0^R 2xe^{-\theta x}dx\right]$$

$$=2\lim_{R\to\infty}\left[\left.\frac{xe^{-\theta x}}{-\theta}\right|_0^R+\frac{1}{\theta}\int_0^R e^{-\theta x}dx\right]=\frac{2}{\theta^2}\lim_{R\to\infty}\left[e^{-\theta x}\right]_0^R=\frac{2}{\theta^2}.$$

$$\therefore\ \mu_2(X)=\frac{2}{\theta^2}-\frac{1}{\theta^2}\text{ as } E(X)=\frac{1}{\theta}$$

$$=\frac{1}{\theta^2}.\ \therefore\ \sigma(X)=\frac{1}{\theta}.$$

9. Uniform distribution

$$\alpha_2(X)=\int_a^b x^2\cdot\frac{1}{b-a}\,dx=\frac{1}{3(b-a)}\left(b^3-a^3\right)=\frac{b^2+ba+a^2}{3}.$$

$$\therefore\ \mu_2(X)=\alpha_2(X)-\{E(X)\}^2=\frac{b^2+ba+a^2}{3}-\left(\frac{b+a}{2}\right)^2=\frac{1}{12}\left(4b^2+4ba+4a^2-3b^2-3a^2-6ba\right)$$

$$=\frac{1}{12}\left(b^2+a^2-2ba\right)=\frac{1}{12}(b-a)^2.\ \therefore\ \sigma(X)=\frac{b-a}{2\sqrt{3}}.$$

10. Cauchy distribution

As the mean of the distribution does not exist.

∴ the variance and standard deviation also do not exist.

11. Discrete uniform distribution

$\text{var}(X)=\dfrac{1}{n}\sum_{i=1}^{n}(x_i-\bar{x})^2$, where $\bar{x}=\dfrac{x_1+x_2+\cdots+x_n}{n}$.

In particular, when $x_i=i, i=1,2,\cdots,n, \text{var}(X)=\frac{1}{12}(n^2-1)$.

[Hints. $\bar{x}=\dfrac{1+2+\cdots+n}{n}=\dfrac{n(n+1)}{2n}=\dfrac{n+1}{2}$.

$$\text{var}(X)=\frac{1}{n}\sum_{i=1}^{n}\left(i-\frac{n+1}{2}\right)^2=\frac{1}{n}\left[\sum_{i=1}^{n}i^2-(n+1)\sum_{i=1}^{n}i+\frac{(n+1)^2}{4}\cdot n\right]$$

$$=\frac{1}{n}\left[\frac{n(n+1)(2n+1)}{6}-(n+1)\frac{n(n+1)}{2}+\frac{n(n+1)^2}{4}\right]$$

$$=\frac{1}{12}\left[2\left(2n^2+3n+1\right)-6\left(n^2+2n+1\right)+3\left(n^2+2n+1\right)\right]$$

$$=\frac{1}{12}\left[4n^2+6n+2-3n^2-6n-3\right]=\frac{1}{12}\left(n^2-1\right)$$

and $\sigma_x=\sqrt{\dfrac{n^2-1}{12}}$.]

12. Laplace distribution

$$\alpha_k(X)=E\left(X^k\right)=\frac{1}{2\lambda}\int_{-\infty}^{\infty}x^k e^{-\frac{|x-\mu|}{\lambda}}\,dx,\text{ provided the integral converges absolutely.}$$

Let $y=\frac{x-\mu}{\lambda}$. The integral on simplification

$$\alpha_k(X)=\frac{1}{2}\sum_{r=0}^{k}\binom{k}{r}\lambda^r\mu^{k-r}r!\{1+(-1)^r\}.$$

$\alpha_1(X)=\mu$ (mean), $\alpha_2(X)=\mu^2+2\lambda^2$. ∴ variance $\sigma_x^2=\alpha_2(X)-\alpha_1^2(X)=2\lambda^2$.

6.6 Discussion of third and fourth central moments

Third central moment $\mu_3(X)$

We know that $\mu_1(X)=0$ for every distribution.

We have seen that $\mu_3(X)=\mu_5(X)=\cdots=\mu_{2n+1}(X)=\cdots=0\ (n\in\mathbb{N})$ for a normal distribution.

Normal (m, σ) distribution is symmetrical about the line $x = m$. For any symmetrical distribution of a RV X, the pdf f_x will be symmetrical about the line $x = m_x$ and by the substitution $u = x - m_x$, f_x reduces to f_u, where f_u is an even function of u.

$$\therefore \ \mu_{2n+1}(X) = \int_{-\infty}^{\infty} (x - m_x)^{2n+1} f_x(x)\, dx = \int_{-\infty}^{\infty} u^{2n+1} f_u(u)\, du = 0,$$

as the integrand is an odd function.

$\therefore$ for a distribution $\mu_3(X), \mu_5(X), \cdots$ do not vanish, the distribution is not symmetrical about the line $x = m_x$. The asymmetry or skewness of a probability distribution, therefore, may be measured by the value of $\mu_3(X)$.

To have a dimensionless measure of skewness the natural suggestion is to find $\mu_3(X^*)$, where $X^* = \frac{X - m_x}{\sigma_x}$, the corresponding normalised RV, and such a measure is called *coefficient of skewners* and is denoted by γ_1.

$$\therefore \ \gamma_1 = \mu_3(X^*) = E\left(X^{*^3}\right) = E\left(\frac{X - m_x}{\sigma_x}\right)^3 = \frac{E(X - m_x)^3}{\sigma_x^3} = \frac{\mu_3(X)}{\sigma_x^3} \ [\neq 0, \text{ for an asymmetric distribution}]$$

$\therefore$ $\gamma_1 < 0$ or $\gamma_1 > 0$ for an asymmetric or skew distribution.
When $\gamma_1 < 0$, the distribution is called negatively skew and
when $\gamma_1 > 0$, the distribution is called positively skew.
Hence, for a continuous distribution:
When $\gamma_1 < 0$, longer tail of probability density curve lies on the left of $x = m_x$ and
when $\gamma_1 > 0$, longer tail of probability density curve lies on the right of $x = m_x$.

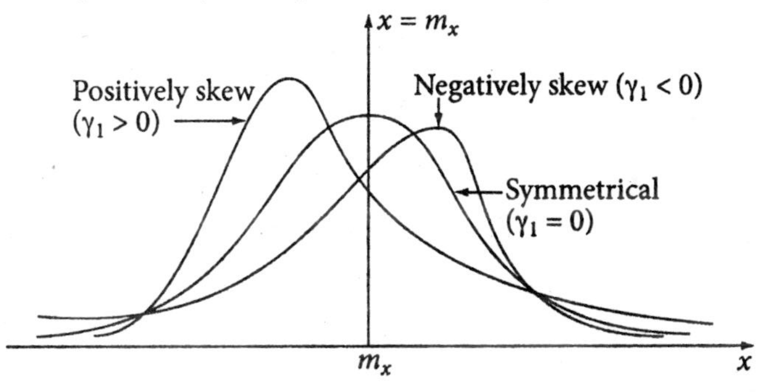

Fig 6.2

For a discrete distribution: $\gamma_1 > 0 \Rightarrow$ most of the spectrum points are on the right of the mean and $\gamma_1 < 0 \Rightarrow$ most of the spectrum point are on left of the mean.

Note: $\gamma_1 \neq 0$ certainly implies that the distribution is not symmetric but $\gamma_1 = 0$ does not necessarily imply that the distribution is symmetric. Sometimes it may so happen that $\mu_5(X) \neq 0$ and in that case asymmetry is measured by $\frac{\mu_5(X)}{\sigma_x^5}$.

Examples of asymmetric distributions and their measures of asymmetry

1. Gamma (l) distribution

We have for gamma (l) distribution

$$\alpha_k(X) = (k + l - 1)(k + l - 2) \cdots (l + 1) l.$$

$\therefore$ $\alpha_1(X) = l, \alpha_2(X) = l(l+1), \alpha_3(X) = l(l+1)(l+2)$.

$$\therefore \ \mu_3(X) = E(X - l)^3 = E(X^3) - 3E(X^2)\, l + 3E(X) l^2 - l^3 = \alpha_3(X) - 3_2\alpha(X) l + 3\alpha_1(X) l^2 - l^3$$
$$= l(l+1)(l+2) - 3l^2(l+1) + 3l^3 - l^3 = 2l.$$

$\mu_2(X) = E(X-l)^2 = \alpha_2(X) - l^2 = l(l+1) - l^2 = l.\ \therefore\ \sigma_x = \sqrt{\mu_2(X)} = \sqrt{l}.$

$\therefore\ \gamma_1 = \dfrac{\mu_3(X)}{\sigma_x^3} = \dfrac{2l}{l\sqrt{l}} = \dfrac{2}{\sqrt{l}}.\ \therefore\ \gamma_1 \propto l^{-l/2}.$

$\therefore$ for large l, asymmetry is small and for small l, asymmetry is high and asymmetry tends to vanish as l tends to infinitely large value. As $\gamma_1 = \frac{2}{\sqrt{l}} > 0$, $\therefore$ gamma distribution is positively skew and the longer tail of probability density curve lies on the right of the line $x = l$.

2. Poisson (μ) distribution

We have $m_x = \mu, \mu_2(x) = \mu.\ \therefore\ \sigma_x = \sqrt{\mu}.$

$$\alpha_3(X) = E\left(X^3\right) = \sum_{i=0}^{\infty} i^3 e^{-\mu}\frac{\mu^i}{i!} = e^{-\mu}\left[\sum_{i=0}^{\infty}\{i(i-1)(i-2)+3i(i-1)+i\}\frac{\mu^i}{i!}\right]$$

$$= e^{-\mu}\left[\mu^3\sum_{i=3}^{\infty}\frac{\mu^{i-3}}{(i-3)!} + 3\mu^2\sum_{i=2}^{\infty}\frac{\mu^{i-2}}{(i-2)!} + \mu\sum_{i=1}^{\infty}\frac{\mu^{i-1}}{(i-1)!}\right]$$

$$= e^{-\mu}\left[\mu^3 e^{\mu} + 3\mu^2 e^{\mu} + \mu e^{\mu}\right], \left[\sum_{i=3}^{\infty}\frac{\mu^{i-3}}{i-3} = 1 + \frac{\mu}{1!} + \frac{\mu^2}{2!} + \cdots = e^{\mu}, \text{ etc.}\right] = \mu^3 + 3\mu^2 + \mu.$$

Similarly, $\alpha_2(X) = \mu^2 + \mu.$

$\mu_3(X) = \alpha_3(X) - 3\alpha_2(X)\mu + 2\mu^3\ [\because\ m_x = \mu]$

$= \mu^3 + 3\mu^2 + \mu - 3(\mu^2 + \mu)\mu + 2\mu^3 = \mu.$

$\therefore\ \gamma_1 = \dfrac{\mu_3(X)}{\sigma_x^3} = \dfrac{\mu}{\mu\sqrt{\mu}} = \dfrac{1}{\sqrt{\mu}} > 0.$ $\therefore$ Poisson (μ) distribution is also a positively skew distribution.

Following the rule. "For a discrete distribution positively skew means most of the spectrum points are on the right of $x = \mu$," [reverse case for negatively skew discrete distribution] we conclude that for Poisson μ distribution most of the spectrum points lie on the right of μ (mean). Check the situation taking any value of μ, say $\mu = 2$.

3. Binomial (n, p) distribution

$$\gamma_1 = \frac{\mu_3(X)}{\sigma_x^3} = \frac{np(1-p)(1-2p)}{\{np(1-p)\}^{3/2}} = \frac{1-2p}{\sqrt{np(1-p)}}\ \text{[Do yourself]}.$$

See that Binomial (n, p) distribution may be positively skew, negatively skew or may be symmetric.

When (i) $p < \frac{1}{2}, \gamma_1 > 0$ and the distribution is positively skew, (ii) $p > \frac{1}{2}, \gamma_1 < 0$ and the distribution is negatively skew, (iii) $p = \frac{1}{2}, \gamma_1 = 0$ and the distribution is symmetric.

In the first case, most of the step points are on the right of the mean np and in second case, most step points on the left of np and in third case, equal number of points on the left and on the right of np.

Another measure of skewness, as given by Karl Pearson, is $\frac{\text{Mean}-\text{Mode}}{\text{Standard deviation}}$.

Note that for a symmetrical distribution Mean = Mode = Median.

6.7 Fourth central moment $\mu_4(X)$

The fourth central moment of a probability distribution plays an important role in describing the degree peakedness or flatness of the pdf curve near the mean. The fourth central moment of the corresponding standardised variable X^*, i.e., $\frac{\mu^4(X)}{\sigma_x^4}$ is denoted by $\beta_2(X)$ and is called the coefficient of kurtosis. Thus the coefficient of kurtosis $\beta_2(x) = \frac{\mu_4(X)}{\sigma_x^4}$ gives the degree of peakedness near the centre of the probability density curve for a continuous case or of the graph of pmf for the discrete case.

For a normal (m, σ) distribution we have $\mu_4(X) = 3\sigma^4$. $\therefore\ \beta_2(X) = \frac{\mu_4(X)}{\sigma^4} = 3$.

To compare the measure of kurtosis any distribution with that of the normal (m, σ) distribution we introduce $\gamma_2(X) = \beta_2(X) - 3 = \frac{\mu_4(X)}{\sigma_x^4} - 3$.

$\gamma_2(X)$ is called the coefficient of excess of the distribution. It gives the vertical deviation.

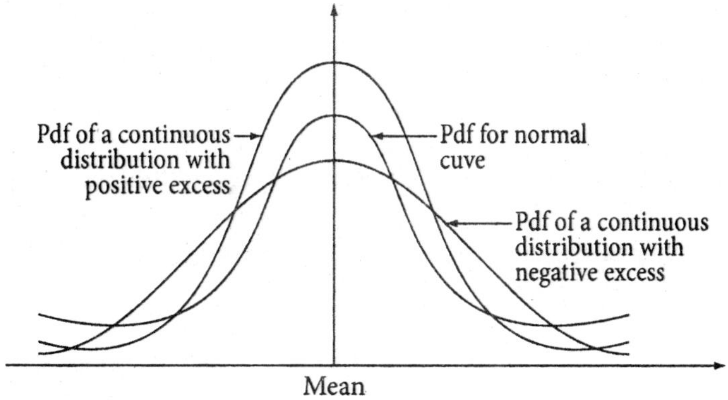

Fig 6.3

Note 1. Variance of a distribution gives the dispersion, skewness gives the horizontal deviation and kurtosis gives the vertical deviation.

2. $\gamma_1(X), \beta_2(X), \gamma_2(X)$ are all pure numbers, i.e., dimensionless. $\beta_2(X)$ and $\gamma_2(X)$ are independent of choice of origin and scale of measurement.

3. If for a distribution $\gamma_2(X) < 0$, the distribution is Platykurtic.
If for a distribution $\gamma_2(X) = 0$, the distribution is Mesokurtic.
If for a distribution $\gamma_2(X) > 0$, the distribution is called *Leptokurtic.*

1. Kurtosis and coefficient of excess for Poisson (μ) distribution

For the Poisson (μ) distribution we have $\alpha_1(X) = \mu, \mu_2(X) = \mu, \mu_3(X) = \mu, \mu_4(X) = \alpha_4(X) - 4\alpha_3(X)\mu + 6\alpha_2(X)\mu^2 - 3\mu^4$. (1)

$$\alpha_4(X) = E\left(X^4\right) = \sum_{i=0}^{\infty} i^4 e^{-\mu}\frac{\mu^i}{i!} = \sum_{i=0}^{\infty}\{i(i-1)(i-2)(i-3) + 6i(i-1)(i-2) + 7i(i-1) + i\} e^{-\mu}\frac{\mu^i}{i!}$$

$$= e^{-\mu}\mu^4\sum_{i=4}^{\infty}\frac{\mu^{i-4}}{(i-4)!} + 6e^{-\mu}\mu^3\sum_{i=3}^{\infty}\frac{\mu^{i-3}}{(i-3)!} + 7e^{-\mu}\mu^2\sum_{i=2}^{\infty}\frac{\mu^{i-2}}{(i-2)!} + e^{-\mu}\mu\sum_{i=1}^{\infty}\frac{\mu^{i-1}}{(i-1)!}$$

$$= \mu^4 + 6\mu^3 + 7\mu^2 + \mu \left[\because \sum_{i=4}^{\infty}\frac{\mu^{i-4}}{(i-4)!} = \sum_{i=3}^{\infty}\frac{\mu^{i-3}}{(i-3)!} = \cdots = e^{\mu}\right]$$

$\alpha_3(X) = \mu^3 + 3\mu^2 + \mu, \alpha_2(X) = \mu^2 + \mu$.

$\therefore\ \mu_4(X) = \mu^4 + 6\mu^3 + 7\mu^2 + \mu - 4\left(\mu^4 + 3\mu^3 + \mu^2\right) + 6\left(\mu^4 + \mu^3\right) - 3\mu^4 = 3\mu^2 + \mu$.

$$\therefore\ \beta_2(X) = \frac{\mu_4(X)}{\mu^2}\ \left[\because \sigma_x = \sqrt{\mu}\right] = 3 + \frac{1}{\mu}.\ \therefore\ \gamma_2(X) = \beta_2(X) - 3 = \frac{1}{\mu}.$$

2. Kurtosis and coefficient of excess for a gamma (l) distribution

Ordinary moment of order n is given by $\alpha_n(X) = (n+l-1)(n+l-2)\cdots(l+1)l, n \geq 1$ and $\alpha_0(X) = E\left(X^0\right) = 1$.

$$\begin{aligned}\therefore \mu_4(X) &= E\left\{(X-l)^4\right\}, \text{ as mean} = l \\ &= E\left(X^4\right) - 4E\left(X^3\right) l + 6E\left(X^2\right) l^2 - 4E(X) l^3 + l^4 \\ &= \alpha_4(X) - 4\alpha_3(X) l + 6\alpha_2(X) l^2 - 4\alpha_1(X) l^3 + l^4 \\ &= (l+3)(l+2)(l+1)l - 4(l+2)(l+1)l^2 + 6(l+1)l^3 - 4l^4 + l^4 \\ &= \left(l^2+5l+6\right)\left(l^2+l\right) - (4l+8)\left(l^3+l^2\right) + (6l+6)l^3 - 3l^4 \\ &= 3l^2 + 6l = 3l(l+2).\end{aligned}$$

$$\therefore \beta_2(X) = \frac{\mu_4(X)}{\{(l+1)l-l^2\}^2} = \frac{3l(l+2)}{l^2} = \frac{6}{l} + 3. \therefore \gamma_2(X) = \beta_2(X) - 3 = \frac{6}{l}.$$

3. Kurtosis and coefficient of excess of binomial (n, p) distribution

Outlines: $\alpha_2(X) = n(n-1)p^2 + np$

$\alpha_3(X) = n(n-1)(n-2)p^3 + 3n(n-1)p^2 + np$

$\alpha_4(X) = n(n-1)(n-2)(n-3)p^4 + 6n(n-1)(n-2)p^3 + 7n(n-1)p^2 + np$

$\mu_2(X) = np(1-p), \mu_3(x) = np(1-p)(1-2p)$

$\mu_4(X) = np(1-p)\{1+3p(1-p)(1-p)(n-2)\}$ [Calculate all those].

$$\therefore \beta_2(X) = \frac{\mu_4(X)}{\sigma_x^4} = \frac{np(1-p)\{1+3p(1-p)(n-2)\}}{n^2p^2(1-p)^2} = 3 + \frac{1+6p(p-1)}{np(1-p)}.$$

$$\therefore \gamma_2(X) = \beta_2(X) - 3 = \frac{1-6p(1-p)}{np(1-p)} \text{ and } \gamma_1(X) = \frac{\mu_3(x)}{\sigma_x^3} = \frac{np(1-p)(1-2p)}{np(1-p)\sqrt{np(1-p)}} = \frac{1-2p}{\sqrt{np(1-p)}}.$$

$\therefore$ when $1-2p=0$, i.e., $p=1/2$, the binomial distribution is symmetric.

It is positively skewed, when $\gamma_1(X) > 0$, i.e., when $p < 1/2$ and it is negatively skewed, when $\gamma_1(X) < 0$, i.e., $p > 1/2$.

6.8 Mathematical preliminaries

Linear transformation. A transformation T is said to be linear if for every pair of functions f_1 and f_2 and for each pair of parametric constants c_1 and c_2, T satisfies the relation

$$T\{c_1 f_1(t) + c_2 f_2(t)\} = c_1 T\{f_1(t)\} + c_2 T\{f_2(t)\}.$$

Thus a linear transformation of a linear combination of two functions is the *same* linear combination of the transformed functions.

In general, for a set of functions $f_1, f_2, \cdots, f_n$ and a set scalars $c_1, c_2, \cdots, c_n$ a linear transformation satisfies $T\{c_1 f_1(t) + c_2 f_2(t) + \cdots + c_n f_n(t)\} = c_1 T\{f_1(t)\} + c_2 T\{f_2(t)\} + \cdots + c_n T\{f_n(t)\}$, i.e.,

$$T\left\{\sum_{i=1}^{n} c_i f_i(t)\right\} = \sum_{i=1}^{n} c_i T\{f_i(t)\}.$$

Integral transformation

It is a linear transformation. Let k be a prescribed function of a variable t and a parameter s, called *Kernel.* A linear integral transformation T of a function F w.r.t. to some kernel k is defined by

$$T\{F(t)\} = \int_a^b k(t,s) F(t)\, dt = f(s), \text{ say,}$$

where (a, b) is the domain of definition of the function F. $f(s)$ is called the *image* of F.

6.8.1 Moment generating function

Moment generating function of a RV X about the point a is a function of a real variable t to be denoted by $M_{x(a)}$ and is defined by $M_{x(a)}(t)=E\left(e^{t(X-a)}\right)$, in case when it exists

$$\therefore\ M_{x(a)}(t)=\int_{-\infty}^{\infty} e^{t(x-a)} f_x(x)\,dx \text{ for a continuous RV}$$

$$\text{and } M_{x(a)}(t)=\sum_{-\infty}^{\infty} e^{t(x_i-a)} f_{xi} \text{ for a discrete RV.}$$

In particular, if the point a is the origin, $M_{x(0)(t)}$ or simply $M_x(t)=E\left(e^{tX}\right)$, whenever exists, i.e., whenever the infinite sum or the improper integral converges absolutely.

$$\therefore\ M_x(t)=E\left(e^{tX}\right)=E\left(1+tX+\frac{t^2}{2!}X^2+\frac{t^3}{3!}X^3+\cdots+\frac{t^r}{r!}X^r+\cdots\right)$$

$$=1+\frac{t}{1!}E(X)+\frac{t^2}{2!}E\left(X^2\right)+\frac{t^3}{3!}E\left(X^3\right)+\cdots+\frac{t^r}{r!}E(X^r)+\cdots$$

$$=1+\frac{t}{1!}\alpha_1(X)+\frac{t^2}{2!}\alpha_2(X)+\cdots+\frac{t^r}{r!}\alpha_r(X)+\cdots=\sum_{r=0}^{\infty}\frac{t^r}{r!}\alpha_r(X).$$

Also, $M_x^r(0)=\alpha_r(X)$, when M_x is continuously differentiable in some neighbourhood of the origin. Thus the rth order ordinary moment (moment about the origin) is the coefficient of $t^r/r!$ in the expansion of $M_x(t)$, whenever $\alpha_r(X)$ exists and hence the name MGF.

Again, $M_{x(a)}(t)=E\left(e^{t(X-a)}\right)=e^{-at}E\left(e^{tX}\right)=e^{-at}M_x(t)$.

Now, $M_{x(m)}(t)=E\left(e^{t(X-m)}\right)$, where m = mean of X, i.e., m_x

$$=E\left\{1+\frac{t}{1!}(X-m)+\frac{t^2}{2!}(X-m)^2+\cdots+\frac{t^r}{r!}(X-m)^r+\cdots\right\}$$

$$=1+\frac{t}{1!}E(X-m)+\frac{t^2}{2!}E(X-m)^2+\cdots+\frac{t^r}{r!}E(X-m)^r+\cdots$$

$$=1+\frac{t}{1!}\mu_1(X)+\frac{t^2}{1!}\mu_2(X)+\cdots+\frac{t^r}{r!}\mu_r(X)+\cdots=\sum_{r=0}^{\infty}\frac{t^r}{r!}\mu_r(X)=e^{-mt}M_x(t).$$

Also, $M^r_{x(m)}(0)=\mu_r(X)$, when $M_{x(m)}$ is continuously differentiable in some nbd of origin. Hence, the central moment of order r is the coefficient of $\frac{t^r}{r!}$ in the expansion of $M_{x(m)}(t)$.

Note: The MGF of a distribution, if it exists, uniquely determines the distribution.

Moment generating functions for some standard distributions

1. Binomial (n,p) distribution

$$M_x(t)=E\left(e^{tX}\right)=\sum_{k=0}^{n} e^{tk}\binom{n}{k}p^k q^{n-k}=\sum_{k=0}^{n}\binom{n}{k}\left(pe^t\right)^k q^{n-k}=\left(pe^t+q\right)^n.$$

$M_x(t)$ exists $\forall\ t\in\mathbb{R}$ as e^t is absolutely convergent $\forall\ t\in\mathbb{R}$.

2. Poisson (μ) distribution

$$M_x(t)=E\left(e^{tX}\right)=\sum_{k=0}^{\infty} e^{tk}\frac{e^{-\mu}\mu^k}{k!}=e^{-\mu}\sum_{k=0}^{\infty}\frac{\left(\mu e^t\right)^k}{k!}=e^{-\mu}\cdot e^{\mu e^t}=e^{\mu\left(e^t-1\right)}.$$

$M_x(t)$ exists $\forall\ t\in\mathbb{R}$ as e^t is absolutely convergent $\forall\ t\in\mathbb{R}$.

3. Normal (m,σ) distribution

Let a RV X follows normal (m,σ) distribution. The MGF of X is given by

$$M_x(t)=E\left(e^{tX}\right)=\frac{1}{\sqrt{2\pi}\sigma}\int_{-\infty}^{\infty}e^{tx}\cdot e^{-\frac{(x-m)^2}{2\sigma^2}}dx=\frac{1}{\sqrt{2\pi}\sigma}\int_{-\infty}^{\infty}e^{-\frac{(x-m)^2-2tx\sigma^2}{2\sigma^2}}dx$$

$$=\frac{1}{\sqrt{2\pi}\sigma}\int_{-\infty}^{\infty}e^{\frac{(x-m)^2-2(x-m)t\sigma^2-2mt\sigma^2+\left(t\sigma^2\right)^2-t^2\sigma^4}{2\sigma^2}}dx=e^{mt+\frac{1}{2}\sigma^2t^2}\cdot\frac{1}{\sqrt{2\pi}\sigma}\int_{-\infty}^{\infty}e^{-\frac{\left(x-m-t\sigma^2\right)^2}{2\sigma^2}}dx$$

$$=e^{mt+\frac{1}{2}\sigma^2t^2}\left[\because\ \frac{1}{\sqrt{2\pi}\sigma}\int_{-\infty}^{\infty}e^{-\frac{\left(x-m-t\sigma^2\right)}{2\sigma^2}}dx=\frac{1}{\sqrt{2\pi}\sigma}\int_{-\infty}^{\infty}e^{-\frac{\left(x-m^1\right)^2}{2\sigma^2}}dx=1,\text{ when } m^1=m+t\sigma^2\right]$$

$M_x(t)$ exists $\forall\ t\in\mathbb{R}$, since e^x is absolutely convergent $\forall\ x\in\mathbb{R}$.

Note: The existence of MGF of a RV X does not guarantee the existence of moments of X.

Consider the following example of a discrete distribution.

$X=x_k=k, k\in\mathbb{N}$ is the spectrum
and $f_{xk}=P(X=k)=\frac{1}{k(k+1)}$, so that $f_{xk}>0$, $\forall\ k\in\mathbb{N}$ and $\sum_{k\in\mathbb{N}}f_{xk}=1$.

[$\sum_{k=1}^{\infty}\frac{1}{k(k+1)}$ be the series. If $\{S_n\}_n$ be the sequence of partial sums, then $S_n=\frac{1}{1\cdot2}+\frac{1}{2\cdot3}+\cdots+\frac{1}{n(n+1)}=\left(1-\frac{1}{2}\right)+\left(\frac{1}{2}-\frac{1}{3}\right)+\cdots+\left(\frac{1}{n}-\frac{1}{n+1}\right)=1-\frac{1}{n+1}$. $\therefore\ \lim S_n=1$. $\therefore$ the series converges and the sum is 1.]

Now, $\alpha_r(X)=E(X^r)=\sum_{k=1}^{\infty}\frac{k^r}{k(k+1)}$, provided the series is absolutely convergent.

Now, note that $\frac{k^r}{k(k+1)}=\frac{k^{r-1}}{k+1}\geq\frac{1}{k+1}$ for $k=1,2,3,\cdots$ and the series $\sum_{k=1}^{\infty}\frac{1}{k+1}=\frac{1}{2}+\frac{1}{3}+\cdots$ is the well-known harmonic series (neglecting the first term) and is divergent and diverges to ∞.

Hence, by comparison $\sum_{k=1}^{\infty}\frac{k^r}{k(k+1)}$ is divergent $\forall\ r\in\mathbb{N}$.

$\therefore$ excepting the ordinary moment of order 0 (which exists for all distributions), all other ordinary moments does not exist.

Though the ordinary moments $\alpha_r(x), r\in\mathbb{N}$ does not exist we show that its MGF exists. [In such case, should we say the function is moment generating?]

$$M_x(t)=E\left(e^{tX}\right)=\sum_{k=1}^{\infty}e^{tk}f_{xk}=\sum_{k=1}^{\infty}e^{tk}\cdot\frac{1}{k(k+1)}=\lim_{n\to\infty}\sum_{k=1}^{n}e^{tk}\left(\frac{1}{k}-\frac{1}{k+1}\right)$$

$$=\lim_{n\to\infty}\left[e^t\left(1-\frac{1}{2}\right)+e^{2t}\left(\frac{1}{2}-\frac{1}{3}\right)+e^{3t}\left(\frac{1}{3}-\frac{1}{4}\right)+\cdots+e^{nt}\left(\frac{1}{n}-\frac{1}{n+1}\right)\right]$$

$$=\lim_{n\to\infty}\left[\left\{e^t+\frac{(e^t)^2}{2}+\frac{(e^t)^3}{3}+\cdots+\frac{(e^t)^n}{n}\right\}-e^{-t}\left\{\frac{(e^t)^2}{2}+\frac{(e^t)^3}{3}+\cdots+\frac{(e^t)^{n+1}}{n+1}\right\}\right]$$

$$=-\ln\left(1-e^t\right)-e^{-t}\left\{-\ln\left(1-e^t\right)-e^t\right\}$$

$$=1-\left(1-e^{-t}\right)\ln\left(1-e^t\right),\ \text{when } e^t<1,\ \text{i.e., } t<\ln 1,\ \text{i.e., } t<0.$$

When $t=0, M_x(t)=E\left(e^0\right)=E(1)=1$. $\therefore$ MGF of X is given by

$$M_x(t)=\begin{cases}1-(1-e^{-t})\ln(1-e^t), & t<0\\ 1, & \text{when } t=0.\end{cases}$$

6.8.2 Characteristic function

The characteristic function of a RV X is a function of a real variable t to be denoted by $K_x(t)$ and is defined by the mean value $E\left(e^{itX}\right)$, i.e.,

$$K_x(t)=E\left(e^{itX}\right)=\begin{cases}\displaystyle\int_{-\infty}^{\infty} e^{itx} f_x(x)\,dx, & \text{for continuous RV } X\\ \displaystyle\sum_{k=-\infty}^{\infty} e^{itk} f_{xk}, & \text{for discrete RV } X.\end{cases}$$

It is important to note that characteristic function of all distributions exist as the integral or the infinite sum converges absolutely for every probability distribution as shown below:

$$\int_{-\infty}^{\infty}\left|e^{itx} f_x(x)\right| dx=\int_{-\infty}^{\infty}\left|e^{itx}\right|\left|f_x(x)\right| dx=\int_{-\infty}^{\infty} f_x(x)\,dx,$$

$$\because \left|e^{itx}\right|=|\cos tx+i\sin tx|=\sqrt{\cos^2 tx+\sin^2 tx}=1 \text{ and } f_x(x)\geq 0$$

converges, as $\displaystyle\int_{-\infty}^{\infty} f_x(x)\,dx=1.$

Again,

$$\sum_{k=-\infty}^{\infty}\left|e^{itk} f_{xk}\right|=\sum_{k=-\infty}^{\infty}\left|e^{itk}\right|\left|f_{xk}\right|=\sum_{k=-\infty}^{\infty} f_{xk},\ \because \left|e^{itk}\right|=1 \text{ and } f_{xk}>0,$$

which converges, as $\displaystyle\sum_{k=-\infty}^{\infty} f_{xk}=1.$

The important role of a characteristic function is to produce moments. However, the existence of characteristic function of a distribution does not guarantee the existence of moments of the distribution. This is due to the fact that the characteristic function exists for all distribution but each distribution does not possess moments of all orders. Beside the existence of characteristic function for every distribution it has the excellence of uniqueness, i.e., a distribution function of a distribution is uniquely determined by its characteristic function.

That is, whenever we know the characteristic function of an unknown distribution we can find the distribution (uniquely). Following is the rule of finding the distribution function F_x, whenever its characteristic function $K_x(t)$ is known.

$$F_x(x)=\frac{1}{2\pi i}\lim_{c\to-\infty}\left\{\lim_{d\to\infty}\int_{-d}^{d}\frac{e^{-itc}-e^{-itx}}{t}K_x(t)\,dt\right\}.$$

The following is the theorem (without proof) that gives the pdf of a continuous distribution when its characteristic function is known.

Theorem If X be a RV whose characteristic function $K_x(t)$ is known, then its pdf f_x is given by

$$f_x(x)=F_x'(x)=\frac{1}{2\pi}\int_{-\infty}^{\infty} e^{-itx} K_x(t)\,dt,\ x\subset(-\infty,\infty),$$

provided $\displaystyle\int_{-\infty}^{\infty}|K_x(t)|\,dt$ converges.

Role of characteristic function for finding moments

From the definition of characteristic function we have

$$K_x(t) = E\left(e^{itX}\right) = E\left\{1 + \frac{itX}{1!} + \frac{(it)^2}{2!}X^2 + \frac{(it)^3}{3!}X^3 + \cdots + \frac{(it)^r}{r!}X^r + \cdots\right\}$$

$$= 1 + \frac{it}{1!}E(X) + \frac{(it)^2}{2!}E\left(X^2\right) + \cdots + \frac{(it)^r}{r!}E(X^r) + \cdots$$

$$= 1 + \frac{it}{1!}\alpha_1(X) + \frac{(it)^2}{2!}\alpha_2(X) + \cdots + \frac{(it)^r}{r!}\alpha_r(X) + \cdots$$

$$= \sum_{r=0}^{\infty}\frac{(it)^r}{r!}\alpha_r(X).$$

$\therefore\ \alpha_r(X)$ **is the coefficient of** $\frac{(it)^r}{r!}$ **in the expansion of the characteristic function** $K_x(t)$**, whenever** $\alpha_r(X)$ **exists.**

Properties of characteristic function

1. $K_{aX+b}(t) = E\left(e^{it(aX+b)}\right) = e^{ibt}E\left(e^{iatX}\right) = e^{ibt}K_x(at).$

   In particular, $K_{x^*}(t) = E\left(e^{itX^*}\right) = E\left(e^{it\left(\frac{1}{\sigma_x}X - \frac{m_x}{\sigma_x}\right)}\right) = e^{-i\frac{m_x}{\sigma_x}t}K_x\left(\frac{t}{\sigma_x}\right).$

2. $K_x(-t) = E\left(e^{-itX}\right) = E(\cos tX - i\sin tX) = E(\cos tX) - iE(\sin tX) = \overline{E(\cos tX) + iE(\sin tX)} = \overline{K_x(t)}.$

Character function of some well-known distributions

1. Poisson (μ) distribution

$$K_x(t) = E\left(e^{itX}\right) = \sum_{k=0}^{\infty} e^{itk}\frac{e^{-\mu}\mu^k}{k!} = e^{-\mu}\sum_{k=0}^{\infty}\frac{\left(\mu e^{it}\right)^k}{k!} = e^{-\mu}e^{\mu e^{it}} = e^{\mu\left(e^{it}-1\right)}.$$

2. Binomial (n, p) distribution

$$K_x(t) = \sum_{k=0}^{n} e^{itk}\binom{n}{k}p^kq^{n-k},\quad q = 1-p$$

$$= \sum_{k=0}^{n}\binom{n}{k}\left(pe^{it}\right)^kq^{n-k} = \left(pe^{it}+q\right)^n.$$

3. Normal (m, σ) distribution

$$K_x(t) = \int_{-\infty}^{\infty} e^{itx}\cdot\frac{1}{\sqrt{2\pi}\sigma}\exp\left[-\frac{(x-m)^2}{2\sigma^2}\right]dx = \int_{-\infty}^{\infty}\frac{1}{\sqrt{2\pi}\sigma}\exp\left[-\frac{(x-m)^2 - 2it\sigma^2x}{2\sigma^2}\right]dx$$

$$= \frac{1}{\sqrt{2\pi}\sigma}\int_{-\infty}^{\infty}\exp\left[-\frac{\left(x-m-it\sigma^2\right)^2 - 2itm\sigma^2 + t^2\sigma^4}{2\sigma^2}\right]dx$$

$$= e^{imt-\frac{1}{2}\sigma^2t^2}\cdot\frac{1}{\sqrt{2\pi}\sigma}\int_{-\infty}^{\infty}\exp\left[-\frac{(x-u)^2}{2\sigma^2}\right]dx,\ \text{when } u = m + it\sigma^2$$

$$= e^{imt-\frac{1}{2}\sigma^2t^2}\quad\left[\because\ \frac{1}{\sqrt{2\pi}\sigma}\int_{-\infty}^{\infty}\exp\left[-\frac{(x-u)^2}{2\sigma^2}\right]dx = 1\right].$$

4. Gamma (l) distribution

$$K_x(t) = \int_0^{\infty} e^{itx}\frac{e^{-x}\cdot x^{l-1}}{\Gamma(l)}dx = \frac{1}{\Gamma(l)}\lim_{\substack{\delta\to 0+\\ R\to\infty}}\int_{\delta}^{R} e^{-(1-it)x}x^{l-1}\,dx.$$

Let $(1-it)x=z$. $\therefore\ (1-it)dx=dz$,

x	δ	R
z	$(1-it)\delta$	$(1-it)R$

$$\therefore\ K_x(t)=\frac{1}{\Gamma(l)}\lim_{\substack{\delta\to0+\\R\to\infty}}\int_{(1-it)\delta}^{(1-it)R} e^{-z}\frac{z^{l-1}}{(1-it)^{l-1}}\cdot\frac{dz}{1-it}=\frac{1}{\Gamma(l)}\cdot\frac{1}{(1-it)^l}\int_0^\infty e^{-z}z^{l-1}dz$$

$$=\frac{1}{(1-it)^l\Gamma(l)}\int_0^\infty e^{-x}x^{l-1}dx=\frac{1}{(1-it)^l\Gamma(l)}\cdot\Gamma(l)=(1-it)^{-l}.$$

6.9 Median

Median is a point on the line of probability distribution up to which the probability is $\frac{1}{2}$. If X be a RV with probability distribution function F_x, then the median μ is such a point on the line of probability distribution such that $F_x(\mu)=\frac{1}{2}$.

For any continuous distribution, F_x is continuous, monotone increasing and increases from 0 to 1. So, the distribution curve $y=F_x(x)$ cuts the line $y=\frac{1}{2}$ at a definite point $\left(\mu,\frac{1}{2}\right)$, so that $F_x(\mu)=\frac{1}{2}$. Therefore, for any continuous distribution, median in uniquely defined.

For a discrete distribution either of the two situations will arise

(i) $y=F_x(x)$, which is a step curve, does not cut the line $y=\frac{1}{2}$ at any point

or, (ii) $y=\frac{1}{2}$ coincides with $y=F_x(x)$, $\forall\ x\in[x_i,x_{i-1})$.

To make the median unique we modify the definition for the case (i) as:

μ is called the median if $\lim\limits_{x\to\mu-}F_x(x)\le\frac{1}{2}$ and $F_x(\mu)\ge\frac{1}{2}$.

For case (ii) as every point in $[x_i,x_{i+1})$ by definition claims to be a median. Therefore, in such case we take the midpoint of the interval, viz., $\frac{x_i+x_{i+1}}{2}$ as the median, i.e., $\mu=\frac{x_i+x_{i+1}}{2}$.

Theorem

For any distribution, continuous or discrete, the first absolute moment about any point is minimum, when taken about the median.

Proof. First we consider continuous distribution. Let a be a point on the line of probability distribution and $a>\mu$ [$\mu=$ median]

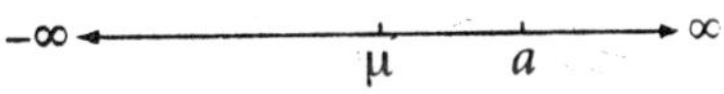

Line of probability distribution

Fig 6.4

$$\therefore\ E(|X-a|)=\int_{-\infty}^{\infty}|x-a|f_x(x)dx,\ f_x(x)=\text{pdf}$$

$$=\int_{-\infty}^{a}(a-x)f_x(x)dx+\int_a^\infty(x-a)f_x(x)dx$$

$$=\int_{-\infty}^{\mu}(a-x)f_x(x)dx+\int_\mu^a(a-x)f_x(x)dx+\int_\mu^\infty(x-a)f_x(x)dx-\int_\mu^a(x-a)f_x(x)dx$$

$$=\int_{-\infty}^{\mu}(\mu-x)f_x(x)dx+\int_{-\infty}^{\mu}(a-\mu)f_x(x)dx+\int_\mu^\infty(x-\mu)f_x(x)dx+\int_\mu^\infty(\mu-a)f_x(x)dx$$
$$+2\int_\mu^a(a-x)f_x(x)dx$$

$$=\int_{-\infty}^{\mu}(\mu-x)f_x(x)dx+\int_\mu^\infty(x-\mu)f_x(x)dx+(a-\mu)F_x(\mu)+(\mu-a)(1-F_x(u))+2\int_\mu^a(a-x)f_x(x)dx$$

$$= \int_{-\infty}^{\infty} |x-\mu| f_X(x)\,dx + (a-\mu)\tfrac{1}{2} + (\mu-a)\cdot\tfrac{1}{2} + 2\int_{\mu}^{a} (a-x) f_X(x)\,dx \quad \left[\because F_X(\mu)=\tfrac{1}{2}\right]$$

$$= E(|x-\mu|) + 2\int_{\mu}^{a} (a-x) f_X(x)\,dx$$

$$\geq E(|X-\mu|), \text{ as } \int_{\mu}^{a} (a-x) f_X(x)\,dx \geq 0, \text{ since } (a-x) f_X(x) \geq 0.$$

$$\therefore\ E(|x-a|) \geq E(|x-\mu|).$$

Now, consider $a < \mu$.

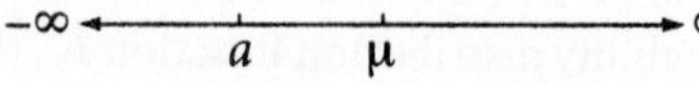

Line of probability distribution

Fig 6.5

$$\therefore\ E(|X-a|) = \int_{-\infty}^{\infty} |x-a| f_X(x)\,dx$$

$$= \int_{-\infty}^{a} (a-x) f_X(x)\,dx + \int_{a}^{\infty} (x-a) f_X(x)\,dx$$

$$= \int_{-\infty}^{\mu} (a-x) f_X(x)\,dx - \int_{a}^{\mu} (a-x) f_X(x)\,dx + \int_{a}^{\mu} (x-a) f_X(x)\,dx + \int_{\mu}^{\infty} (x-a) f_X(x)\,dx$$

$$= \int_{-\infty}^{\mu} (\mu-x) f_X(x)\,dx + \int_{-\infty}^{\mu} (a-\mu) f_X(x)\,dx + 2\int_{a}^{\mu} (x-a) f_X(x)\,dx$$
$$+ \int_{\mu}^{\infty} (x-\mu) f_X(x)\,dx + \int_{\mu}^{\infty} (\mu-a) f_X(x)\,dx$$

$$= \int_{-\infty}^{\mu} (\mu-x) f_X(x)\,dx + \int_{\mu}^{\infty} (x-\mu) f_X(x)\,dx + (a-\mu)F_X(\mu) + 2\int_{a}^{\mu} (x-a) f_X(x)\,dx$$
$$+ (\mu-a)\,[1-F_X(\mu)]$$

$$= \int_{-\infty}^{\infty} |x-\mu| f_X(x)\,dx + (a-\mu)\tfrac{1}{2} + 2\int_{a}^{\mu} (x-a) f_X(x)\,dx + (\mu-a)\left(1-\tfrac{1}{2}\right), \text{ as } F_X(\mu)=\tfrac{1}{2}$$

$$= E(|x-\mu|) + 2\int_{a}^{\mu} (x-a) f_X(x)\,dx$$

$$\geq E(|x-\mu|) \text{ as } \int_{a}^{\mu} (x-a) f_X(x)\,dx \geq 0, \text{ since } (x-a) f_X(x) \geq 0.$$

When X is a discrete RV with spectrum $x_i, i = 0, \pm 1, \pm 2, \cdots$.
Let μ be the median and a be any other point and $a > \mu$ [for $a < \mu$ proof is similar].

$$\therefore\ E(|X-a|) = \sum_{i=-\infty}^{\infty} |x_i - a|\, f_{xi} = \sum_{x_i \leq a} |x_i - a|\, f_{xi} + \sum_{x_i > a} (x_i - a)\, f_{xi}$$

$$= \sum_{x_i \leq a} (a - x_i)\, f_{xi} + \sum_{x_i > a} (x_i - a)\, f_{xi}$$

$$= \sum_{x_i \leq \mu} (a - x_i)\, f_{xi} + \sum_{\mu < x_i \leq a} (a - x_i)\, f_{xi} + \sum_{x_i > \mu} (x_i - a)\, f_{xi} - \sum_{\mu < x_i \leq a} (x_i - a)\, f_{xi}$$

$$= \sum_{x_i \le \mu} (\mu - x_i) f_{xi} + \sum_{x_i \le \mu} (a-\mu) f_{xi} + \sum_{x_i > \mu} (x_i - \mu) f_{xi} + \sum_{x_i > \mu} (\mu - a) f_{xi} + 2 \sum_{\mu < x_i \le a} (a - x_i) f_{xi}$$

$$= \sum_{i=-\infty}^{\infty} |x_i - \mu| f_{xi} + (a-\mu) \sum_{x_i \le \mu} f_{xi} + (\mu - a) \sum_{x_i > \mu} f_{xi} + 2 \sum_{\mu < x_i \le a} (a - x_i) f_{xi}$$

$$= E(|X-\mu|) + (a-\mu) F_x(\mu) + (\mu - a)(1 - F_x(\mu)) + 2 \sum_{\mu < x_i \le a} (a - x_i) f_{xi}$$

$$= E(|X-\mu|) + 2 \sum_{\mu < x_i \le a} (a - x_i) f_{xi}, \because F_x(\mu) = \tfrac{1}{2}$$

$$\ge E(|X-\mu|), \text{ as } \sum_{\mu < x_i \le a} (a - x_i) f_{xi} \ge 0, \text{ since } a - x_i \ge 0 \text{ and } f_{xi} > 0.$$

$\therefore\ E(|X-a|) \ge E(|X-\mu|)$.

Note: $E(|X-\mu|)$ is the measure of dispersion of the distribution of X from the median μ. If $E(|X-\mu|)$ is small, the spectrum points x_i lie close to μ.

6.10 Quartiles

Let $p \in (0,1)$. The quartile of order p is a point ζ_p on the line of probability distribution such that

$$F_x(\zeta_p) = p.$$

In case of continuous distribution, ζ_p exists uniquely.

In discrete case, the definition is taken as

$$F_x(\zeta_p - 0) \le p \text{ and } F_x(\zeta_p) \ge p.$$

When $F_x(x) = p$ in $x_i \le x < x_{i+1}, \zeta_p = \frac{x_i + x_{i+1}}{2}$.

In particular, $\zeta_{1/4}$ and $\zeta_{3/4}$ are upper and lower quartiles. $\frac{1}{2}(\zeta_{3/4} - \zeta_{1/4})$ is called the semi-interquartile range. This is also treated as a measure of dispersion.

The quartile of order $\frac{k}{10}$ is called a decile of order k. $\therefore$ 5th decile of the median.

The quartile of order $\frac{k}{100}$ is called a percentile of order k.

$\therefore$ percentile of order 25, 50 and 75 are respectively the lower quartile, median and the upper quartile.

Percentile of order 10 is the decile of 1, percentile of order 20 is the decile of order 2, etc.

Distribution will be prominent if we are given decile of all orders and will be much more prominent if we are given percentile of all orders.

Already we have obtained two measures of central tendency, viz., the mean and median. Now, we are going to have another measure of central tendency, viz., the mode.

6.11 Mode

For a discrete distribution the spectrum point x_r will be a mode if $f_{r-1} \le f_r \ge f_{r+1}$, where $f_k - p(X - x_k)$. According to this definition, there may have more than one mode. When there is only one mode the distribution is called *unimodal*, for two modes the distribution is bimodal and for more than two it is multimodal. If $f_{r-1} < f_r = f_{r+1} > f_{r+2}, f_r, f_{r+1}$ are both modes, etc.

For a continuous distribution, a point within the range of the RV X is a mode if the pdf f_x is maximum there. That is $x = a$ is a mode if $f'(a) = 0$, (where $f'(x)$ exists at $x = a$) and $f''(a) < 0$ or in general $f'(a) = f''(a) = \cdots = f^{n-1}(a) = 0$ and $f^n(a) < 0$, when n is a positive even integer.

In a continuous case also, the distribution may be unimodal, bimodal or multimodal. It is important to note that mode may not exist for a distribution, discrete or continuous [follows from definition].

For a unimodal distribution, the mode M may be taken as a measure of central tendency of the distribution. For a symmetrical unimodal distribution with mode M and with mean m, it can be shown that $m-M=0$, i.e., $M=m$. So, $m-M$ or rather the dimensionless quantity $\frac{m-M}{\sigma}(\sigma=\text{SD})$ may be used as a measure of skewness of the distribution.

1. Mode of binomial (n,p) distribution

According to definition, a point $x_r = r$ of the spectrum is a mode if $f_{r-1} \le f_r \ge f_{r+1}$, where $f_k = P(X=x_k)$.

$$\therefore \binom{n}{r-1}p^{r-1}q^{n-r+1} \le \binom{n}{r}p^r q^{n-r} \ge \binom{n}{r+1}p^{r+1}q^{n-r-1}, q=1-p$$

$$\text{or, } \frac{n!}{(r-1)!\,(n-r+1)!}p^{r-1}q^{n-r+1} \le \frac{n!}{r!\,(n-r)!}p^r q^{n-r} \ge \frac{n!}{(r+1)!\,(n-r-1)!}p^{r+1}q^{n-r-1}.$$

Taking the first two members:

$$\frac{q}{n-r+1} \le \frac{p}{r} \text{ or, } qr \le np-pr+p \text{ or, } r \le np+p, \because p+q=1.$$

Taking the last two members:

$$\frac{q}{n-r} \ge \frac{p}{r+1} \text{ or, } qr+q \ge np-pr \text{ or, } r \ge np-q.$$

$\therefore x_r$ will be a mode provided r satisfies

$$np-q \le r \le np+p \text{ or, } (n+1)p-1 \le r \le (n+1)p. \tag{1}$$

As $(n+1)p$ and $(n+1)p-1$ differ by 1, so a unique value of r (positive integer) is there to satisfy (1), when $(n+1)p$ is not an integer and so the distribution is unimodal and $x_r = r$ is the mode, when r satisfies (1).

When $(n+1)p$ is a positive integer, $(n+1)p-1$ is a nonnegative integer, then the distribution is not unimodal; $x_{(n+1)p-1}$ and $x_{(n+1)p}$ are the modes.

2. Mode of Poisson (μ) distribution

According to definition, $x_r = r$ will be a mode, if

$$f_{r-1} \le f_r \ge f_{r+1} \text{ or, } e^{-\mu}\frac{\mu^{r-1}}{(r-1)!} \le e^{-\mu}\frac{\mu^r}{r!} \ge e^{-\mu}\frac{\mu^{r+1}}{(r+1)!} \text{ or, } 1 \le \frac{\mu}{r} \ge \frac{\mu^2}{(r+1)r} \Rightarrow \mu \ge r \text{ and } \mu \le r+1,$$

$$\therefore \mu-1 \le r \le \mu. \tag{2}$$

When μ is not an integer, only one value (positive integer) of r satisfies (1) and $x_r = r$ is the mode, where $r=[\mu]$, integral part of μ.

When μ is an integer (positive), there are two modes, viz., $\mu-1$ and μ.

3. Mode of Gamma (l) distribution

$$\text{Here the pdf is } f_x(x) = \begin{cases} \frac{e^{-x}x^{l-1}}{\Gamma(l)}, & 0<x<\infty \\ 0, & \text{elsewhere.} \end{cases}$$

$$\therefore f_x'(x) = \frac{1}{\Gamma(l)}\left\{-e^{-x}x^{l-1}+(l-1)e^{-x}x^{l-2}\right\} = \frac{e^{-x}x^{l-2}}{\Gamma(l)}(l-1-x).$$

$\therefore\ f_x'(x)=0 \Rightarrow x=l-1$ and $x=0$, when $l>2$.

$$f_x''(x)=\frac{1}{\Gamma(l)}\left\{-e^{-x}x^{l-2}(l-1-x)+(l-2)e^{-x}x^{l-3}(l-1-x)-e^{-x}x^{l-2}\right\}$$

$$=\frac{e^{-x}x^{l-3}}{\Gamma(l)}\{-x(l-1-x)+(l-2)(l-1-x)-1\}=\frac{e^{-x}x^{l-3}}{\Gamma(l)}\{(l-1-x)(l-2-x)-1\}.$$

$$\therefore\ f_x''(l-1)=-\frac{e^{-(l-1)}(l-1)^{l-3}}{\Gamma(l)}<0.$$

$\therefore\ (l-1)$ is a mode of the distribution.

At $x=0, f_X(x)=0$ for $l>2$ and so $f_X(x)$ cannot be maximum at $x=0$. $\therefore$ the distribution is unimodal and $(l-1)$ is the only mode.

4. Mode of normal (m,σ) distribution

Here the pdf is $f_X(x)=\frac{1}{\sqrt{2\pi}\sigma}e^{-\frac{(x-m)^2}{2\sigma^2}}$, $-\infty<x<\infty$.

$\therefore\ f_x'(x)=\frac{1}{\sqrt{2\pi}\sigma}e^{-\frac{(x-m)^2}{2\sigma^2}}\left[-\frac{x-m}{\sigma^2}\right]$ exists $\forall\ x\in(-\infty,\infty)$. $\therefore\ f_x'(x)=0$, only when $x=m$.

$$f_x''(x)=\frac{1}{\sqrt{2\pi}\sigma}e^{-\frac{(x-m)^2}{2\sigma^2}}\left[-\frac{x-m}{\sigma^2}\right]^2-\frac{1}{\sqrt{2\pi}\sigma^3}e^{-\frac{(x-m)^2}{2\sigma^2}}.\ \therefore\ f_x''(m)=-\frac{1}{\sqrt{2\pi}\sigma^3}<0.$$

$\therefore\ f_X(x)$ attains local maximum only at $x=m$. $\therefore\ m$ is the only mode of the normal (m,σ) distribution. Hence, the normal (m,σ) is unimodal.

Note: As this distribution is symmetrical about $x=m$, $\therefore$ median = mode = mean = m.

Ex. 5. Prove that first absolute moment about the mean cannot exceed the standard deviation for any distribution.

Solution. Let X be a RV.

$$\therefore\ \text{var}(X)=E(X-m_x)^2,\ m_x=E(X)=\text{mean of } X$$
$$=E\left(X^2\right)-2m_xE(X)+m_x^2=E\left(X^2\right)-2m_x^2+m_x^2$$
$$=E\left(X^2\right)-m_x^2=E\left(X^2\right)-\{E(X)\}^2.$$

For any distribution of X, $\text{var}(X)\geq 0$. $\therefore\ E\left(X^2\right)\geq\{E(X)\}^2$ for every distribution.

Let us now apply the above inequality for the RV $|X-m_x|$. The inequality for the RV $|X-m_x|$ becomes

$$E\left(|X-m_x|^2\right)\geq\{E(|X-m_x|)\}^2$$
$$\text{or, } E(X-m_x)^2\geq\{E(|X-m_x|)\}^2$$
$$\text{or, } \{E(|X-m_x|)\}^2\leq\sigma_x^2,\ \text{where } \sigma_x=\text{SD of } X$$
$$\text{or, } E(|X-m_x|)\leq\sigma_x,\ \text{as } \sigma_x \text{ and } E(|X-m_x|) \text{ both nonnegative.}$$

Hence, the result.

6. Find the mode and median of the continuous distribution given by the following pdf f_x :

$$f_x(x)=\begin{cases}\sin x, & 0\leq x\leq\frac{\pi}{2}\\ 0, & \text{elsewhere.}\end{cases}$$

Find also its variance.

Solution. f_x is everywhere differentiable. $\therefore$ for a mode, if exist, we have $f_x'(x)=0$, i.e., $\cos x=0\Rightarrow x=\frac{\pi}{2}$.

$$f_x''(x)\Big]_{x=\pi/2}=-\sin x\Big]_{x=\pi/2}=-1<0.\ \therefore\ f_x \text{ is maximum at } x=\pi/2.$$

$\therefore$ the distribution is unimodal and mode is $\frac{\pi}{2}$.

If λ be the median, then $\int_{-\infty}^{\lambda} f_X(x)\,dx = \frac{1}{2}$, i.e., $\int_0^{\lambda} \sin x\,dx = \frac{1}{2}$ or, $1-\cos\lambda = \frac{1}{2}$. $\therefore$ $\cos\lambda = \frac{1}{2} \Rightarrow \lambda = \frac{\pi}{3}$.

$$\text{Mean} = E(X) = \int_0^{\pi/2} x\sin x\,dx = -x\cos x\Big|_0^{\pi/2} + \int_0^{\pi/2} \cos x\,dx = \sin\frac{\pi}{2} - \sin 0 = 1.$$

$$\therefore \text{ variance} = E(X-1)^2 = \int_0^{\pi/2} (x-1)^2 \sin x\,dx = -(x-1)^2\cos x\Big|_0^{\pi/2} + 2\int_0^{\pi/2} (x-1)\cos x\,dx$$

$$= 1 + 2\left[(x-1)\sin x\Big|_0^{\pi/2} - \int_0^{\pi/2} \sin x\,dx\right] = 1 + 2\left(\frac{\pi}{2} - 1\right) + 2\Big[\cos x\Big]_0^{\pi/2}$$

$$= \pi - 1 + 2(-1) = \pi - 3.$$

7. Show that the mean deviation from the median is the minimum.

Proof. Let mean deviation be minimum from a point $x = \lambda$. Mean deviation of a distribution from the point λ

$$= E(|X-\lambda|) = \int_{-\infty}^{\infty} |x-\lambda| f_X(x)\,dx = \int_{-\infty}^{\lambda} (\lambda - x) f_X(x)\,dx + \int_{\lambda}^{\infty} (x-\lambda) f_X(x)\,dx.$$

$\therefore$ $\frac{d}{d\lambda}E(|X-\lambda|) = \int_{-\infty}^{\lambda} f_X(x)\,dx - \int_{\lambda}^{\infty} f_X(x)\,dx$, by the differential under the sign of integration

$= F_X(\lambda) - (1 - F_X(\lambda))$, where $f_X(x) = F_x'(x)$, F_X is the distribution function

$= 2F_X(\lambda) - 1.$

For extremum value of $E(|X-\lambda|)$, we get $2F_X(\lambda) - 1 = 0$, i.e., $F_X(\lambda) = \frac{1}{2} \Rightarrow \lambda$ is the median.

Again, $\frac{d^2}{d\lambda^2}E(|X-\lambda|) = 2F_\lambda'(\lambda) = 2f_X(\lambda) > 0$.

$\therefore$ $E(|X-\lambda|)$ is minimum, when λ is the median.

8. Find the mean, median and mode of the continuous probability distribution defined by the pdf f_X given by $f_X(x) = \begin{cases} kx(1-x), & 0 \le x \le 1 \\ 0, & \text{elsewhere.} \end{cases}$

Is it a symmetrical distribution?

Solution. First, we find k. We have $\int_{-\infty}^{\infty} f_X(x)\,dx = 1$.

$$\therefore k\int_0^1 x(1-x)\,dx = 1 \text{ or, } \frac{1}{k} = \left[\frac{x^2}{2} - \frac{x^3}{3}\right]_6^1. \therefore k = 6.$$

$$m_X = E(X) = 6\int_0^1 x^2(1-x)\,dx = 6\left[\frac{x^3}{3} - \frac{x^4}{4}\right]_0^1 = 6 \times \frac{1}{12} = \frac{1}{2}.$$

Let λ be the median.

$$\therefore 6\int_0^{\lambda} \left(x - x^2\right)dx = \frac{1}{2} \text{ or, } \frac{1}{12} = \frac{\lambda^2}{2} - \frac{\lambda^2}{3} \text{ or, } \lambda^2(3-2\lambda) = \frac{1}{2} \text{ or, } 6\lambda^2 - 4\lambda^3 - 1 = 0$$

$$\text{or, } 4\lambda^3 - 6\lambda^2 + 1 = 0 \Rightarrow \lambda = \frac{1}{2}, \frac{1}{2}\left(1 \pm \sqrt{3}\right).$$

$\therefore$ median $\lambda = \frac{1}{2}$ as the other values are outside the range of nonzero probability density.

$f_x'(x) = 6(1-2x), f_x''(x) = -12 < 0$. As f_x is differentiable everywhere, $\therefore$ at the mode $f_x'(x) = 0$, i.e., $x = \frac{1}{2}$. Also, $f_x''\left(\frac{1}{2}\right) = -12 < 0$. $\therefore$ f_x is maximum at $x = \frac{1}{2}$. $\therefore$ $\frac{1}{2}$ is the mode.

$\therefore$ mean = median = mode = $\frac{1}{2}$. $\therefore$ the distribution is symmetrical about the point $\frac{1}{2}$.

WORKED-OUT EXAMPLES

Ex. 1. Find the mean and variance of the hypergeometric distribution.

Solution. The hypergeometric distribution is defined by the pmf

$$f_i = P(X=i) = \frac{\binom{M}{i}\binom{N-M}{n-i}}{\binom{N}{n}}, \quad i = 0, 1, 2, \cdots, \min\{n, M\}$$

with the set of parameters $\{N, M, n\} \subset \mathbb{N}$ satisfying $M \le N, n \le N$ [$\mathbb{N}$ = set of natural numbers]

$$\text{Mean} = E(X) = \sum_{i=0}^{n} i \frac{\binom{M}{i}\binom{N-M}{n-i}}{\binom{N}{n}}, \text{ whenever } n \le M$$

$$= \sum_{i=1}^{n} \frac{M\binom{M-1}{i-1}\binom{N-M}{n-i}}{\binom{N}{n}\binom{N-1}{n-1}} = \frac{nM}{N}\sum_{j=0}^{n-1} \frac{\binom{M-1}{j}\binom{N-M}{n-j-1}}{\binom{N-1}{n-1}}, \text{ when } i = j+1$$

$$= \frac{nM}{N} \cdot \frac{1}{\binom{N-1}{n-1}} \sum_{j=0}^{n-1} \binom{M-1}{j}\binom{N-M}{n-j-1} = \frac{nM}{N} \cdot \frac{1}{\binom{N-1}{n-1}} \cdot \binom{N-1}{n-1} = \frac{nM}{N}.$$

When $n > M$,

$$\text{Mean} = \sum_{i=0}^{M} i \frac{\binom{M}{i}\binom{N-M}{n-i}}{\binom{N}{n}} = \sum_{i=1}^{M} M \frac{\binom{M-1}{i-1}\binom{N-M}{n-i}}{\frac{N}{n}\binom{N-1}{n-1}} = \frac{nM}{N} \cdot \frac{1}{\binom{N-1}{n-1}} \sum_{j=0}^{M-1} \binom{M-1}{j}\binom{N-M}{n-j-1}$$

$$= \frac{nM}{N} \cdot \frac{1}{\binom{N-1}{n-1}} \binom{N-1}{n-1} = \frac{nM}{N}.$$

To have the variance we use the formula $\text{var}(X) = E\{X(X-1)\} - E(X)\{E(X)-1\}$.

Now,

$$E\{X(X-1)\} = \sum_{i=0}^{n} i(i-1) \frac{\binom{M}{i}\binom{N-M}{n-i}}{\binom{N}{n}}, \text{ whenever } n \le M$$

$$= \sum_{i=2}^{n} M(M-1) \frac{\binom{M-2}{i-2}\binom{N-M}{n-i}}{\frac{N(N-1)}{n(n-1)}\binom{N-2}{n-2}} = \frac{M(M-1)n(n-1)}{N(N-1)} \sum_{j=0}^{n-2} \frac{\binom{M-2}{j}\binom{N-M}{n-j-2}}{\binom{N-2}{n-2}}, \ i = j+2$$

$$= \frac{n(n-1)M(M-1)}{N(N-1)} \frac{1}{\binom{N-2}{n-2}} \sum_{j=0}^{n-2} \binom{M-2}{j}\binom{N-M}{n-j-2}$$

$$= \frac{n(n-1)M(M-1)}{N(N-1)} \cdot \frac{1}{\binom{N-2}{n-2}} \binom{N-2}{n-2} = \frac{n(n-1)M(M-1)}{N(N-1)}.$$

$$\therefore \text{var} X = \frac{n(n-1)M(M-1)}{N(N-1)} - \frac{nM}{N}\left(\frac{nM}{N} - 1\right) = \frac{nM}{N}\left\{\frac{(n-1)(M-1)}{N-1} - \frac{nM-N}{N}\right\}$$

$$= \frac{nM}{N} \frac{(n-N)(M-N)}{N(N-1)}.$$

Also, $\operatorname{var} X = \frac{nM}{N} \cdot \frac{(n-N)(M-N)}{N(N-1)}$, when $n > M$.

$$\therefore \sigma(X) = \sqrt{\frac{nM}{N} \cdot \frac{(n-N)(M-N)}{N(N-1)}}.$$

Ex. 2. Show that the first absolute moment about the mean for normal (m, σ) distribution is $\sqrt{\frac{2}{\pi}}\sigma$.[CH. 07]

Solution. First absolute moment about the mean m

$$= E(|X-m|) = \int_{-\infty}^{\infty} |x-m| \cdot \frac{1}{\sqrt{2\pi}\sigma} e^{-\frac{(x-m)^2}{2\sigma^2}} dx$$

$$= -\int_{-\infty}^{m} (x-m) \cdot \frac{1}{\sqrt{2\pi}\sigma} e^{-\frac{(x-m)^2}{2\sigma^2}} dx + \int_{m}^{\infty} (x-m) \frac{1}{\sqrt{2\pi}\sigma} e^{-\frac{(x-m)^2}{2\sigma^2}} dx$$

Put $\frac{(x-m)^2}{2\sigma^2} = u$, $\therefore \frac{2(x-m)}{2\sigma^2} dx = du$. $\therefore (x-m)dx = \sigma^2 du$, $u \to \infty$ as $x \to \pm\infty$

$$= -\int_{\infty}^{0} \frac{1}{\sqrt{2\pi}\sigma} e^{-u} \cdot \sigma^2 du + \int_{0}^{\infty} \frac{1}{\sqrt{2\pi}\sigma} e^{-u} \cdot \sigma^2 du = 2\int_{0}^{\infty} \frac{\sigma}{\sqrt{2\pi}} e^{-u} du = \sqrt{\frac{2}{\pi}}\sigma[-e^{-u}]_0^{\infty}$$

$$= \sqrt{\frac{2}{\pi}}\sigma.$$

Ex. 3. A point P is chosen at random on a line segment AB of length $2b$. Find the expected values of (i) the area of the rectangle with AP and PB as adjacent sides, (ii) the absolute difference of AP and PB and (iii) $\min\{AP, PB\}$.

[Recall of coordinate in one dimension]:

Direct line. A straight line is called a *directed line* when one direction is considered as positive and the other as negative direction.

Directed line segment. Any two points on a directed line constitute what is known as a *direct line segment.* In Fig 6.6 $\overline{AB}$ is a directed line segment.

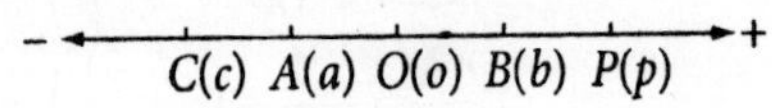

Fig 6.6

Value of a directed line segment. The direction of $\overline{AB}$ is from A towards B. If the direction coincides with the positive direction of the directed line, then the value of $\overline{AB}$ is denoted by AB and is defined by $AB = |\overline{AB}|$ = length of the segment. On the other hand, if the direction coincides with the negative direction of directed line then, $AB = -|\overline{AB}|$.

Identity of values. For any three points A, B, C on a directed line we have $AB + BC = AC$, which is the identity of values.

Coordinate in one dimension. Take a point O on a directed line g as origin. Let P be a point on g, then $OP(=p)$ is the coordinate of P.

Distance between two points on a directed line. Let A and B be two points on the directed line g and a, b be respectively their coordinates, then distance between A and B is $|a-b|$]

Solution. Let O, the midpoint of $\overline{AB}$, as origin of coordinate. Let P be a point chosen at random on $\overline{AB}$ with coordinate x. Let X be the RV denoting the value OP, i.e., x is the real value of X. X follows uniform distribution in $(-b, b)$.

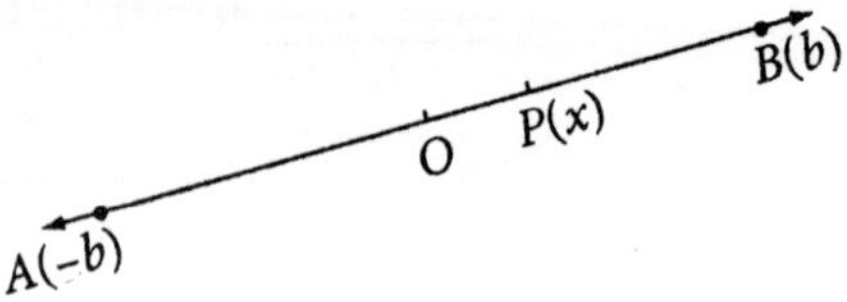

Fig 6.7

$$\therefore f_X(x) = \begin{cases} \frac{1}{2b}, & -b < x < b \\ 0, & \text{elsewhere.} \end{cases}$$

(i) Area of the rectangle with AP and PB as adjacent sides $= AP \cdot PB$. We are to find $E(AP \cdot PB)$.

$$E(AP \cdot PB) = E\{(X+b)(b-X)\} = \frac{1}{2b}\int_{-b}^{b}(x+b)(b-x)\,dx = \frac{1}{2b}\int_{-b}^{b}\left(b^2-x^2\right)dx$$

$$= \frac{1}{2b}\left[b^2x - \frac{x^3}{3}\right]_{-b}^{b} = \frac{1}{2b}\left(2b^3 - \frac{2b^3}{3}\right) = b^2 - \frac{1}{3}b^2 = \frac{2b^2}{3}.$$

(ii) Absolute difference of AP and PB is $|AP - PB|$.

$$E(|AP-PB|) = E(|X+b-b+X|) = E(|2X|) = \frac{1}{2b}\int_{-b}^{b} 2|x|\,dx = \frac{1}{b}\left[\int_{-b}^{0}(-x)\,dx + \int_{0}^{b} x\,dx\right]$$

$$= \frac{1}{b}\left[-\frac{x^2}{2}\Big|_{-b}^{0} + \frac{x^2}{2}\Big|_{0}^{b}\right] = b.$$

(iii) $$E[\min\{AP, PB\}] = E[\min\{X+b, b-X\}] = \frac{1}{2b}\int_{-b}^{b}\min\{x+b, b-x\}\,dx$$

$$= \frac{1}{2b}\int_{-b}^{0}\min\{x+b, b-x\}\,dx + \frac{1}{2b}\int_{0}^{b}\min\{x+b, b-x\}\,dx$$

$$= \frac{1}{2b}\int_{-b}^{0}(x+b)\,dx + \frac{1}{2b}\int_{0}^{b}(b-x)\,dx = \frac{1}{2b}\left[\frac{x^2}{2}+bx\right]_{-b}^{0} + \frac{1}{2b}\left[bx - \frac{x^2}{2}\right]_{0}^{b}$$

$$= \frac{1}{2b}\left(-\frac{b^2}{2}+b^2\right) + \frac{1}{2b}\left(b^2 - \frac{b^2}{2}\right) = 2\cdot\frac{1}{2b}\frac{1}{2}b^2 = \frac{b}{2}.$$

Ex. 4. Find the mean and variance of the discrete distribution: $x_i = i, i = 0, 1, 2, \cdots$ and $f_{xi} = P(X = x_i) = P(X = i) = \frac{1}{1+\lambda}\left(\frac{\lambda}{1+\lambda}\right)^i, \lambda (> 0)$ is the only parameter. [This is known as *Pascal distribution*.]

Solution. $m_x = E(X) = \sum_{i=0}^{\infty} i \frac{1}{1+\lambda}\left(\frac{\lambda}{1+\lambda}\right)^i = \frac{1}{1+\lambda}\sum_{i=1}^{\infty} i\left(\frac{\lambda}{1+\lambda}\right)^i = \frac{1}{1+\lambda}\left(k+2k^2+3k^3+\cdots\right)$,

when $k = \dfrac{\lambda}{1+\lambda}$.

$$\text{Let} \quad s = k + 2k^2 + 3k^3 + 4k^4 + \cdots$$

$$sk = \quad k^2 + 2k^3 + 3k^4 + \cdots \tag{1}$$

Subtracting, $s(1-k) = k + k^2 + k^3 + \cdots = \frac{k}{1-k}$ as $k < 1$

$$\therefore s = \frac{k}{(1-k)^2} = \frac{\lambda}{1+\lambda}(1+\lambda)^2 = \lambda(1+\lambda). \; \therefore m_x = \frac{1}{1+\lambda}\lambda(1+\lambda) - \lambda.$$

$$E\{X(X-1)\} = \frac{1}{1+\lambda}\sum_{i=2}^{\infty} i(i-1)\left(\frac{\lambda}{1+\lambda}\right)^i$$

$$= \frac{1}{1+\lambda}\left[2\cdot1\cdot\left(\frac{\lambda}{1+\lambda}\right)^2 + 3\cdot2\cdot\left(\frac{\lambda}{1+\lambda}\right)^3 + 4\cdot3\cdot\left(\frac{\lambda}{1+\lambda}\right)^4 + \cdots \text{ to } \infty\right]$$

$$= \frac{2\lambda^2}{(1+\lambda)^3}\left[1 + 3\frac{\lambda}{1+\lambda} + 6\left(\frac{\lambda}{1+\lambda}\right)^2 + 10\left(\frac{\lambda}{1+\lambda}\right)^3 + \cdots \text{ to } \infty\right]$$

$$\text{Let} \quad s' = 1+3k+6k^2+10k^3+\cdots, \text{when } k = \tfrac{\lambda}{1+\lambda}$$

$$ks' = \quad k+3k^2+6k^3+\cdots$$

$$\text{Subtracting, } (1-k)s' = 1+2k+3k^2+4k^3+\cdots \text{ to } \infty$$

$$= \frac{s}{k} \text{ [by (1)]} = \frac{\lambda(1+\lambda)}{\lambda/(1+\lambda)} = (1+\lambda)^2 .$$

$$\therefore \left(1-\frac{\lambda}{1+\lambda}\right)s' = (1+\lambda)^2. \therefore s' = (1+\lambda)^3. \therefore E\{X(X-1)\} = \frac{2\lambda^2}{(1+\lambda)^3}\cdot(1+\lambda)^3 = 2\lambda^2.$$

$$\therefore \operatorname{var} X = E\{X(X-1)\} - E(X)\{E(X)-1\} = 2\lambda^2 - \lambda(\lambda-1) = \lambda^2+\lambda = \lambda(\lambda+1).$$

Ex. 5. The distribution of a continuous RV X defined by the pdf: $f_X(x) = \begin{cases} \frac{1}{\sqrt{2\pi}x} e^{-\frac{1}{2}(\ln x)^2}, & x>0 \\ 0, & x<0 \end{cases}$ is called ln-normal distribution. Find the mean, mode, variance and coefficient of skewness, coefficient of kurtosis and coefficient of excess of kurtosis.

Solution.

$$\text{Mean} = E(X) = \frac{1}{\sqrt{2\pi}}\int_0^\infty e^{-\frac{1}{2}(\ln x)^2}\,dx$$

Putting $\frac{\ln x}{\sqrt{2}} = u$, $dx = \sqrt{2}e^{\sqrt{2}u}\,du$. As $x \to 0+, u \to -\infty$ and as $x \to \infty, u \to \infty$

$$= \frac{1}{\sqrt{2\pi}}\int_{-\infty}^{\infty} e^{-u^2}\sqrt{2}e^{\sqrt{2}u}\,du = \frac{1}{\sqrt{\pi}}\int_{-\infty}^{\infty} e^{-\left(u-\frac{1}{\sqrt{2}}\right)^2+\frac{1}{2}}\,du = \frac{\sqrt{e}}{\sqrt{\pi}}\int_{-\infty}^{\infty} e^{-v^2}\,dv, \text{ when } v = u-\tfrac{1}{\sqrt{2}}$$

$$= \frac{2\sqrt{e}}{\sqrt{\pi}}\int_0^\infty e^{-v^2}\,dv, \text{ as integrand is an even function}$$

$$= \frac{2\sqrt{e}}{\sqrt{\pi}}\cdot\frac{\sqrt{\pi}}{2} = \sqrt{e}.$$

$$\text{Now, } f'_X(x) = -\frac{1}{x^2}e^{-\frac{1}{2}(\ln x)^2} + \frac{1}{x}e^{-\frac{1}{2}(\ln x)^2}\left[-\frac{1}{2}\cdot 2\ln x\cdot\frac{1}{x}\right] = -\frac{1}{x^2}e^{-\frac{1}{2}(\ln x)^2}(1+\ln x), \text{ exists } \forall\, x>0.$$

$\therefore$ at a mode $f'_X(x) = 0 \Rightarrow 1+\ln x = 0 \Rightarrow x = \frac{1}{e}$.

$$f'_X\left(\tfrac{1}{e}-0\right) > 0 \text{ and } f'_X\left(\tfrac{1}{e}+0\right) < 0.$$

$\therefore$ $f_X(x)$ is maximum, when $x = \frac{1}{e}$. Hence, mode $= \frac{1}{e}$.

$$E\left(X^2\right) = \frac{1}{\sqrt{2\pi}}\int_0^\infty x e^{-\frac{1}{2}(\ln x)^2}\,dx$$

Putting $\frac{\ln x}{\sqrt{2}} = u, dx = \sqrt{2}e^{\sqrt{2}u}\,du$. $\therefore$ $u \to -\infty$ as $x \to 0+$ and $u \to \infty$ as $x \to \infty$

$$= \frac{1}{\sqrt{2\pi}}\int_{-\infty}^{\infty} e^{\sqrt{2}u}e^{-u^2}\sqrt{2}e^{\sqrt{2}u}\,du = \frac{1}{\sqrt{\pi}}\int_{-\infty}^{\infty} e^{2\sqrt{2}u-u^2}\,du$$

$$= \frac{1}{\sqrt{\pi}}\int_{-\infty}^{\infty} e^{-(u-\sqrt{2})^2+2}\,du = \frac{e^2}{\sqrt{\pi}}\int_{-\infty}^{\infty} e^{-v_1^2}\,dv_1, \text{ when } v_1 = u-\sqrt{2}$$

$$= \frac{2e^2}{\sqrt{\pi}}\int_0^\infty e^{-v_1^2}\,dv_1, \text{ as integrand is an even function.}$$

$$= \frac{2e^2}{\sqrt{2\pi}}\cdot\frac{\sqrt{\pi}}{2} = e^2.$$

$\therefore$ var$X = E(X^2) - \{E(X)\}^2 = e^2 - (\sqrt{e})^2 = e^2 - e = e(e-1)$. $\therefore$ SD $= \sigma(X) = \sqrt{e}\sqrt{e-1}$.

$$\mu_3(X) = E\left(X - \sqrt{e}\right)^3 = E\left(X^3\right) - 3\sqrt{e}E\left(X^2\right) + 3eE(X) - e\sqrt{e}$$
$$= E\left(X^3\right) - 3\sqrt{e}\,e^2 + 2e\sqrt{e}, \because E(X) = \sqrt{e} \text{ and } E\left(X^2\right) = e^2.$$

$$E\left(X^3\right) = \frac{1}{\sqrt{2\pi}}\int_0^\infty x^2 e^{-\frac{1}{2}(\ln x)^2}\,dx = \frac{1}{\sqrt{2\pi}}\int_{-\infty}^\infty e^{2\sqrt{2}u}\,e^{-u^2}\sqrt{2}\,e^{\sqrt{2}u}\,du, \text{ when } u = \frac{\ln x}{\sqrt{2}}$$
$$= \frac{1}{\sqrt{\pi}}\int_{-\infty}^\infty e^{-\left(u^2 - 3\sqrt{2}u\right)}\,du = \frac{1}{\sqrt{\pi}}\int_{-\infty}^\infty e^{-\left(u-\frac{3}{\sqrt{2}}\right)^2 + 9/2}\,du = 2\frac{e^{9/2}}{\sqrt{\pi}}\int_0^\infty e^{-v_2^2}\,dv_2,\ v_2 = u - \frac{3}{\sqrt{2}}$$
$$= \frac{2e^{9/2}}{\sqrt{\pi}}\cdot\frac{\sqrt{\pi}}{2} = e^{9/2}.$$

$\therefore \mu_3(X) = e^{9/2} - 3\sqrt{e}\,e^2 + 2e\sqrt{e} = e\sqrt{e}\left(e^3 - 3e + 2\right)$.

$\therefore$ coefficient of skewness $= \dfrac{\mu_3(X)}{\sigma^3(X)} = \dfrac{e\sqrt{e}\left(e^3 - 3e + 2\right)}{\sqrt{e}\sqrt{e-1}\cdot e(e-1)} = \dfrac{(e-1)^2(e+2)}{(e-1)\sqrt{e-1}} = (e+2)\sqrt{e-1}$.

$$\mu_4(X) = E\left(X - \sqrt{e}\right)^4 = E\left(X^4\right) - 4\sqrt{e}E\left(X^3\right) + 6eE\left(X^2\right) - 4e\sqrt{e}E(X) + e^2.$$

$$E\left(X^4\right) = \frac{1}{\sqrt{2\pi}}\int_0^\infty x^3 e^{-\frac{1}{2}(\ln x)^2}\,dx = \frac{1}{\sqrt{2\pi}}\int_{-\infty}^\infty e^{3\sqrt{2}u}\,e^{-u^2}\sqrt{2}\,e^{\sqrt{2}u}\,du, \text{ when } u = \frac{\ln x}{\sqrt{2}}$$
$$= \frac{1}{\sqrt{\pi}}\int_{-\infty}^\infty e^{-\left(u^2 - 4\sqrt{2}u\right)}\,du = \frac{1}{\sqrt{\pi}}\int_{-\infty}^\infty e^{-\left(u - 2\sqrt{2}\right)^2 + 8}\,du = \frac{2e^8}{\sqrt{\pi}}\int_0^\infty e^{-v_3^2}\,dv_3,\ v_3 = u - 2\sqrt{2}$$
$$= \frac{2e^8}{\sqrt{\pi}}\frac{\sqrt{\pi}}{2} = e^8.$$

$\therefore \mu_4(X) = e^8 - 4\sqrt{e}\cdot e^{9/2} + 6e\cdot e^2 - 4e\sqrt{e}\cdot\sqrt{e} + e^2 = e^8 - 4e^5 + 6e^3 - 4e^2 + e^2 = e^2\left(e^6 - 4e^3 + 6e - 3\right)$.

$\therefore$ coefficient of kurtosis $\beta_2 = \dfrac{\mu_4(X)}{\sigma^4(X)} = \dfrac{e^2\left(e^6 - 4e^3 + 6e - 3\right)}{e^2(e-1)^2} = e^4 + 2e^3 + 3e^2 - 3$.

$\therefore$ coefficient of excess (of kurtosis) $\gamma_2 = \beta_2 - 3 = e^4 + 2e^3 + 3e^2 - 6 = (e-1)\left(e^3 + 3e^2 + 6e + 6\right)$.
Note that $\gamma_2 > 0$.

Ex. 6. If X is a binomial (n, p) variate, then prove that $\mu_{k+1} = p(1-p)\left(nk\mu_{k-1} + \frac{d\mu_k}{dp}\right)$, where μ_k is the kth central moment. [C.H. 2007]

Solution. The pmf for the binomial (n, p) distribution is

$$f_{xi} = P(X = i) = \tbinom{n}{i}p^i q^{n-i}, (q = 1-p) \text{ for } i = 0, 1, 2, \cdots, n.$$

$$\therefore \mu_k = \mu_k(X) = E(X - np)^k, \because E(X) = np$$
$$= \sum_{i=0}^n (i - np)^k \binom{n}{i} p^i q^{n-i}.$$

$$\therefore \frac{d\mu_k}{dp} = \sum_{i=0}^n \frac{d}{dp}\left\{(i - np)^k \binom{n}{i} p^i (1-p)^{n-i}\right\}$$
$$= \sum_{i=0}^n \left\{k(-n)(i - np)^{k-1}\binom{n}{i}p^i(1-p)^{n-i} + (i - np)^k\binom{n}{i} i p^{i-1}(1-p)^{n-i}\right.$$
$$\left. + (i - np)^k\binom{n}{i}p^i(i-n)(1-p)^{n-i-1}\right\}$$

$$=-nk\sum_{i=0}^{n}(i-np)^{k-1}\binom{n}{i}p^i(1-p)^{n-i}+\sum_{i=0}^{k}(i-np)^k\binom{n}{i}p^i(1-p)^{n-i}\left\{\frac{i}{p}-\frac{n-i}{1-p}\right\}$$

$$=-nk\mu_{k-1}+\sum_{i=0}^{k}(i-np)^{k+1}\binom{n}{i}p^i(1-p)^{n-i}/p(1-p)$$

$$=-nk\mu_{k-1}+\frac{1}{p(1-p)}\mu_{k+1}.$$

$$\therefore\ \mu_{k+1}=p(1-p)\left[nk\mu_{k-1}+\frac{d\mu_k}{dp}\right]\ \textbf{(Proved).}$$

Ex. 7. Let λ and μ_r denote the mean and rth moment about the mean respectively of a Poisson distribution. Obtain the following recurrence relation: $\mu_{r+1}=\lambda\left[r\mu_{r-1}+\frac{d\mu_r}{d\lambda}\right]$. [C.H. 2005]

Solution. Let X follow Poisson distribution with mean λ. So the distribution is given by

$$f_{xi}=P(X=i)=e^{-\lambda}\frac{\lambda^i}{i!},\ i=0,1,2,\cdots$$

$$\therefore\ \mu_r=\mu_r(X)=E(X-\lambda)^r=\sum_{i=0}^{\infty}(i-\lambda)^r\frac{e^{-\lambda}\lambda^i}{i!},$$

where the infinite sum is absolutely convergent.

$$\therefore\ \frac{d\mu_r}{d\lambda}=\sum_{i=0}^{\infty}\frac{d}{d\lambda}\left[(i-\lambda)^r\frac{e^{-\lambda}\lambda^i}{i!}\right],$$ the process of term by term differentiation holds as the series converges absolutely.

$$=\sum_{i=0}^{\infty}\left\{-r(i-\lambda)^{r-1}e^{-\lambda}\frac{\lambda^i}{i!}+(i-\lambda)^r\cdot\frac{-e^{\lambda}\lambda^i}{i!}+(i-\lambda)^r\frac{e^{-\lambda}i\lambda^{i-1}}{i!}\right\}$$

$$=-r\mu_{r-1}+\sum_{i=0}^{\infty}\frac{(i-\lambda)^r}{i!}e^{-\lambda}\cdot\lambda^{i-1}(i-\lambda)=-r\mu_{r-1}+\frac{1}{\lambda}\sum_{i=0}^{\infty}(i-\lambda)^{r+1}\frac{e^{-\lambda}\lambda^i}{i!}$$

$$=-r\mu_{r-1}+\frac{1}{\lambda}\mu_{r+1}.$$

$$\therefore\ \mu_{r+1}=\lambda\left(\frac{d\mu_r}{d\lambda}+r\mu_{r-1}\right)\ \textbf{(proved).}$$

Ex. 8. If X is a normal (m,σ) variate, prove that $\mu_{2(r+1)}=\sigma^2\mu_{2r}+\sigma^3\frac{d\mu_{2r}}{d\sigma}$.

Solution. X is normal (m,σ).

$$\therefore\ \mu_{2r}=\mu_{2r}(X)=E(X-m)^{2r}$$

$$\text{or, }\mu_{2r}=\int_{-\infty}^{\infty}(x-m)^{2r}\frac{1}{\sqrt{2\pi}\sigma}e^{-\frac{(x-m)^2}{2\sigma^2}}dx,$$

where the integral converges absolutely.

$$\therefore\ \frac{d\mu_{2r}}{d\sigma}=\int_{-\infty}^{\infty}(x-m)^{2r}\left\{-\frac{1}{\sqrt{2\pi}\sigma^2}e^{-\frac{(x-m)^2}{2\sigma^2}}+\frac{1}{\sqrt{2\pi}\sigma}e^{-\frac{(x-m)^2}{2\sigma^2}}\cdot\frac{(x-m)^2}{\sigma^3}\right\}dx,$$

(assuming the validity of differentiation under the sign of integration)

$$\therefore\ \sigma^3\frac{du_{2r}}{d\sigma}=-\sigma^2\int_{-\infty}^{\infty}(x-m)^{2r}\cdot\frac{1}{\sqrt{2\pi}\sigma}e^{-\frac{(x-m)^2}{2\sigma^2}}dx+\int_{-\infty}^{\infty}(x-m)^{2r+2}\cdot\frac{1}{\sqrt{2\pi}\sigma}e^{-\frac{(x-m)^2}{2\sigma^2}}dx$$

$$=-\sigma^2\mu_{2r}+\mu_{2(r+1)}$$

$$\therefore\ \mu_{2(r+1)}=\sigma^2\mu_{2r}+\sigma^3\frac{d\mu_{2r}}{d\sigma},\ \textbf{proved.}$$

Ex. 9. Find the median and mode of a distribution having the pdf $\lambda e^{-\lambda x}(x>0)$. [C.H. 2006]

Solution. Obviously, $\lambda > 0$ as density function ≥ 0. Let μ be the median.

$$\therefore\ \lambda\int_0^{\mu} e^{-\lambda x}\,dx=\frac{1}{2} \text{ or, } -\left[e^{-\lambda x}\right]_0^{\mu}=\frac{1}{2} \text{ or, } 1-e^{-\lambda\mu}=\frac{1}{2}.\ \therefore\ e^{-\lambda\mu}=\frac{1}{2}.$$

$$\therefore\ -\lambda\mu=\ln\frac{1}{2}=-\ln 2.\ \therefore\ \mu=\frac{\ln 2}{\lambda}.$$

Let $f_x(x)=\begin{cases}\lambda e^{-\lambda x}, & x>0\ [\lambda>0]\\ 0, & \text{elsewhere.}\end{cases}$

$\therefore\ f_x'(x)=-\lambda^2 e^{-\lambda x}$ exists for all $x>0$. But, $f_x'(x)$ never vanishes as $f_x'(x)<0,\ \forall\ x>0$. $\therefore\ f_x$ has no local maxima for $x>0$. Again, $f_x(x)=0,\ \forall\ x\leq 0$. So, f_x has no local maxima for $x<0$. Again, $f_x(0)=0,\ f_x(x)>0$ for $x>0$.

$\therefore$ there does not exist any interval $(-\delta,\delta),(\delta>0)$ such that $f_x(0)>f_x(x),\ \forall\ x\in(-\delta,\delta)$. $\therefore\ x=0$ is also not a point of local maxima. $\therefore$ the distribution has no mode.

Ex. 10. A RV X is distributed with probability density $0.5\cos x$ for $x\in\left(-\frac{\pi}{2},\frac{\pi}{2}\right)$. Find the mean value and variance of the RV $y=\sin X$. [C.H. 2003]

Solution. Given: $f_x(x)=\begin{cases}\frac{1}{2}\cos x, & -\frac{\pi}{2}<X<\frac{\pi}{2}\\ 0, & \text{elsewhere.}\end{cases}$

First, we find the distribution of $Y(=\sin X)$.

In real values of X and Y we have $y=\sin x$.

$$\therefore\ \frac{dy}{dx}=\cos x>0\ \forall\ x\in\left(-\frac{\pi}{2},\frac{\pi}{2}\right)$$

$\therefore$ by the law of transformation of RV the pdf f_y of y is given by

$$f_y(y)=f_x(x)\left|\frac{dx}{dy}\right|=\frac{1}{2}\cos x\cdot\frac{1}{\cos x}=\frac{1}{2},\ -1<y<1.$$

$\therefore\ y$ is uniformly distributed in $(-1,1)$.

$\therefore$ mean value $=E(Y)=\int_{-1}^{1}\frac{1}{2}y\,dx=\frac{1}{4}\left[y^2\right]_{-1}^{1}=0$ and variance $=E\{(Y-E(Y))\}^2=E\left(Y^2\right)=\frac{1}{2}\int_{-1}^{1}y^2\,dy=\frac{1}{6}\left[y^3\right]_{-1}^{1}=\frac{1}{3}.$

Ex. 11. The pdf of a RV X is given by

$$f_x(x)=ce^{-(x^2+2x+3)},-\infty<x<\infty.$$

Find the value of the constant c, the expectation and variance of the distribution. [C.H. 2000]

Solution. We have $\int_{-\infty}^{\infty}f_x(x)\,dx=1.\ \therefore\ c\int_{-\infty}^{\infty}e^{-(x^2+2x+3)}\,dx=1$

$$\text{or, } \frac{1}{c}=\int_{-\infty}^{\infty}e^{-(x+1)^2-2}\,dx=\frac{1}{e^2}\int_{-\infty}^{\infty}e^{-(x+1)^2}\,dx=\frac{1}{e^2}\int_{-\infty}^{\infty}e^{-u^2}\,du,\ u=x+1$$

$$=\frac{2}{e^2}\int_0^{\infty}e^{-u^2}\,du=\frac{1}{e^2}\int_0^{\infty}e^{-v}v^{\frac{1}{2}-1}\,dv, \text{ when } u^2=v$$

$$=\frac{1}{e^2}\Gamma\left(\frac{1}{2}\right)=\frac{\sqrt{\pi}}{e^2}.\ \therefore\ c=\frac{e^2}{\sqrt{\pi}}.$$

$$\therefore\ f_X(x) = \frac{e^2}{\sqrt{\pi}} e^{-(x^2+2x+3)}, -\infty < x < \infty.$$

$$E(X) = \frac{e^2}{\sqrt{\pi}} \int_{-\infty}^{\infty} x e^{-(x+1)^2-2}\,dx = \frac{1}{\sqrt{\pi}} \int_{-\infty}^{\infty} x e^{-(x+1)^2}\,dx$$

Put $x+1=z$, $\therefore\ dx = dz$ and as $x \to -\infty, z \to -\infty$ and $x \to \infty, z \to \infty$

$$= \frac{1}{\sqrt{\pi}} \int_{-\infty}^{\infty} (z-1) e^{-z^2}\,dz = \frac{1}{\sqrt{\pi}} \left[\int_{-\infty}^{\infty} z e^{-z^2}\,dz - \int_{-\infty}^{\infty} e^{-z^2}\,dz \right]$$

$$= -\frac{1}{\sqrt{\pi}} \cdot 2 \int_0^{\infty} e^{-z^2}\,dz,$$ as first integrand is an odd function and second integrand is an even function

$$= -\frac{1}{\sqrt{\pi}} \cdot 2 \frac{\sqrt{\pi}}{2} = -1.$$

$$E\left(X^2\right) = \frac{e^2}{\sqrt{\pi}} \int_{-\infty}^{\infty} x^2 e^{-(x+1)^2-2}\,dx = \frac{1}{\sqrt{\pi}} \int_{-\infty}^{\infty} (u-1)^2 e^{-u^2}\,du,$$ where $u = x+1$

$$= \frac{1}{\sqrt{\pi}} \int_{-\infty}^{\infty} \left(u^2 - 2u + 1\right) e^{-u^2}\,du$$

$$= \frac{1}{\sqrt{\pi}} \int_{-\infty}^{\infty} u^2 e^{-u^2}\,du - \frac{2}{\sqrt{\pi}} \int_{-\infty}^{\infty} u e^{-u^2}\,du + \frac{2}{\sqrt{\pi}} \int_0^{\infty} e^{-u^2}\,du$$

$$= \frac{2}{\sqrt{\pi}} \int_0^{\infty} u^2 e^{-u^2}\,du + \frac{2}{\sqrt{\pi}} \cdot \frac{\sqrt{\pi}}{2},$$ as the second integrand is an odd function and the first and third integrands are even functions

Put $u^2 = v$, $\therefore\ 2u\,du = dv$, the limits remain the same.

$$= \frac{2}{\sqrt{\pi}} \int_0^{\infty} v e^{-v} \frac{dv}{2\sqrt{v}} + 1 = \frac{1}{\sqrt{\pi}} \int_0^{\infty} e^{-v} v^{\frac{3}{2}-1}\,dv + 1 = \frac{1}{\sqrt{\pi}} \Gamma\left(\frac{3}{2}\right) + 1$$

$$= \frac{1}{\sqrt{\pi}} \cdot \frac{1}{2} \Gamma\left(\frac{1}{2}\right) + 1 = \frac{1}{2} + 1, \text{ as } \Gamma\left(\frac{1}{2}\right) = \sqrt{\pi}$$

$$= \frac{3}{2}.$$

$$\therefore\ \text{var}(X) = E\left(X^2\right) - \{E(X)\}^2 = \frac{3}{2} - (-1)^2 = \frac{3}{2} - 1 = \frac{1}{2}\ \left[\text{SD} = \frac{1}{\sqrt{2}}\right].$$

Ex. 12. A coin is tossed repeatedly until a head appears. If the tosses are independent and probability of head is p for each toss, find the expected number of tosses. [B.H. 2003]

Solution. Let X denotes the number of tosses required until head appears. Then, the spectrum of X is i, when $i = 1, 2, \cdots$

Now, $P(X=1) = p, P(X=2) = qp, \cdots, P(X=n) = q^{n-1}p$, where $q = 1-p$ = probability of tail for each trial.

$$\therefore\ E(X) = \sum_{i=1}^{\infty} i f_{xi},\ f_{xi} = P(X=i)$$

$$= p + 2qp + 3q^2p + \cdots + nq^{n-1}p + \cdots \infty = p\left[1 + 2q + 3q^2 + \cdots\right] = p(1-q)^{-2} = \tfrac{1}{p}.$$

EXERCISE

1. k unbiased dice rolled at a time. Show that the expected value of the products of number of faces appeared is $\left(\frac{7}{2}\right)^k$.
2. The distribution of a RV X is given by the pmf: $f_{xk} = P(X = k) = 2^{-k}$, $k = 1, 2, 3, \cdots$. Show that $E(X) = 2$ and $\text{var}(X) = 2$. [B.H. 1986]
3. Find the variance of a continuous RV X with pdf: $f_x(x) = \begin{cases} 1-|1-x|, & 0 < x \leq 2 \\ 0, & \text{elsewhere.} \end{cases}$ [B.H. 1992]

   (*Ans.* var$(X) = 1/6$)
4. From a point on a circle of radius r, chords are drawn at random. Find the expected value of the length of chords and the variance of the length of chords.

   [**Hints.** $AP = X = 2r\cos\theta$. Here r is a constant and θ varies uniformly in $\left[-\frac{\pi}{2}, \frac{\pi}{2}\right]$. So, the distribution of θ is given by $f_\theta(\theta) = \begin{cases} \frac{1}{\pi}, & -\frac{\pi}{2} \leq \theta \leq \frac{\pi}{2} \\ 0, & \text{elsewhere.} \end{cases}$

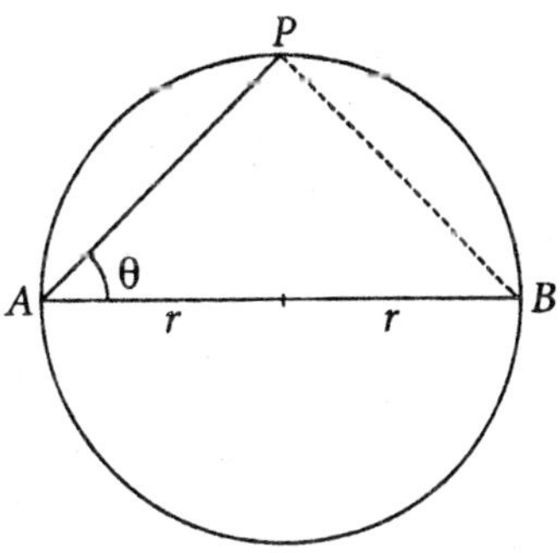

$$\therefore\ E(X) = 2r\int_{-\pi/2}^{\pi/2} \cos\theta\, f_\theta(\theta)\, d\theta = \frac{2r}{\pi}\int_{-\pi/2}^{\pi/2} \cos\theta\, d\theta$$

$$= \frac{2r}{\pi}\left[\sin\theta\right]_{-\pi/2}^{\pi/2} = \frac{4r}{\pi}.$$

$$E\left(X^2\right) = 4r^2\int_{-\pi/2}^{\pi/2} \cos^2\theta \cdot \frac{1}{\pi}\, d\theta = \frac{2r^2}{\pi}\int_{-\pi/2}^{\pi/2} (1+\cos 2\theta)\, d\theta$$

$$= \frac{2r^2}{\pi}[\pi] = 2r^2.$$

$$\therefore\ \text{var}(X) = E\left(X^2\right) - (E(X))^2 = 2r^2 - \frac{16r^2}{\pi^2} = 2r^2\left(1 - \frac{8}{\pi^2}\right).]$$

5. If X is a binomial (n, p) variate, then prove that $\mu_{k+1} = p(1-p)\left(nk\mu_{k-1} + \frac{d\mu_k}{dp}\right)$, where μ_k is the kth order central moment. [C.H. 2007]
6. Let λ and μ_r denote the mean and the central moment of order r respectively. Obtain the following recurrence relation: $\mu_{r+1} = \lambda\left[r\mu_{r-1} + \frac{d\mu_r}{d\lambda}\right]$. [C.H. 2005]
7. Prove that the MGF of a uniform distribution over the interval $[-a, a]$ is $\frac{\sinh at}{at}$. Hence, calculate the central moments.
8. For the discrete uniform distribution with spectrum $1, 2, \cdots, n$, find mean, median, mode and variance.

   [**Hints.** Let the RV follows discrete uniform distribution. $\therefore\ f_{xi} = \frac{1}{n}$, $i = 1, 2, \cdots, n$.

$$m_x = E(X) = \sum_{i=1}^{n} i f_{xi} = \frac{1}{n}(1+2+\cdots+n) = \frac{n+1}{2}.$$

$$E\left(X^2\right) = \sum_{i=1}^{n} i^2 f_{xi} = \frac{1}{n}\left(1^2+2^2+\cdots+n^2\right) = \frac{(n+1)(2n+1)}{6}.$$

$$\therefore\ \text{var}(X) = E\left(X^2\right) - \{E(X)\}^2 = \frac{(n+1)(2n+1)}{6} - \left(\frac{n+1}{2}\right)^2 = \frac{1}{12}\left(n^2-1\right).$$

m will be median, if $\frac{m}{n} \le \frac{1}{2} < \frac{m+1}{n}$. $\therefore$ when $n=4$, $m=2$
when $n=5$, $m=2$
when $n=6$, $m=3$
when $n=7$, $m=3$
when $n=8$, $m=4$, etc.

As probability at every node is the same, so there is no distinct mode.

9. Find the mean, variance and mode of binomial $(6, \frac{1}{3})$ distribution.

(*Ans.* median = 2, mode =2, variance = 4/3)

10. Let $X \sim B(n,p)$, n is fixed and p is a parameter. For what value of p, var(X) is the maximum.

(*Ans.* $p = 1/2$)

11. Find the geometric mean of the RV X, when its pdf is given by $f_X(x) = \begin{cases} 6(2-x)(x-1), & 1 \le x \le 2 \\ 0, & \text{elsewhere.} \end{cases}$

[**Hints.** If X be a RV and G and H denote respectively geometric mean and harmonic mean, then by definition, $\ln G = E(\ln x)$ and $\frac{1}{H} = E\left(\frac{1}{X}\right)$.

$$\therefore \ln G = 6\int_1^2 \ln x \cdot (2-x)(x-1)\,dx = 6\int_1^2 \left(-x^2+3x-2\right)\ln x\,dx = -\ln 16 + \frac{19}{6}. \therefore G = \frac{1}{16}e^{19/6}.]$$

12. Find the harmonic mean of a RV X, when its pdf is given by $f_X(x) = \begin{cases} e^{-x}\frac{x^n}{n!}(n \in \mathbb{N}), & 0 < x < \infty \\ 0, & \text{elsewhere.} \end{cases}$

[**Hints.** If H be the harmonic mean then $\frac{1}{H} = E\left(\frac{1}{X}\right) = \frac{1}{n!}\int_0^\infty e^{-x}x^{n-1}\,dx = \frac{\Gamma(n)}{n!} = \frac{(n-1)!}{n!} = \frac{1}{n}$.]

13. If a continuous RV X follows uniform distribution in $[a,b]$, find mean, variance, coefficient of skewness and coefficient of kurtosis.

Hints. mean $= \int_a^b x \cdot \frac{1}{b-a}\,dx = \frac{1}{2(b-a)}\left(b^2-a^2\right) = \frac{b+a}{2}$.

$$E(X^2) = \int_a^b x^2 \frac{1}{b-a}\,dx = \frac{1}{3(b-a)}\left(b^2-a^3\right) = \frac{1}{3}\left(b^2+ba+a^2\right).$$

$\therefore$ var$(X) = E(X^2) - \{E(X)\}^2 = \frac{1}{2}(b-a)^2$.

It is a symmetrical distribution (symmetry about mean). $\therefore$ coefficient of skewness $(\beta_1) = 0$; coefficient of kurtosis $(\beta_2) = 1.8$ (try yourself).

14. The radius X of a circle is uniformly distributed in $(3,4)$. Find the mean and variance of the circumference of the circle.

(*Ans.* mean = 7π, variance = $\pi^2/3$.)

15. If $\phi_X(t)$ is a real characteristic function of a RV X, prove that $1 + \phi_X(2t) \ge 2\{\phi_X(t)\}^2$.

Chapter 7

Mathematical Expectation II

7.1 Introduction

Let $g : \mathbb{R} \times \mathbb{R} \to \mathbb{R}$ be a continuously differentiable mapping and (X, Y) be a two-dimensional RV. Then, the distribution of the RV $g(X, Y)$ is uniquely determined by the joint probability distribution of X and Y.

The expectation or the mean value (MV) of the distribution of $g(X, Y)$ is defined by

$$E\{g(X,Y)\} = \sum_{i=-\infty}^{\infty}\sum_{j=-\infty}^{\infty} g\left(x_i, y_j\right) f_{ij}, \text{ for a discrete distribution}$$

$$= \int_{-\infty}^{\infty}\int_{-\infty}^{\infty} g(x,y) f_{x,y}(x,y)\,dx\,dy, \text{ for a continuous distribution,}$$

provided the infinite sum or the improper integral converges absolutely. The infinite series or improper integral does not converges or even they converge but not absolutely, the expectation does not exist.

Some information about absolute convergence and the reason for defining expectation by absolute convergence:

Definition. An infinite series Σu_n is said to be absolutely convergent if the series $\Sigma |u_n|$ is convergent.

Theorem 7.1.1 An absolutely convergent series is convergent.

Theorem 7.1.2 If Σa_n and Σb_n be two absolutely convergent series with a and b as their sums, then their product $\Sigma a_n \Sigma b_n$ [i.e., $a_0b_0+(a_0b_1+a_1b_0)+(a_0b_2+a_1b_1+a_2b_0)+\cdots$, the Cauchy product] is absolutely convergent and converges to ab.

Theorem 7.1.3 If the series Σa_n and Σb_n are convergent with their sums a and b respectively and one of them converges absolutely, then the series Σc_n, where $c_n = a_0b_n + a_1b_{n-1} + a_2b_{n-2} + \cdots + a_nb_0$ is convergent and its sum is ab [Σc_n is the Cauchy product of Σa_n and Σb_n].

Note. The product of two non-absolutely convergent series may not be convergent but if converges, then it converges to the product of the individual sums of the series.

If $g(X, Y) = u(X)v(Y)$ and RVs $u(X)$ and $v(Y)$ are independent, then $E[g(X, Y)] = E(u(X))\, E(v(Y))$ cannot be written if we define expectation by ordinary convergence, as the ordinary convergence (non-absolutely convergence) of the series $E\{u(X)\}$ and $E\{v(Y)\}$ do not guarantee the convergence of their product series. Here is an example from real analysis. In this example, we show that the product of two non-absolutely series is a divergent series.

Ex. Consider the series: $\sum_{n=0}^{\infty} \frac{(-1)^n}{\sqrt{n+1}} = 1 - \frac{1}{\sqrt{2}} + \frac{1}{\sqrt{3}} - \frac{1}{\sqrt{4}} + \cdots$.

The series is convergent by Dirichlet's test, since the sequence of partial sums of the series $\sum_{n=0}^{\infty}(-1)^n$, i.e., $\{1,-1,1,-1,\cdots\}$ is bounded and the sequence $\left\{\frac{1}{\sqrt{n+1}}\right\}_n$ is monotone decreasing and converges to 0. The series is not absolutely convergent as

$$\sum_{n=0}^{\infty}\left|\frac{(-1)^n}{\sqrt{n+1}}\right|=\sum_{n=0}^{\infty}\frac{1}{\sqrt{n+1}}=1+\frac{1}{\sqrt{2}}+\frac{1}{\sqrt{3}}+\cdots,$$

which is a p-series with $p=1/2\leq 1$.

Now, consider the product of this series with itself, the product series $=1-\left(\frac{1}{\sqrt{2}}+\frac{1}{\sqrt{2}}\right)+\left(\frac{1}{\sqrt{3}}+\frac{1}{\sqrt{2}\sqrt{2}}+\frac{1}{\sqrt{3}}\right)-\left(\frac{1}{\sqrt{4}}+\frac{1}{\sqrt{3}\sqrt{2}}+\frac{1}{\sqrt{2}\sqrt{3}}+\frac{1}{\sqrt{4}}\right)+\cdots=\Sigma c_n$, say.

[**Definition.** The product Σa_n and Σb_n is Σc_n, where $c_n=\sum_{r=0}^{n}a_r b_{n-r}\,(n=0,1,2,\cdots)$],

where $c_n=(-1)^n\sum_{r=0}^{n}\frac{1}{\sqrt{(n-r+1)(r+1)}}$ is divergent as shown below:

$(n-r+1)(r+1)=\left(\frac{n}{2}+1\right)^2-\left(\frac{n}{2}-r\right)^2\leq\left(\frac{n}{2}+1\right)^2.$

$$\therefore\ |c_n|\geq\sum_{r=0}^{n}\frac{2}{n+2}=\frac{2(n+1)}{n+2}.$$

$\therefore$ c_n does not tend to 0 as $n\to\infty$ and so Σc_n is not convergent. [necessary condition of convergence of Σa_n is $a_n\to 0$ as $n\to\infty$]

Theorem 7.1.4 (Due to Mertens). Let $\sum_{n=0}^{\infty}a_n$ converges absolutely to the sum a and $\sum_{n=0}^{\infty}b_n$ converges to the sum b.

Then, $\sum_{n=0}^{\infty}c_n$, where $c_n=\sum_{r=0}^{n}a_r b_{n-r}\,(n=0,1,2,\cdots)$ is convergent and converges to ab, i.e., the product of two convergent series converges to the product of their sums if at least one of the two series converges absolutely.

Moreover, most of the test of convergence of a series like comparison test, root test, ratio test are really a test for absolute convergence.

The definition of expectation by the absolute convergence is due to the fact that the convergence crisis may not occur in any situation and the rearrangement of terms does not affect the convergence and its sum for a absolutely convergent series.

Note 1: $E\{g(X,Y)\}=\sum_{i=-\infty}^{\infty}\sum_{j=-\infty}^{\infty}g(x_i y_j)\,f_{ij}$, provided the series on the RHS is absolutely convergent, i.e., the series

$$\sum_{i=-\infty}^{\infty}\sum_{j=-\infty}^{\infty}\left|g(x_i,y_j)\,f_{ij}\right|,\text{ i.e., }\sum_{i=-\infty}^{\infty}\sum_{j=-\infty}^{\infty}\left|g(x_i,y_j)\right|f_{ij}\text{ as }f_{ij}>0,\ \forall\, i,j$$

is convergent, but $E\{g(X,Y)\}=\sum_{i=-\infty}^{\infty}\sum_{j=-\infty}^{\infty}g(x_i,y_j)\,f_{ij}$ and $\neq\sum_{i=-\infty}^{\infty}\sum_{j=-\infty}^{\infty}\left|g(x_i,y_j)\right|f_{ij}$.

The convergence of $\sum_{i=-\infty}^{\infty}\sum_{j=-\infty}^{\infty}\left|g(x_i,y_j)\right|f_{ij}$ is the sufficient condition for the convergence of $\sum_{i=-\infty}^{\infty}\sum_{j=-\infty}^{\infty}g(x_i,y_j)\,f_{ij}$.

Note 2: The expectation of a RV $g(X)$ may be calculated from the distribution of X above or from the joint distribution of the RVs X and Y, i.e., from the distribution of the two-dimensional RV (X, Y) or from the distribution of three-dimensional RV (X, Y, Z), etc.

Verification. $E\{g(X)\} = \sum_{i=-\infty}^{\infty} \sum_{j=-\infty}^{\infty} g(x_i) f_{ij} = \sum_{i=-\infty}^{\infty} g(x_i) \sum_{j=-\infty}^{\infty} f_{ij} = \sum_{i=-\infty}^{\infty} g(x_i) f_{xi}$, for discrete case.

$$E\{g(X)\} = \int_{-\infty}^{\infty}\int_{-\infty}^{\infty} g(x) f_{x,y}(x,y)\,dx\,dy = \int_{-\infty}^{\infty} g(x)\left\{\int_{-\infty}^{\infty} f_{x,y}(x,y)\,dy\right\} dx = \int_{-\infty}^{\infty} g(x) f_x(x)\,dx,$$

for continuous case.

From the distribution of the RV (X, Y, Z):

$$E\{g(X)\} = \sum_{i=-\infty}^{\infty}\sum_{j=-\infty}^{\infty}\sum_{k=-\infty}^{\infty} g(x_i) f_{ijk} = \sum_{i=-\infty}^{\infty} g(x_i)\left[\sum_{j=-\infty}^{\infty}\sum_{k=-\infty}^{\infty} f_{ijk}\right] = \sum_{i=-\infty}^{\infty} g(x_i) f_{xi}$$

and $$E\{g(X)\} = \int_{-\infty}^{\infty}\int_{-\infty}^{\infty}\int_{-\infty}^{\infty} g(x) f_{x,y,z}(x,y,z)\,dx\,dy\,dz = \int_{-\infty}^{\infty} g(x)\left[\int_{-\infty}^{\infty}\int_{-\infty}^{\infty} f_{x,y,z}(x,y,z)\,dy\,dz\right] dx$$

$$= \int_{-\infty}^{\infty} g(x) f_x(x)\,dx, \text{ etc.}$$

[assuming the existence of $E\{g(X)\}$].

7.2 Linearity of expectation

For any set of scalars $\{c_1, c_2, \cdots, c_n\}$ and for any set of RVs $\{g_1(X,Y), g_2(X,Y), \cdots, g_n(X,Y)\}$ such that $E\{g_i(X,Y)\}$ exists $\forall\, i = 1, 2, \cdots, n$, then

$E\{c_1 g_1(X,Y) + c_2 g_2(X,Y) + \cdots + c_n g_n(X,Y)\} = c_1 E\{g_1(X,Y)\} + c_2 E\{g_2(X,Y)\} + \cdots + c_n E\{g_n(X,Y)\}$,

i.e., $E\left\{\sum_{k=1}^{n} c_k g_k(X,Y)\right\} = \sum_{k=1}^{n} c_k E\{g_k(X,Y)\}$.

Proof. (*Discrete case*). Let (X, Y) is a discrete RV in two dimensions with spectrum (x_i, y_j), $\forall\, i, j = 0, \pm1, \pm2, \cdots$ and $f_{ij} = P(X = x_i, Y = y_j) > 0$, $\forall\, i, j$.

Since, each $E\{g_k(X,Y)\}$ exists, $\therefore \sum_{i=-\infty}^{\infty}\sum_{j=-\infty}^{\infty} g_k(x_i, y_i) f_{ij}$ is absolutely convergent for each k. Hence, $\sum_{i=-\infty}^{\infty}\sum_{j=-\infty}^{\infty}\left\{\sum_{k=1}^{n} c_k g_k(x_i, y_i)\right\} f_{ij}$ converges absolutely.

$$\text{Now, } E\left\{\sum_{k=1}^{n} c_k g_k(X,Y)\right\} = \sum_{i=-\infty}^{\infty}\sum_{j=-\infty}^{\infty}\left\{\sum_{k=1}^{n} c_k g_k(x_i, y_i)\right\} f_{ij} = \sum_{k=1}^{n} c_k\left\{\sum_{i=-\infty}^{\infty}\sum_{j=-\infty}^{\infty} g_k(x_i, y_j) f_{ij}\right\}$$

$$= \sum_{k=1}^{n} c_k E\{g_k(X,Y)\}.$$

$$\text{(Continuous case) or, } E\left\{\sum_{k=1}^{n} c_k g_k(X,Y)\right] = \int_{-\infty}^{\infty}\int_{-\infty}^{\infty}\left\{\sum_{k=1}^{n} c_k g_k(x,y)\right\} f_{x,y}(x,y)\,dx\,dy$$

$$= \sum_{k=1}^{n} c_k\left\{\int_{-\infty}^{\infty}\int_{-\infty}^{\infty} g_k(x,y) f_{x,y}(x,y)\,dx\,dy\right\}$$

$$= \sum_{k=1}^{n} c_k E\{g_k(X,Y)\}.$$

Theorem 7.2.1 Let $g:\mathbb{R}\times\mathbb{R}\to\mathbb{R}$ be a function such that $g(x,y)\le 0,\ \forall\,(x,y)\in\mathbb{R}\times\mathbb{R}$.
Then, the MV $E\{g(X,Y)\}\le 0$, provided $E\{g(X,Y)\}$ exists.

Proof. (*Discrete case*). As $E\{g(X,Y)\}$ exists, $\therefore\ \sum_{i=-\infty}^{\infty}\sum_{j=-\infty}^{\infty} g(x_i,y_j)\, f_{ij}$ converges absolutely.

By the given condition $g(x_i,y_j)\le 0$ and $f_{ij}>0$ at all possible spectrum point. $\therefore\ g(x_i,y_i)\, f_{ij}\le 0,\ \forall\, i,j.$
Hence, $\sum_i\sum_j g(x_i,y_j)\, f_{ij}\le 0.\ \therefore\ E\{g(X,Y)\}\le 0.$

(*Continuous case*). As $E\{g(X,Y)\}$ exists, $\therefore\ \int_{-\infty}^{\infty}\int_{-\infty}^{\infty} g(x,y) f_{x,y}(x,y)\,dx\,dy$ converges absolutely, where $f_{x,y}(x,y)$ is the joint pdf.
As $g(x,y)\le 0$ and $f_{x,y}(x,y)\ge 0,\ \forall\,(x,y)\in\mathbb{R}\times\mathbb{R},\ \therefore\ g(x,y) f_{x,y}(x,y)\le 0,\ \forall\,(x,y)\in\mathbb{R}\times\mathbb{R}$.

$$\therefore \int_{-\infty}^{\infty}\int_{-\infty}^{\infty} g(x,y) f_{x,y}(x,y)\,dx\,dy \le 0, \text{ i.e., } E\{g(X,Y)\}\le 0.$$

Note: In particular, when $g(x,y)\le 0,\ \forall\,(x,y)\in\mathbb{R}^2$ and $E\{g(X,Y)\}=0$, then $\sum_i\sum_j g(x_i,y_i)\, f_{ij}=0$. But, $f_{ij}>0$ and $g(x_i,y_j)\le 0,\ \forall\, i,j$, so $g(x_i,y_j)=0,\ \forall\, i,j$.
$\therefore$ the spectrum of the RV $g(X,Y)$ is $\{0\}$.
In continuous case, let g is continuous and $g(x,y)\le 0,\ \forall\,(x,y)\in\mathbb{R}^2$.

$$\therefore \int_{-\infty}^{\infty}\int_{-\infty}^{\infty} g(x,y) f_{x,y}\,dx\,dy = 0.$$

But, $g(x,y)\le 0$ and $f_{x,y}(x,y)\ge 0,\ \forall\,(x,y)\in\mathbb{R}^2$.
This is possible only when $g(x,y)\equiv 0,\ \forall\,(x,y)\in\mathbb{R}^2$.
Hence, the spectrum of the RV $g(X,Y)$ is $\{0\}$.

Note: In case, $g(x,y)\ge 0,\ \forall\,(x,y)$, then $E\{g(X,Y)\}\ge 0$, whenever exists.

Theorem 7.2.2 If X and Y are independent RVs, then show that $E\{g_1(X)\cdot g_2(Y)\}=E\{g_1(X)\}\cdot E\{g_2(X)\}$, provided all the MVs exist.

Proof.
$$E\{g_1(X)\cdot g_2(Y)\} = \sum_{i=-\infty}^{\infty}\sum_{j=-\infty}^{\infty} g_1(x_i)\, g_2(y_j)\, f_{ij}$$

$$= \sum_{i=-\infty}^{\infty}\sum_{j=-\infty}^{\infty} g_1(x_i)\, g_2(y_j)\, f_{xi} f_{yj}, \text{ since } X \text{ and } Y \text{ are independent}$$

$$= \sum_{i=-\infty}^{\infty} g_1(x_i)\, f_{xi} \sum_{j=-\infty}^{\infty} g_2(y_j)\, f_{yj}, \text{ the rearrangement is justified since the series converges absolutely}$$

$$= E\{g_1(X)\}\cdot E\{g_2(Y)\}.$$

When (X, Y) is a continuous RV:

$$E\{g_1(X)\cdot g_2(Y)\} = \int_{-\infty}^{\infty}\int_{-\infty}^{\infty} g_1(x)g_2(y)f_{x,y}(x,y)\,dx\,dy$$

$$= \int_{-\infty}^{\infty}\int_{-\infty}^{\infty} g_1(x)g_2(y)f_x(x)f_y(y)\,dx\,dy, \text{ since } X, Y \text{ are independent}$$

$$= \int_{-\infty}^{\infty} g_1(x)f_x(x)\,dx \int_{-\infty}^{\infty} g_2(y)f_y(y)\,dy = E\{g_1(X)\}\cdot E\{g_2(Y)\}.$$

Note: The proof in reverse way is done later.

7.3 Moments of a probability distribution

Moments play an important role in every distribution pattern. In the one-dimensional probability distribution we have seen moments of a distribution help us to measure dispersion of the distribution from its mean, the measure of skewness, Curtosis, etc. For the two-dimensional probability distribution of the RV (X, Y), the moment defines the jointless of X and Y, i.e., their covariance, correlation coefficient, etc.

Moment about the origin $(0,0)$:

Let $r, s \in \mathbb{N}\cup\{0\}$. The moment of the distribution of (X, Y) about the origin $(0,0)$ of order $(r+s)$ is denoted by α_{rs} and is defined by the MV $\alpha_{rs} = E(X^r Y^s)$.

This is also called *ordinary moment* of order $(r+s)$.

In particular, when $s = 0, \alpha_{r0} = E(X^r) = \alpha_{xr}$, which is the ordinary moment (moment about the origin) for the one-dimensional distribution of X of order r. Similarly, $r = 0$ gives $\alpha_{os} = E(Y^s) = \alpha_{ys}$.

(1) *When* $r = 0$, $s = 0$; $\alpha_{00} = E(1) = 1$ (2) *When* $r = 1$, $s = 0$; $\alpha_{10} = E(X) = m_x$ (3) *When* $r = 0$, $s = 1$; $\alpha_{01} = E(Y) = m_y$, *whenever they exist.*

So, every ordinary moment of the one-dimensional case is obtained from the case of two dimensions as a particular case.

Moment about the centre of mass (m_x, m_y):

Let $r, s \in \mathbb{N}\cup\{0\}$.

The moment of the distribution of (X, Y) about the centre of mass (m_x, m_y) of order (r, s) is denoted by μ_{rs} and is defined by the MV

$$\mu_{rs} = E\left\{(X - m_x)^r\left(Y - m_y\right)^s\right\}.$$

This is also called *central moment of order* $(r+s)$.

Particular cases. $\mu_{10} = E(X - m_x) = 0, \mu_{01} = 0, \mu_{00} = E(1) = 1; \mu_{r0} = E(X - m_x)^r = \mu_{xr}; \mu_{0s} = E\left(Y - m_y\right)^s = \mu_{ys}; \mu_{20} = \mu_{x2} = \sigma_x^2; \mu_{02} = \mu_{y2} = \sigma_y^2$, etc., whenever they exist.

So, every central moment of one dimensional distribution is obtained from the central moments of two-dimensional distribution as particular cases.

Covariance, correlation coefficient

The second order mixed central moment μ_{11} is defined as the covariance of two RVs, say, X and Y. The name itself suggests that it gives some relation of variations of the RVs, whether the variation of X depends on the variation of Y, if depends how? Whether the increase of Y implies that X also increases or the increase of Y implies that X decreases, etc. It is denoted by $\text{cov}(X, Y)$.

The property of such a jointness of the RVs may be provided in a dimensionless measure by the covariance of their standardised forms X^* and Y^*, where $X^* = \frac{X-m_x}{\sigma_x}$ and $Y^* = \frac{Y-m_y}{\sigma_y}$.

The dimensionless measure of the covariance of X and Y is known as correlation coefficient between X, Y and is denoted by ρ_{xy} or $\rho(X, Y)$ or simply by ρ, i.e.,

$$\rho_{x,y} = \text{cov}\,(X^*, Y^*) = E\,(X^* Y^*) \text{ as } m_{x^*} = 0,\ m_{y^*} = 0$$

$$= E\left(\frac{X - m_x}{\sigma_x} \cdot \frac{Y - m_y}{\sigma_y}\right) = \frac{E(X - m_x)\big(Y - m_y\big)}{\sigma_x \sigma_y} = \frac{\text{cov}(X, Y)}{\sigma_x \sigma_y}, \text{ provided } \sigma_x \sigma_y \neq 0.$$

Properties of correlation coefficient

1. Correlation coefficient of two RVs is independent of the choice of origin and scale of measurement. Let the constant $a_1, a_2, b_1, b_2 \in \mathbb{R}$ such that $a_1 a_2 \neq 0$.

Then, $\rho\,(a_1X + b_1, a_2Y + b_2) = \dfrac{a_1 a_2}{|a_1|\,|a_2|}\rho(X, Y)$ and $\rho\,(a_1X + b_1, a_2Y + b_2) = \rho(X, Y)$, when $a_1 a_2 > 0$.

Proof. $E(a_1X + b_1) = a_1\, E(X) + b_1 = a_1 m_x + b_1;\ E(a_2Y + b_2) = a_2\, E(Y) + b_2 = a_2 m_y + b_2$.

$\sigma^2_{a_1x+b_1} = E\{a_1X + b_1 - E(a_1X + b_1)\}^2 = E\{a_1X + b_1 - a_1m_x - b_1\}^2 = a_1^2\, E(X - m_x)^2 = a_1^2 \sigma_x^2$.

$\therefore\ \sigma_{a_1x+b_1} = |a_1|\,\sigma_x\ [\because\ \sigma_x \geq 0]$. Similarly, $\sigma_{a_2y+b_2} = |a_2|\,\sigma_y$.

$$\therefore\ (a_1X + b_1)^* = \frac{a_1X + b_1 - a_1 m_x - b_1}{|a_1|\,\sigma_x} = \frac{a_1}{|a_1|}\,\frac{X - m_x}{\sigma_x} = \frac{a_1}{|a_1|}X^*.$$

Similarly, $(a_2Y + b_2)^* = \frac{a_2}{|a_2|}Y^*$.

$$\therefore\ \rho\,(a_1X + b_1, a_2Y + b_2) = E\,\{(a_1X + b_1)^*(a_2Y + b_2)^*\} = E\left\{\frac{a_1}{|a_1|}X^* \cdot \frac{a_2}{|a_2|}Y^*\right\}$$

$$= \frac{a_1 a_2}{|a_1|\,|a_2|}\,E\,(X^* Y^*) = \frac{a_1 a_2}{|a_1 a_2|}\,\rho(X, Y).$$

When $a_1 a_2 > 0, |a_1 a_2| = a_1 a_2$,

$$\rho\,(a_1X + b_1, a_2Y + b_2) = \rho(X, Y).$$

2. Correlation coefficient between any two RVs cannot exceed 1 numerically, whenever it exists.

That is, $|\rho(X, Y)| \leq 1$ or equivalently $-1 \leq \rho(X, Y) \leq 1$.

Proof. Obviously, $\left(\dfrac{x - m_x}{\sigma_x} + \dfrac{y - m_y}{\sigma_y}\right)^2 \geq 0,\ \forall\,(x, y) \in \mathbb{R} \times \mathbb{R}$.

$$\therefore\ E\left(\frac{X - m_x}{\sigma_x} + \frac{Y - m_y}{\sigma_y}\right)^2 \geq 0, \text{ i.e., } E(X^* + Y^*)^2 \geq 0$$

or, $E\left\{(X^*)^2\right\} + E\left\{(Y^*)^2\right\} + 2E(X^*Y^*) \geq 0$

or, $1 + 1 + 2\rho(X, Y) \geq 0$ or, $2\rho(X, Y) \geq -2$, or, $-1 \leq \rho(X, Y)$. (1)

Again, $\left(\dfrac{x - m_x}{\sigma_x} - \dfrac{y - m_y}{\sigma_y}\right)^2 \geq 0,\ \forall\,(x, y) \in \mathbb{R} \times \mathbb{R}$.

$$\therefore\ E\left(\frac{X - m_x}{\sigma_x} - \frac{Y - m_y}{\sigma_y}\right)^2 \geq 0 \text{ or, } E(X^* - Y^*)^2 \geq 0 \text{ or, } E\left\{(X^*)^2\right\} + E\left\{(Y^*)^2\right\} - 2E(X^*Y^*) \geq 0$$

or, $1 + 1 - 2\rho(X, Y) \geq 0$ or, $\rho(X, Y) \leq 1$. (2)

(1) and (2) $\Rightarrow -1 \leq \rho(X, Y) \leq 1$.

Significance of $\rho(X, Y)$

Correlation coefficient gives the measure of linear dependence between X and Y. If $\rho(X, Y)=0$; X, Y are said to be uncorrelated.

When $\rho(X, Y) \simeq \pm 1$, the RVs are approximately linear. When $\rho(X, Y)=\pm 1$; the RVs will be perfectly linear, i.e., the whole two-dimensional probability distribution of (X, Y) lies on a straight line and conversely.

Verification. From (1) and (2), $\rho(X, Y)=\pm 1 \Rightarrow X^* \mp Y^*=0$ or, $Y^*=\pm X^*$, i.e., $\frac{Y-m_y}{\sigma_y}=\pm\frac{X-m_x}{\sigma_x}$.

To have the converse: Let $Y=aX+b, a\neq 0$ and a, b are real constants.

$\therefore\ E(Y)=aE(X)+b$ or, $m_y=am_x+b$.

$\therefore\ \sigma_y^2=E\left(Y-m_y\right)^2=E(aX+b-am_x-b)^2=a^2E(X-m_x)^2=a^2\sigma_x^2$. $\therefore\ \sigma_y=|a|\sigma_x$ $[\because\ \sigma_x\geq 0]$.

$\therefore\ Y^*=(aX+b-am_x-b)/|a|\sigma_x=\frac{a}{|a|}\frac{X-m_x}{\sigma_x}=\frac{a}{|a|}X^*$.

$\therefore\ \rho(X, Y)=\operatorname{cov}(X^*, Y^*)=E(X^*Y^*)=E\left(X^*\cdot\frac{a}{|a|}X^*\right)=\frac{a}{|a|}E\left\{(X^*)^2\right\}=\frac{a}{|a|}=\pm 1.$

Relation between variance and covariance

The variance of the RV $X+Y$ does not depend only on the variances of X and Y but also depends on their covariance, i.e., $\operatorname{var}(X+Y)=\operatorname{var}(X)+\operatorname{var}(Y)+2\operatorname{cov}(X, Y)$.

Verification

$E(X+Y)=E(X)+E(Y)=m_x+m_y$.

$$\begin{aligned}\therefore\ \operatorname{var}(X+Y) &= E\left\{X+Y-\left(m_x+m_y\right)\right\}^2=E\left\{(X-m_x)+\left(Y-m_y\right)\right\}^2\\ &= E\left\{(X-m_x)^2+\left(Y-m_y\right)^2+2(X-m_x)\left(Y-m_y\right)\right\}\\ &= E(X-m_x)^2+E\left(Y-m_y\right)^2+2E(X-m_x)\left(Y-m_y\right)\\ &= \operatorname{var}(X)+\operatorname{var}(Y)+2\operatorname{cov}(X, Y).\end{aligned}$$

$$\therefore\ \sigma_{x+y}^2=\sigma_x^2+\sigma_y^2+2\sigma_x\sigma_y\rho(X, Y).$$

However, when X and Y are uncorrelated, i.e., $\rho(X, Y)=0$, $\sigma_{x+y}^2=\sigma_x^2+\sigma_y^2$.

In a similar way: $\operatorname{var}(X-Y)=\operatorname{var}(X)+\operatorname{var}(Y)-2\operatorname{cov}(X, Y)$,

i.e., $\sigma_{x-y}^2=\sigma_x^2+\sigma_y^2-2\sigma_x\sigma_y\rho(X, Y)=\sigma_x^2+\sigma_y^2$, when $\rho(X, Y)=0$.

Note: $\operatorname{var}(X+Y)=\operatorname{var}(X-Y)$, if $\rho(X, Y)=0$.

Generalisation. For a set of RVs $\{X_1, X_2, \cdots, X_n\}$ and a set of real constants $\{a_1, a_2, \cdots, a_n\}$,

$$E(a_1X_1+a_2X_2+\cdots+a_nX_n)=a_1E(X_1)+a_2E(X_2)+\cdots+a_nE(X_n)=a_1m_{x_1}+a_2m_{x_2}+\cdots+a_nm_{x_n}.$$

$$\begin{aligned}&\therefore\ \operatorname{var}(a_1X_1+a_2X_2+\cdots+a_nX_n)\\ &= E\{(a_1X_1+a_2X_2+\cdots+a_nX_n)-(a_1m_{x_1}+a_2m_{x_2}+\cdots+a_nm_{x_n})\}^2\\ &= E\{a_1(X_1-m_{x_1})+a_2(X_2-m_{x_2})+\cdots+a_n(X_n-m_{x_n})\}^2\\ &= a_1^2E(X_1-m_{x_1})^2+a_2^2E(X_2-m_{x_2})^2+\cdots+a_n^2E(X_n-m_{x_n})^2\\ &\quad +2a_1a_2E(X_1-m_{x_1})(X_2-m_{x_2})+2a_2a_3E(X_2-m_{x_2})(X_3-m_{x_3})+\cdots\\ &\quad +2a_{n-1}a_nE(X_{n-1}-m_{x_{n-1}})(X_n-m_{x_n})\end{aligned}$$

$$= a_1^2\sigma_{x_1}^2 + a_2^2\sigma_{x_2}^2 + \cdots + a_n^2\sigma_{x_n}^2 + 2a_1a_2\text{cov}(X_1,X_2) + 2a_2a_3\text{cov}(X_2,X_3) + \cdots$$
$$+ 2a_{n-1}a_n\text{cov}(X_{n-1},X_n)$$
$$= a_1^2\sigma_{x_1}^2 + a_2^2\sigma_{x_2}^2 + \cdots a_n^2\sigma_{x_n}^2 + 2a_1a_2\sigma_{x_1}\sigma_{x_2}\rho(X_1,X_2)$$
$$+ 2a_2a_3\sigma_{x_2}\sigma_{x_3}\rho(X_2,X_3) + \cdots + 2a_{n-1}a_n\sigma_{x_{n-1}}\sigma_{x_n}\rho(X_{n-1},X_n)$$
$$= \sum_{i=1}^{n} a_i^2\sigma_{x_i}^2 + 2\sum_{\substack{i,j=1\\(i<j)}}^{n} a_ia_j\sigma_{x_i}\sigma_{x_j}\rho\left(X_i,X_j\right).$$

In particular, when $X_1, X_2, \cdots, X_n$ are pairwise uncorrelated, i.e., $\rho\left(X_i,X_j\right)=0, i\neq j$, then

$$\text{var}\,(a_1X_1 + a_2X_2 + \cdots + a_nX_n) = \sum_{i=1}^{n} a_i^2\sigma_{x_i}^2 = \sum_{i=1}^{n} a_i^2\,\text{var}\,X_i.$$

In such a case, $\text{var}\,(a_1X_1 \pm a_2X_2 \pm \cdots \pm a_nX_n) = \sum_{i=1}^{n} a_i^2\,\text{var}\,X_i$.

Theorem 7.3.1 Central moment or $k+l$ can be expressed as a linear combination of ordinary moments of order $\leq k+l$ for two RVs X and Y.

Proof.
$$\mu_{kl} = E\left\{(X-m_x)^k\left(Y-m_y\right)^l\right\} = E\left[\left\{\sum_{r=0}^{k}(-1)^r\binom{k}{r}X^{k-r}m_x^r\right\}\left\{\sum_{s=0}^{l}(-1)^s\binom{l}{s}Y^{l-s}m_y^s\right\}\right]$$
$$= E\left[\sum_{r=0}^{k}\sum_{s=0}^{l}(-1)^{r+s}\binom{k}{r}\binom{l}{s}X^{k-r}Y^{l-s}m_x^rm_y^s\right]$$
$$= \sum_{r=0}^{k}\sum_{s=0}^{l}(-1)^{r+s}\binom{k}{r}\binom{l}{s}E\left(X^{k-r}Y^{l-s}\right)m_x^rm_y^s, \text{ by the linearity property of expectation}$$
$$= \sum_{r=0}^{k}\sum_{s=0}^{l}(-1)^{r+s}\binom{k}{r}\binom{l}{s}\alpha_{k-r\ l-s}m_x^rm_y^s.$$

In particular,

$$\mu_{11} = \sum_{r=0}^{1}\sum_{s=0}^{1}(-1)^{r+s}\binom{1}{r}\binom{1}{s}\alpha_{1-r\,1-s}m_x^rm_y^s = \alpha_{11} - \alpha_{10}m_y - \alpha_{01}m_x + \alpha_{00}m_xm_y$$
$$= \alpha_{11} - m_xm_y - m_ym_x + m_xm_y\ [\because \alpha_{00}=1]$$
$$= \alpha_{11} - m_xm_y.$$

Now, we show that two correlated RVs can be made uncorrelated by a linear transformation when the transformation is a suitable rotation by the following theorem.

Theorem 7.3.2 Let X, Y be two correlated RVs and U, V be two RVs obtained from RVs X, Y by the rotation of coordinate axes through an angle α, i.e.,

$$U = X\cos\alpha + Y\sin\alpha$$
$$V = -X\sin\alpha + Y\cos\alpha,$$

then U and V will be uncorrelated, if $\alpha = \frac{1}{2}\tan^{-1}\left(\frac{2\sigma_x\sigma_y\rho(X,Y)}{\sigma_x^2-\sigma_y^2}\right)$.

Proof. $m_u = E(U) = m_x \cos\alpha + m_y \sin\alpha$; $m_v = E(V) = -m_x \sin\alpha + m_y \cos\alpha$.

$$\begin{aligned}
\therefore\ \operatorname{cov}(U,V) &= E(U-m_u)(V-m_v)\\
&= E\left[\{(X-m_x)\cos\alpha + \left(Y-m_y\right)\sin\alpha\}\{-(X-m_x)\sin\alpha + \left(Y-m_y\right)\cos\alpha\}\right]\\
&= E\left\{-\tfrac{1}{2}(X-m_x)^2\sin 2\alpha + \tfrac{1}{2}\left(Y-m_y\right)^2\sin 2\alpha + (X-m_x)\left(Y-m_y\right)\cos 2\alpha\right\}\\
&= -\tfrac{1}{2}E\left\{(X-m_x)^2\right\}\sin 2\alpha + \tfrac{1}{2}E\left\{\left(Y-m_y\right)^2\right\}\sin 2\alpha + E\left\{(X-m_x)\left(Y-m_y\right)\right\}\cos 2\alpha\\
&= -\tfrac{1}{2}\sigma_x^2\sin 2\alpha + \tfrac{1}{2}\sigma_y^2\sin 2\alpha + \operatorname{cov}(X,Y)\cos 2\alpha.
\end{aligned}$$

$\rho(U,V)=0$, if $\operatorname{cov}(U,V)=0$, i.e., if $\operatorname{cov}(X,Y)=\frac{1}{2}\left(\sigma_x^2-\sigma_y^2\right)\tan 2\alpha$, provided $\cos 2\alpha \neq 0$

or, $\tan 2\alpha = \frac{2\operatorname{cov}(X,Y)}{\sigma_x^2-\sigma_y^2} = \frac{2\sigma_x\sigma_y\rho(X,Y)}{\sigma_x^2-\sigma_y^2}$, i.e., $\alpha = \frac{1}{2}\tan^{-1}\left(\frac{2\sigma_x\sigma_y\rho(X,Y)}{\sigma_x^2-\sigma_y^2}\right)$.

When $\cos 2\alpha = 0, \alpha = \frac{\pi}{4}, \operatorname{cov}(U,V) = \frac{1}{2}\left(\sigma_y^2-\sigma_x^2\right)$, then U, V will be uncorrelated, if $\sigma_y = \sigma_x$.

7.4 Correlation coefficient of a bivariate normal distribution

The bivariate normal distribution of the RVs X and Y is defined by the following pdf $f_{x,y}(x,y)$:

$$f_{x,y}(x,y) = \frac{1}{2\pi\sigma_x\sigma_y\sqrt{1-\rho^2}}\exp\left[-\frac{1}{2(1-\rho^2)}\left\{\left(\frac{x-m_x}{\sigma_x}\right)^2 - 2\rho\frac{x-m_x}{\sigma_x}\cdot\frac{y-m_y}{\sigma_y} + \left(\frac{y-m_y}{\sigma_y}\right)\right\}\right],$$

$$(-\infty < x < \infty, -\infty < y < \infty)$$

with five parameters $m_x, m_y, \sigma_x (>0), \sigma_y (>0)$ and $\rho(-1<\rho<1)$.

Now,

$$\begin{aligned}
\operatorname{cov}(X,Y) &= E\left\{(X-m_x)\left(Y-m_y\right)\right\}\\
&= \frac{1}{2\pi\sigma_x\sigma_y\sqrt{1-\rho^2}}\int_{-\infty}^{\infty}\int_{-\infty}^{\infty}(x-m_x)\left(y-m_y\right)\\
&\qquad \exp\left[-\frac{1}{2(1-\rho^2)}\left\{\left(\frac{x-x_{m_x}}{\sigma_x}\right)^2 - 2\rho\frac{x-m_x}{\sigma_x}\cdot\frac{y-m_y}{\sigma_y} + \left(\frac{y-m_y}{\sigma_y}\right)^2\right\}\right]dx\,dy.
\end{aligned}$$

Let us put $\frac{x-m_x}{\sigma_x} = x^*$ and $\frac{y-m_y}{\sigma_y} = y^*$. $\therefore\ dx = \sigma_x dx^*, dy = \sigma_y dy^*$.

The limits of x, y and of x^*, y^* are the same, $\because\ m_x, m_y \in \mathbb{R}$ and $\sigma_x > 0, \sigma_y > 0$.

$$\begin{aligned}
\therefore\ \operatorname{cov}(X,Y) &= \frac{\sigma_x\sigma_y}{2\pi\sqrt{1-\rho^2}}\int_{-\infty}^{\infty}\int_{-\infty}^{\infty}x^*y^*\exp\left[-\frac{1}{2(1-\rho^2)}\left\{x^{*2}-2\rho x^*y^*+y^{*2}\right\}\right]dx^*dy^*\\
&= \frac{\sigma_x\sigma_y}{2\pi\sqrt{1-\rho^2}}\int_{-\infty}^{\infty}\int_{-\infty}^{\infty}xy\exp\left[-\frac{1}{2(1-\rho^2)}\left\{x^2-2\rho xy+y^2\right\}\right]dx\,dy,
\end{aligned}$$

by the property of the definite integral

$$= \frac{\sigma_x\sigma_y}{2\pi\sqrt{1-\rho^2}}\int_{-\infty}^{\infty}\int_{-\infty}^{\infty}xy\exp\left[-\frac{1}{2(1-\rho^2)}\left\{(x-\rho y)^2+\left(1-\rho^2\right)y^2\right\}\right]dx\,dy.$$

$$\therefore \rho(X,Y) = \frac{\text{cov}(x,y)}{\sigma_x \sigma_y} = \frac{1}{\sqrt{2\pi}} \int_{-\infty}^{\infty} y e^{-y^2/2} \left\{ \frac{1}{\sqrt{2\pi}\sqrt{1-\rho^2}} \int_{-\infty}^{\infty} x e^{-\frac{(x-\rho y)^2}{2(1-\rho^2)}} dx \right\} dy$$

$$= \frac{1}{\sqrt{2\pi}} \int_{-\infty}^{\infty} y e^{-y^2/2} \left\{ \frac{1}{\sqrt{2\pi}\sqrt{1-\rho^2}} \int_{-\infty}^{\infty} (x-\rho y+\rho y) e^{-\frac{(x-\rho y)^2}{2(1-\rho^2)}} dx \right\} dy$$

$$= \frac{1}{\sqrt{2\pi}} \int_{-\infty}^{\infty} y e^{-y^2/2} \left\{ \frac{1}{\sqrt{2\pi}\sqrt{1-\rho^2}} \int_{-\infty}^{\infty} (x-\rho y) e^{-\frac{(x-\rho y)^2}{2(1-\rho^2)}} dx + \right.$$

$$\left. \rho y \cdot \frac{1}{\sqrt{2\pi}\sqrt{1-\rho^2}} \int_{-\infty}^{\infty} e^{-\frac{(x-\rho y)^2}{2(1-\rho^2)}} dx \right\} dy.$$

Now,

$$\int_{-\infty}^{\infty} (x-\rho y) e^{-\frac{(x-\rho y)^2}{2(1-\rho^2)}} dx = \int_{-\infty}^{\infty} u e^{-\frac{u^2}{2(1-\rho^2)}} du, \text{ when } u = x-\rho y \text{ and } y \text{ is treated as constant}$$

$$= 0, \text{ as the integrand is an odd function}$$

and $\frac{1}{\sqrt{2\pi}\sqrt{1-\rho^2}} \int_{-\infty}^{\infty} e^{-\frac{(x-\rho y)^2}{2(1-\rho^2)}} dx = 1$, as $\frac{1}{\sqrt{2\pi}\sqrt{1-\rho^2}} e^{-\frac{(x-\rho y)^2}{2(1-\rho^2)}}$ is the pdf for a normal $\left(\rho y, \sqrt{1-\rho^2}\right)$ distribution [treating y as constant for such an integral].

$$\therefore \rho(X,Y) = \rho \cdot \frac{1}{\sqrt{2\pi}} \int_{-\infty}^{\infty} y^2 e^{-y^2/2} dy = \frac{2\rho}{\sqrt{2\pi}} \int_{0}^{\infty} y^2 e^{-y^2/2} dy, \text{ as the integral is an even function}$$

$$= \frac{2\rho}{\sqrt{2\pi}} \lim_{R\to\infty} \int_{0}^{R} y^2 e^{-y^2/2} dy.$$

Let $y = \sqrt{2v}$, $\therefore y^2 = 2v$, $\therefore 2y\,dy = 2dv$, $\therefore y\,dy = dv$

$$= \frac{2\rho}{\sqrt{2\pi}} \lim_{R\to\infty} \int_{0}^{\frac{1}{2}R^2} 2v e^{-v} \frac{dv}{\sqrt{2v}} = \frac{2\rho}{\sqrt{\pi}} \int_{0}^{\infty} e^{-v} v^{\frac{3}{2}-1} dv = \frac{2\rho}{\sqrt{\pi}} \Gamma\left(\frac{3}{2}\right)$$

$$= \frac{2\rho}{\sqrt{\pi}} \cdot \frac{1}{2} \Gamma\left(\frac{1}{2}\right) = \rho, \text{ as } \Gamma\left(\frac{1}{2}\right) = \sqrt{\pi}.$$

Thus the parameter ρ bears the natural meaning of correlation coefficient between the RVs X and Y.

7.5 Independence of two RVs *X* and *Y*

Theorem 7.5.1 If $h_1 : D_1 \to \mathbb{R}$ and $h_2 : D_2 \to \mathbb{R}$, where $D_1, D_2 \subseteq \mathbb{R}$ be two continuous functions of x and y respectively and X, Y be two independent RVs, then the expectation of the product of two RVs $h_1(X)$ and $h_2(Y)$ is the product of their expectations, i.e., $E\{h_1(X)h_2(Y)\} = E\{h_1(X)\}\,E\{h_2(Y)\}$, if each individual expectation exists.

Proof. (Continuous case). $E\{h_1(X)\} = \int_{-\infty}^{\infty} h_1(x) f_x(x)\,dx$, $E\{h_2(Y)\} = \int_{-\infty}^{\infty} h_2(y) f_y(y)\,dy$, the integrals being absolutely convergent. Their product is also absolutely convergent [Theorem 2].

$$\therefore\ E\{h_1(X)\}\,E\{h_2(Y)\} = \int_{-\infty}^{\infty} h_1(x) f_x(x)\,dx \int_{-\infty}^{\infty} h_2(y) f_y(y)\,dy = \int_{-\infty}^{\infty} h_1(x) h_2(y) f_x(x) f_y(y)\,dx\,dy$$

$$= \int_{-\infty}^{\infty} h_1(x) h_2(y) f_{x,y}(x,y)\,dx\,dy \quad \because X, Y \text{ are independent} \quad \therefore f_{x,y}(x,y) = f_x(x) f_y(y)$$

$$= E\{h_1(X) h_2(Y)\}.$$

(**Discrete case**). $E\{h_1(X)\} = \sum_{i=-\infty}^{\infty} h_1(x_i) f_{xi}$, $E\{h_2(Y)\} = \sum_{j=-\infty}^{\infty} h_2(y_j) f_{yj}$, the infinite sums are assumed to be absolutely convergent.

So, their product is also converges absolutely.

$$\therefore\ E\{h_1(X)\}\,E\{h_2(Y)\} = \sum_{i=-\infty}^{\infty} h_1(x_i) f_{xi} \sum_{j=-\infty}^{\infty} h_2(y_j) f_{yj} = \sum_{i=-\infty}^{\infty}\sum_{j=-\infty}^{\infty} h_1(x_i) h_2(y_j) f_{xi} f_{yj},$$

as the series converges absolutely

$$= \sum_{i=-\infty}^{\infty}\sum_{j=-\infty}^{\infty} h_1(x_i) h_2(y_j) f_{ij} \quad \because X, Y \text{ are independent} \therefore f_{ij} = f_{xi} f_{yj}$$

$$= E\{h_1(X) h_2(Y)\}.$$

In particular, if $h_1(X) = X$ and $h_2(Y) = Y$, then $E(XY) = E(X)E(Y)$.

Theorem 7.5.2 If X and Y are two independent RVs, then they are uncorrelated but not conversely.

Proof. The uncorrelation of X and Y implies

$$\text{cov}(X,Y) = E\{(X - m_x)(Y - m_y)\}$$
$$= E(X - m_x)E(Y - m_y) \text{ as } X \text{ and } Y \text{ are independent}$$
$$= 0$$

$\therefore\ \rho(X,Y) = \frac{\text{cov}(X,Y)}{\sigma_x \sigma_y} = 0 \Rightarrow X$ and Y are uncorrelated.

The converse is not necessarily true. See the following example.

Let X is normal $(0,1)$. $\therefore\ m_x = 0$.

Let us consider a RV Y depending on X such that $Y = X^2$, so that X, Y are not independent and Y is a $\chi^2(1)$ distribution.

Now, $\text{cov}(X,Y) = E\{(X-0)(Y - m_y)\} = E[X(X^2 - m_y)] = E(X^3) - m_y E(X) = 0 - m_y \cdot 0 = 0$, as X is normal $(0,1)$ and so it is symmetric about the origin.

$\therefore$ all odd ordered central moments [which is same as ordinary moments for this case as $m_x = 0$] vanish.

$\therefore\ \rho(X,Y) = \frac{\text{cov}(X,Y)}{\sigma_x \sigma_y} = 0$.

Hence, X and Y are uncorrelated although they are dependent by the relation $Y = X^2$. Such a situation arises due to the symmetry of the distribution.

Theorem 7.5.3 If the two-dimensional RV (X,Y) follows a bivariate normal distribution, then X and Y will be independent, if and only if X and Y are uncorrelated.

Proof. For any two independent RVs X, Y their correlation coefficient $\rho(X,Y) = 0$ as $\text{cov}(X,Y) = E(X - m_x)(Y - m_y) = E(X - m_x)\,E(Y - m_y)$, as X, Y are independent

$= 0$.

$\therefore$ in particular, when (X,Y) follows bivariate normal distribution and X, Y are independent, $\rho = \rho(X,Y) = 0$.

Now, let $\rho=0$, we show that X and Y are independent.

[**Note:** This is not so for an arbitrary distribution.]

For a bivariate normal distribution, the pdf $f_{x,y}$ is given by

$$f_{x,y}(x,y)=\frac{1}{2\pi\sigma_x\sigma_y\sqrt{1-\rho^2}}\exp\left[-\frac{1}{2(1-\rho^2)}\left\{\left(\frac{x-m_x}{\sigma_x}\right)^2-2\rho\frac{x-m_x}{\sigma_x}\cdot\frac{y-m_y}{\sigma_y}+\left(\frac{y-m_y}{\sigma_y}\right)^2\right\}\right]$$

$$-\infty<x,y<\infty$$

Let $\rho=\rho(X,Y)=0$.

$$\therefore\ f_{x,y}(x,y)=\frac{1}{2\pi\sigma_x\sigma_y}\exp\left[-\frac{1}{2}\left\{\left(\frac{x-m_x}{\sigma_x}\right)^2+\left(\frac{y-m_y}{\sigma_y}\right)^2\right\}\right]=\frac{1}{\sqrt{2\pi}\sigma_x}e^{-\frac{(x-m_x)^2}{2\sigma_x^2}}\cdot\frac{1}{\sqrt{2\pi}\sigma_y}e^{-\frac{(y-m_y)^2}{2\sigma_y^2}}$$

$=f_X(x)\cdot f_Y(y)=$ product of the pdf of the marginal distributions of X and Y.

Hence, the NASC for the independence of two RVs leads to the conclusion that X and Y are independent.

Note: $f_{x,y}(x,y)=f_X(x)\cdot f_Y(y)=$ product of two functions, one is a function of x alone and the other is a function of y alone. Hence, X and Y are independent.

7.6 Generalisation of expectation in a higher dimension

Let $g:\mathbb{R}^n\to\mathbb{R}$ be a continuous function and $(X_1,X_2,\cdots,X_n)$ be a n-dimensional RV. Then, the distribution of the RVs $g(X_1,X_2,\cdots,X_n)$ is completely determined by the joint distribution of the RVs $X_1,X_2,\cdots,X_n$.

The mean or expected value of the probability distribution determined by the RV $g(X_1,X_2,\cdots,X_n)$ is as usual denoted by $E\{g(X_1,X_2,\cdots,X_n)\}$ and is defined by

$$E\{g(X_1,X_2,\cdots,X_n)\}=\sum_{i_1=-\infty}^{\infty}\sum_{i_2=-\infty}^{\infty}\cdots\sum_{i_n=-\infty}^{\infty}g(x_{i_1},x_{i_2},\cdots,x_{i_n})\,f_{i_1i_2\cdots i_n},$$

when $(x_{i_1},x_{i_2},\cdots,x_{i_n})$ is a point in the spectrum of the RV $(X_1,X_2,\cdots,X_n)$ and $f_{i_1i_2\cdots i_n}=P(X_1=x_{i_1},X_2=x_{i_2},\cdots,X_n=x_{i_n})$, provided the infinite sum converges absolutely, for the discrete case.

In the continuous case:

$$E\{g(X_1,X_2,\cdots,X_n)\}=\int_{-\infty}^{\infty}\int_{-\infty}^{\infty}\cdots\int_{-\infty}^{\infty}g(x_1,x_2,\cdots,x_n)\,f_{X_1,X_2,\cdots,X_n}(x_1,x_2,\cdots,x_n)\,dx_1\,dx_2\cdots dx_n,$$

provided the integral converges absolutely.

Let $g(X_1,X_2,\cdots,X_n)=a_1X_1+a_2X_2+\cdots+a_nX_n$. Then, $E\{g(X_1,X_2,\cdots,X_n)\}=a_1m_1+a_2m_2+\cdots+a_nm_n$, where $m_i=E(X_i), i=1(1)n$.

$$\text{var}\{g(X_1,X_2,\cdots,X_n)=E\{a_1(X_1-m_1)+a_2(X_2-m_2)+\cdots+a_n(X_n-m_n)\}^2$$

$$=\sum_{i=1}^{n}a_i^2\sigma_i^2+2\sum_{\substack{i,j=1\\i<j}}^{n}a_ia_j\,E\{(X_i-m_i)(X_j-m_j)\},$$

$$\sigma_i^2=E(X_i-m_i)^2=\text{var}(X_i)$$

i.e., $$\text{var}(a_1X_1+a_2X_2+\cdots+a_nX_n)=\sum_{i=1}^{n}a_i^2\sigma_i^2+2\sum_{\substack{i,j=1\\i<j}}^{n}a_ia_j\sigma_i\sigma_j\rho(X_i,X_j).$$

When $X_1, X_2, \cdots, X_n$ are pairwise uncorrelated, then $\text{var}\left\{\sum_{i=1}^{n} a_i X_i\right\} = \sum_{i=1}^{n} a_i^2 \sigma_i^2$.

Theorem 7.6.1 If $X_1, X_2, \cdots, X_n$ are mutually independent RVs and $g_1, g_2, \cdots, g_n$ are continuous functions of $X_1, X_2, \cdots, X_n$ respectively such that their MVs exist, then $E = \{g_1(X_1), g_2(X_2), \cdots, g_n(X_n)\} = E\{g_1(X_1)\}\, E\{g_2(X_2)\} \cdots E\{g_n(X_n)\}$.

Proof. (*Discrete case*). Let $f_{1i}, f_{2i}, \cdots, f_{ni}$ be the pmf of the mutually independent RVs $X_1, X_2, \cdots, X_n$ respectively and $f_{i_1 i_2 \cdots i_n}$ be their joint probability mass function, then $f_{i_1 i_2 \cdots i_n} = f_{1i_1} f_{2i_2} \cdots f_{ni_n}$ (1)

$$\therefore\ E\{g_1(X_1)\} = \sum_{i=-\infty}^{\infty} g_1(x_{1i}) f_{1i},\ E\{g_2(X_2)\} = \sum_{i=-\infty}^{\infty} g_2(x_{2i}) f_{2i}, \cdots, E\{g_n(X_n)\} = \sum_{i=-\infty}^{\infty} g_n(x_{ni}) f_{ni}.$$

$$\begin{aligned}
\therefore\ & E\{g_1(X_1)\}\, E\{g_2(X_2)\} \cdots E\{g_n(X_n)\} \\
&= \sum_{i_1=-\infty}^{\infty} g_1(x_{1i_1}) f_{1i_1} \sum_{i_2=-\infty}^{\infty} g_2(x_{2i_2}) f_{2i_2} \cdots \sum_{i_n=-\infty}^{\infty} g_n(x_{ni_n}) f_{ni_n} \\
&= \sum_{i_1=-\infty}^{\infty} \sum_{2_1=-\infty}^{\infty} \cdots \sum_{i_n=-\infty}^{\infty} g_1(x_{1i_1})\, g_2(x_{2i_2}) \cdots g_n(x_{ni_n})\, f_{1i_1} f_{2i_2} \cdots f_{ni_n} \\
&= \sum_{i_1=-\infty}^{\infty} \sum_{i_2=-\infty}^{\infty} \cdots \sum_{i_n=-\infty}^{\infty} g_1(x_{1i_1})\, g_2(x_{2i_2}) \cdots g_n(x_{ni_n})\, f_{i_1 i_2 \cdots i_n} \text{ by (1)} \\
&= E\{g_1(X_1)\, g_2(X_2) \cdots g_{n(X_n)}\} \text{ (Proved).}
\end{aligned}$$

Continuous case. As $X_1, X_2, \cdots, X_n$ are independent,

$\therefore\ f_{X_1, X_2, \cdots, X_n}(x_1, x_2, \cdots, x_n) = f_{X_1}(x_1)\, f_{X_2}(x_2) \cdots f_{X_n}(x_n).$ (2)

$$\begin{aligned}
& E\{g_1(X_1)\} E\{g_2(X_2)\} \cdots E\{g_n(X_n)\} \\
&= \int_{-\infty}^{\infty} g_1(x_1) f_{X_1}(x_1)\, dx_1 \int_{-\infty}^{\infty} g_2(x_2) f_{X_2}(x_2)\, dx_2 \cdots \int_{-\infty}^{\infty} g_n(x_n) f_{X_n}(x_n)\, dx_n \\
&= \int_{-\infty}^{\infty} \int_{-\infty}^{\infty} \cdots \int_{-\infty}^{\infty} g_1(x_1)\, g_2(x_2) \cdots g_n(x_n) f_{X_1}(x_1)\, f_{X_2}(x_2) \cdots f_{X_n}(x_n)\, dx_1\, dx_2 \cdots dx_n \\
&= \int_{-\infty}^{\infty} \int_{-\infty}^{\infty} \cdots \int_{-\infty}^{\infty} g_1(x_1)\, g_2(x_2) \cdots g_n(x_n) f_{X_1, X_2, \cdots, X_n}(x_1, x_2, \cdots, x_n)\, dx_1\, dx_2 \cdots dx_n \text{ [by (2)]} \\
&= E\{g_1(X_1)\, g_2(X_2) \cdots g_n(X_n)\} \text{ (Proved).}
\end{aligned}$$

7.7 Moments

Ordinary moment or moment about the origin of the joint distribution of the random $X_1, X_2, \cdots, X_n$ of order $k_1 + k_2 + \cdots + k_n$ is denoted by $\alpha_{k_1 k_2 \cdots k_n}$ and is defined by the MV $\alpha_{k_1 k_2 \cdots k_n} = E\left(X_1^{k_1} X_2^{k_2} \cdots X_n^{k_n}\right)$, if exists.

The central moment or moment about the centre of mass $(m_1, m_2, \cdots, m_n)$ of the same order $k_1 + k_2 + \cdots + k_n$ is denoted by $\mu_{k_1 k_2 \cdots k_n}$ and is defined by $\mu_{k_1 k_2 \cdots k_n} = E\left\{(X_1 - m_1)^{k_1} (X_2 - m_2)^{k_2} (X_n - m_n)^{k_n}\right\}$ (if exists), where $m_1 = E(X_1), m_2 = E(X_2), \cdots, m_n = E(X_n)$.

The moment about some point $(c_1, c_2 \cdots, c_n)$ of the same order $k_1 + k_2 + \cdots + k_n$ is defined by $E\left\{(X_1 - c_1)^{k_1} (X_2 - c_2)^{k_2} \cdots (X_n - c_n)^{k_n}\right\}$, if exists.

In case of independence of $X_1, X_2, \cdots, X_n$

$$\alpha_{k_1 k_2 \cdots k_n} = E\left(X_1^{k_1}\right) E\left(X_2^{k_2}\right) \cdots E\left(X_n^{k_n}\right)$$

$$\mu_{k_1 k_2 \cdots k_n} = E\left\{(X_1 - m_1)^{k_1}\right\} E\left\{(X_2 - m_2)^{k_2}\right\} \cdots E\left\{(X_n - m_n)^{k_n}\right\}, \text{ etc.}$$

7.7.1 Moment generating function

The MGF of the distribution of a n-dimensional RV $(X_1, X_2, \cdots, X_n)$ is a function $\psi_{x_1,x_2,\cdots,x_n} : \mathbb{R}^n \to \mathbb{R}$ defined by the MV:

$$\psi_{x_1,x_2,\cdots,x_n}(t_1,t_2,\cdots,t_n) = E\left(e^{t_1X_1+t_2X_2+\cdots+t_nX_n}\right), (t_1,t_2,\cdots,t_n) \in \mathbb{R}^n$$

$$= \sum_{i_1=-\infty}^{\infty}\sum_{i_2=-\infty}^{\infty}\cdots\sum_{i_n=-\infty}^{\infty} e^{t_1x_{1i_1}+t_2x_{2i_2}+\cdots+t_nx_{ni_n}} f_{i_1i_2\cdots i_n}, \text{ for a discrete case}$$

$$= \int_{-\infty}^{\infty}\int_{-\infty}^{\infty}\cdots\int_{-\infty}^{\infty} e^{t_1x_1+t_2x_2+\cdots+t_nx_n} f_{x_1,x_2,\cdots,x_n}(x_1,x_2,\cdots,x_n)\, dx_1\, dx_2 \cdots d_{x_n},$$

for a continuous case, provided the infinite sum or the improper integral converges absolutely.

Why moment generating? Consider the RV (X, Y).

$$\therefore \psi_{x,y}(t,u) = E\left\{e^{tX+uY}\right\} = E\left(e^{tX}\cdot e^{uY}\right) = E\left\{\sum_{k=0}^{\infty}\frac{t^kX^k}{k!}\sum_{s=0}^{\infty}\frac{u^sY^s}{s!}\right\} = \sum_{k=0}^{\infty}\sum_{s=0}^{\infty}\frac{t^ku^s}{k!s!}E\left(X^kY^s\right)$$

as the series are absolutely convergent and E is linear (provided $E\left(X^kY^s\right)$ exists)

$$= \sum_{k=0}^{\infty}\sum_{s=0}^{\infty}\frac{t^ku^s}{k!s!}\alpha_{ks}.$$

Thus, the ordinary moment of order $k+s$ is obtained by equating the coefficient of $\frac{t^ku^s}{k!s!}$ in the expansion of the MGF $\psi_{x,y}$ of the parameters t and u.

7.7.2 Moment generating function of a standard normal bivariate distribution

$$\psi_{x_1,x_2}(t_1,t_2) = \int_{-\infty}^{\infty}\int_{-\infty}^{\infty} e^{t_1x_1+t_2x_2} f_{x_1,x_2}(x_1,x_2)\,dx_1\,dx_2,$$

$f_{x_1,x_2}(x_1,x_2)$ = pdf of a standard normal bivariate distribution

$$= \int_{-\infty}^{\infty}\int_{-\infty}^{\infty} e^{t_1x_1+t_2x_2}.\frac{1}{2\pi\sqrt{1-\rho^2}}e^{-\frac{1}{2(1-\rho^2)}\{x_1^2+x_2^2-2\rho x_1x_2\}}\,dx_1\,dx_2$$

$$= \frac{1}{2\pi\sqrt{1-\rho^2}}\int_{-\infty}^{\infty}\int_{-\infty}^{\infty}\exp\left[-\frac{1}{2(1-\rho^2)}\left\{x_1^2+x_2^2-2\rho x_1x_2-2\left(1-\rho^2\right)(t_1x_1+t_2x_2)\right\}\right]dx_1\,dx_2$$

$$= \frac{1}{2\pi\sqrt{1-\rho^2}}\int_{-\infty}^{\infty}\int_{-\infty}^{\infty}\exp\left[-\frac{1}{2(1-\rho^2)}\left\{(x_1-t_1)^2(x_2-t_2)^2-2\rho x_1x_2-t_1^2-t_2^2+2\rho^2(t_1x_1+t_2x_2)\right\}\right]dx_1dx_2$$

$$= \frac{1}{2\pi\sqrt{1-\rho^2}}\int_{-\infty}^{\infty}\int_{-\infty}^{\infty}\exp\left[-\frac{1}{2(1-\rho^2)}\left\{(x_1-t_1-\rho t_2)^2(x_2-t_2-\rho t_1)^2-\left(1+\rho^2\right)\left(t_1^2+t_2^2\right)\right.\right.$$

$$\left.\left.+2\rho t_2(x_1-t_1)+2\rho t_1(x_2-t_2)-2\rho x_1x_2+2\rho^2(t_1x_1+t_2x_2)\right\}\right]dx_1\,dx_2$$

$$= \frac{1}{2\pi\sqrt{1-\rho^2}}\int_{-\infty}^{\infty}\int_{-\infty}^{\infty}\exp\left[-\frac{1}{2(1-\rho^2)}\left\{(x_1-t_1-\rho t_2)^2+(x_2-t_2-\rho t_1)^2-2\rho(x_1-t_1-\rho t_2)(x_2-t_2-\rho t_1)\right.\right.$$

$$\left.\left.-\left(1-\rho^2\right)\left(t_1^2+t_2^2\right)-2\rho t_1t_2+2\rho^3t_1t_2\right\}\right]dx_1dx_2$$

$$= e^{\frac{1}{2}\left(t_1^2+t_2^2+2\rho t_1t_2\right)}\int_{-\infty}^{\infty}\int_{-\infty}^{\infty}\frac{1}{2\pi\sqrt{1-\rho^2}}\exp\left[-\frac{1}{2(1-\rho^2)}\left\{\left(x_1-\overline{t_1+\rho t_2}\right)^2+\left(x_2-\overline{t_2+\rho t_1}\right)^2-2\rho\left(x_1-\overline{t_1+\rho t_2}\right)\left(x_2-\overline{t_2+\rho t_1}\right)\right\}\right]dx_1\,dx_2$$

$$= e^{\frac{1}{2}\left(t_1^2+t_2^2+2\rho t_1t_2\right)},$$

the remaining part is the double integral of a bivariate normal $(t_1+\rho t_2, t_2+\rho t_1, 1, 1, \rho)$ distribution and hence equal to 1.

7.8 Joint characteristic function

The joint characteristic function of the distribution of (X, Y) is a function $\chi_{x,y}:\mathbb{R}^2 \to \mathbb{C}$ defined by $\chi_{x,y}(t,u) = E\left(e^{i(tX+uY)}\right), i=\sqrt{-1}$.

It is very important to note that characteristic function exists for all distribution as

$$\begin{aligned} E\left|e^{i(tx+uy)}\right| &= \int_{-\infty}^{\infty}\int_{-\infty}^{\infty} \left|e^{i(tx+uy)}\right| f_{x,y}(x,y)\,dx\,dy \\ &= \int_{-\infty}^{\infty}\int_{-\infty}^{\infty} |\cos(tx+uy)+i\sin(tx+uy)| f_{x,y}(x,y)\,dx\,dy \\ &= \int_{-\infty}^{\infty}\int_{-\infty}^{\infty} \sqrt{\cos^2(tx+uy)+\sin^2(tx+uy)}\, f_{x,y}(x,y)\,dx\,dy \\ &= \int_{-\infty}^{\infty}\int_{-\infty}^{\infty} f_{x,y}(x,y)\,dx\,dy = 1. \end{aligned}$$

$\therefore$ the integral $\int_{-\infty}^{\infty}\int_{-\infty}^{\infty} e^{i(tx+uy)} f_{x,y}(x,y)\,dx\,dy$ converges absolutely. Hence, $E(e^{i(tx+uy)})$ exists for all distributions, discrete or continuous.

For continuous distribution, characteristic function is an integral transform of the pdf w.r.t. the weight function $e^{i(tx+uy)}$ for a joint distribution of X and Y. So, characteristic function is a unique character of a distribution. Chracteristic function has a one-to-one correspondence with the probability distribution. So, a distribution can be determined from its characteristic function.

The characteristic function of the distribution of a n-dimensional RV $(X_1, X_2, \cdots, X_n)$ is a function $\chi_{x_1,x_2,\cdots,x_n}:\mathbb{R}^n \to \mathbb{C}$ defined by $\chi_{x_1,x_2,\cdots,x_n}(t_1,t_2,\cdots,t_n) = E\left(e^{i(t_1X_1+t_2X_2+\cdots+t_nX_n)}\right)$.

Exactly in a similar manner characteristic function exists for all distribution of $(X_1, X_2, \cdots, X_n)$, discrete or continuous.

$$\begin{aligned} \therefore\ \chi_{x_1,x_2,\cdots,x_n}(t_1,t_2,\cdots,t_n) &= \int_{-\infty}^{\infty}\int_{-\infty}^{\infty}\cdots\int_{-\infty}^{\infty} e^{i(t_1x_1+t_2x_2+\cdots+t_nx_n)} f_{x_1,x_2,\cdots,x_n}(x_1,x_2,\cdots,x_n)\,dx_1\,dx_2\cdots dx_n, \\ &\qquad \text{for a continuous distribution} \\ &= \sum_{i_1=-\infty}^{\infty}\sum_{i_2=-\infty}^{\infty}\cdots\sum_{i_n=-\infty}^{\infty} e^{i\left(t_1x_{1i_1}+t_2x_{2i_2}+\cdots+t_nx_{ni_n}\right)}, \text{ for a discrete distribution,} \end{aligned}$$

where $f_{x_1,x_2,\cdots,x_n}$ is the joint pdf and
$f_{i_1i_2\cdots i_n} = P(X_1 = X_{1i_1}, X_2 = X_{2i_2}, \cdots, X_n = x_{ni_n})$ is the pmf.

Theorem 7.8.1 If two RVs X and Y are independent, then their joint characteristic function is the product of the individual characteristic functions, i.e., $\chi_{x,y}(t,u) = \chi_x(t)\chi_y(u)$ and conversely.

Proof. (*Condition necessary.*) Let X, Y are independent.

$$\therefore\ \chi_{x,y}(t,u)=\int_{-\infty}^{\infty}\int_{-\infty}^{\infty} e^{i(tx+uy)} f_{x,y}(x,y)\,dx\,dy$$

$$=\int_{-\infty}^{\infty} e^{i(tx+uy)} f_x(x) f_y(y)\,dx\,dy, \text{ as } X, Y \text{ are independent}$$

$$=\int_{-\infty}^{\infty} e^{itx} f_x(x)\,dx \int_{-\infty}^{\infty} e^{iuy} f_y(y)\,dy, \text{ where the integrals are absolutely convergent}$$

$$=\chi_x(t)\chi_y(u).$$

Conversely, let $\chi_{x,y}(t,u)=\chi_x(t)\chi_y(u)$. (1)

$$\therefore\ \int_{-\infty}^{\infty}\int_{-\infty}^{\infty} e^{i(tx+my)} f_{x,y}(x,y)\,dx\,dy=\int_{-\infty}^{\infty} e^{itx} f_x(x)\,dx \int_{-\infty}^{\infty} e^{iuy} f_y(y)\,dy$$

$$=\int_{-\infty}^{\infty}\int_{-\infty}^{\infty} e^{i(tx+uy)} f_x(x) f_y(y)\,dx\,dy$$

$$\text{or, } \int_{-\infty}^{\infty}\int_{-\infty}^{\infty} e^{i(tx+uy)}\left\{f_{x,y}(x,y)-f_x(x)f_y(y)\right\}dx\,dy\equiv 0,$$

for all distributions satisfying (1).

$\therefore\ f_{x,y}(x,y)=f_x(x)f_y(y)$, whenever (1) is true $\Rightarrow X$ and Y are independent.

Discrete case. Let X and Y are independent.

$$\chi_{x,y}(t,u)=\sum_{j=-\infty}^{\infty}\sum_{k=-\infty}^{\infty} e^{i(tx_j+uy_k)} f_{jk}=\sum_{j=-\infty}^{\infty}\sum_{k=-\infty}^{\infty} e^{itx_j+ity_k} f_{xj} f_{yk} \text{ as } X \text{ and } Y \text{ are independent}$$

$$=\sum_{j=-\infty}^{\infty} e^{itx_j} f_{xj}\sum_{k=-\infty}^{\infty} e^{ity_k} f_{yk}, \text{ where the infinite series are absolutely convergent}$$

$$=\chi_x(t)\chi_y(u),$$

where $f_{jk}=P(X=x_j, Y=y_k)=P(X=x_j)P(Y=y_k)=f_{xj}f_{yk}$ in case of independence.

Conversely, let $\chi_{x,y}(t,u)=\chi_x(t)\chi_y(u)$. (2)

$$\therefore\ \sum_{j=-\infty}^{\infty}\sum_{k=-\infty}^{\infty} e^{i(tx_j+uy_k)} f_{jk}=\sum_{j=-\infty}^{\infty} e^{itx_j} f_{xj}\sum_{k=-\infty}^{\infty} e^{iuy_k} f_{yk}=\sum_{j=-\infty}^{\infty}\sum_{k=-\infty}^{\infty} e^{i(tx_j+uy_k)} f_{xj} f_{yk}$$

$$\text{or, } \sum_{j=-\infty}^{\infty}\sum_{k=-\infty}^{\infty} e^{i(tx_j+uy_k)}\left\{f_{jk}-f_{xj}f_{yk}\right\}=0,$$

for all distributions of (X, Y) satisfying (2).

$\therefore\ f_{jk}=f_{xj}f_{yk}\Rightarrow X$ and Y are independent.

Theorem 7.8.2 The RVs $X_1, X_2, \cdots, X_n$ will be mutually independent, if and only if their joint characteristic function can be expressed as the product of their individual characteristic functions, i.e., $\chi_{x_1,x_2,\cdots,x_n}(t_1,t_2,\cdots,t_n)=\chi_{x_1}(t_1)\chi_{x_2}(t_2)\cdots\chi_{x_n}(t_n)$.

Proof. (*Continuous case*). Condition necessary.

Let $X_1 X_2, \cdots, X_n$ are mutually independent.

$$\therefore\ f_{x_1,x_2,\cdots,x_n}(x_1, x_2, \cdots, x_n) = f_{x_1}(x_1)\, f_{x_2}(x_2) \cdots f_{x_n}(x_n). \qquad (1)$$

$$\therefore\ \chi_{x_1,x_2,\cdots,x_n}(t_1, t_2, \cdots, t_n) = \int_{-\infty}^{\infty}\int_{-\infty}^{\infty}\cdots\int_{-\infty}^{\infty} e^{i(t_1x_1+t_2x_2+\cdots+t_nx_n)} f_{x_1,x_2,\cdots,x_n}(x_1, x_2, \cdots, x_n)\, dx_1\, dx_2 \cdots dx_n$$

$$= \int_{-\infty}^{\infty}\int_{-\infty}^{\infty}\cdots\int_{-\infty}^{\infty} e^{it_1x_1+it_2x_2+\cdots+it_nx_n} f_{x_1}(x_1)\, f_{x_2}(x_2) \cdots f_{x_n}(x_n)\, dx_1\, dx_2 \cdots dx_n$$

[using (1)]

$$= \int_{-\infty}^{\infty} e^{it_1x_1} f_{x_1}(x_1) dx_1 \int_{-\infty}^{\infty} e^{it_2x_2} f_{x_2}(x_2)\, dx_2 \cdots \int_{-\infty}^{\infty} e^{it_nx_n} f_{x_n}(x_n)\, dx_n,$$

where every series absolutely converges

$$= \chi_{x_1}(t_1)\chi_{x_2}(t_2) \cdots \chi_{x_n}(t_n).$$

Condition sufficient. Let $\chi_{x_1,x_2,\cdots,x_n}(t_1, t_2, \cdots, t_n) = \chi_{x_1}(t_1)\chi_{x_2}(t_2) \cdots \chi_{x_n}(t_n)$.

$$\therefore\ \int_{-\infty}^{\infty}\int_{-\infty}^{\infty}\cdots\int_{-\infty}^{\infty} e^{i(t_1x_1+t_2x_2+\cdots+t_nx_n)} f_{x_1,x_2,\cdots,x_n}(x_1, x_2, \cdots, x_n)\, dx_1\, dx_2 \cdots dx_n$$

$$= \int_{-\infty}^{\infty} e^{it_1x_1} f_{x_1}(x_1)\, dx_1 \int_{-\infty}^{\infty} e^{it_2x_2} f_{x_2}(x_2)\, dx_2 \cdots \int_{-\infty}^{\infty} e^{it_nx_n} f_{x_n}(x_n)\, dx_n$$

$$= \int_{-\infty}^{\infty}\int_{-\infty}^{\infty}\cdots\int_{-\infty}^{\infty} e^{i(t_1x_1+t_2x_2+\cdots+t_nx_n)} f_{x_1}(x_1) f_{x_2}(x_2) \cdots f_{x_n}(x_n)\, dx_1\, dx_2 \cdots dx_n$$

or, $$\int_{-\infty}^{\infty}\int_{-\infty}^{\infty}\cdots\int_{-\infty}^{\infty} e^{i(t_1x_1+t_2x_2+\cdots+t_nx_n)} \{f_{x_1,x_2,\cdots,x_n}(x_1, x_2, \cdots, x_n) - f_{x_1}(x_1) f_{x_2}(x_2) \cdots f_{x_n}(x_n)\}\, dx_1\, dx_2 \cdots dx_n = 0,$$

for every distribution of $(X_1, X_2, \cdots, X_n)$.

$\therefore\ f_{x_1,x_2,\cdots,x_n}(x_1, x_2, \cdots, x_n) = f_{x_1}(x_1)\, f_{x_2}(x_2) \cdots f_{x_n}(x_n)$. Hence, $X_1, X_2, \cdots, X_n$ are mutually independent.

(*Discrete case.*) Condition necessary. Let $X_1, X_2, \cdots, X_n$ are mutually independent, i.e., $f_{i_1i_2\cdots i_n} = f_{1i_1} f_{2i_2} \cdots f_{ni_n}$.

$$\therefore\ \chi_{x_1,x_2,\cdots,x_n}(t_1, t_2, \cdots, t_n) = \sum_{i_1=-\infty}^{\infty}\sum_{i_2=-\infty}^{\infty}\cdots\sum_{i_n=-\infty}^{\infty} e^{i(t_1x_{1i_1}+t_2x_{2i_2}+\cdots+t_nx_{ni_n})} f_{i_1i_2\cdots i_n}$$

$$= \sum_{i_1=-\infty}^{\infty}\sum_{i_2=-\infty}^{\infty}\cdots\sum_{i_n=-\infty}^{\infty} e^{i(t_1x_{1i_1}+t_2x_{2i_2}+\cdots+t_nx_{ni_n})} f_{1i_1} f_{2i_2} \cdots f_{ni_n}$$

$$= \sum_{i_1=\infty}^{\infty} e^{it_1x_{1i_1}} f_{1i_1} \sum_{i_2=\infty}^{\infty} e^{it_2x_{2i_2}} f_{2i_2} \cdots \sum_{i_n=\infty}^{\infty} e^{it_nx_{ni_n}} f_{ni_n}$$

$$= \chi_{x_1}(t_1)\chi_{x_2}(t_2) \cdots \chi_{x_n}(t_n).$$

Condition sufficient. Let $\chi_{x_1,x_2,\cdots,x_n}(t_1, t_2, \cdots, t_n) = \chi_{x_1}(t_1)\chi_{x_2}(t_2) \cdots \chi_{x_n}(t_n)$.

$$\therefore \sum_{i_1=-\infty}^{\infty}\sum_{i_2=-\infty}^{\infty}\cdots\sum_{i_n=-\infty}^{\infty} e^{i(t_1x_{1i_1}+t_2x_{2i_2}+\cdots+t_nx_{ni_n})} f_{i_1i_2\cdots i_n}$$

$$=\sum_{i_1=-\infty}^{\infty} e^{it_1x_{1i_1}} f_{1i_1}\sum_{i_2=-\infty}^{\infty} e^{it_2x_{1i_2}} f_{2i_2}\cdots\sum_{i_n=-\infty}^{\infty} e^{it_nx_{1in}} f_{ni_n}$$

$$=\sum_{i_1=-\infty}^{\infty}\sum_{i_2=-\infty}^{\infty}\cdots\sum_{i_n=-\infty}^{\infty} e^{i(t_1x_{1i_1}+t_2x_{2i_2}+\cdots+t_nx_{ni_n})} f_{1i_1}f_{2i_2}\cdots f_{ni_n}$$

$$\text{or, } \sum_{i_1=-\infty}^{\infty}\sum_{i_2=-\infty}^{\infty}\cdots\sum_{i_n=-\infty}^{\infty} e^{i(t_1x_{1i_1}+t_2x_{2i_2}+\cdots+t_nx_{ni_n})}\{f_{i_1i_2\cdots i_n}-f_{1i_1}f_{2i_2}\cdots f_{ni_n}\}=0,$$

for every distribution of $(X_1, X_2, \cdots, X_n)$.

$\therefore\ f_{i_1i_2\cdots i_n}=f_{1i_1}f_{2i_2}\cdots f_{ni_n} \Rightarrow X_1, X_2, \cdots, X_n$ are mutually independent.

Note. If a set of RVs $\{X_1, X_2, \cdots, X_n\}$ are mutually independent then the RV of any subset of this set are also mutually independent.

Theorem 7.8.3 If the RVs $X_1, X_2, \cdots, X_n$ are mutually independent and $g_i : S_i \to \mathbb{R}$, $i = 1(1)n$, be continuous functions, such that $E\{g_i(X_i)\}$ exists $\forall\ i$, then

$$E\{g_1(X_1)\, g_2(X_2)\cdots g_n(X_n)\} = E\{g_1(X_1)\}\, E\{g_2(X_2)\}\cdots E\{g_n(X_n)\}.$$

$[S_i$ = event space of $X_i]$.

Proof. (Continuous case)

As $X_1, X_2, \cdots, X_n$ are mutually independent.

$$\therefore\ f_{x_1,x_2,\cdots,x_n}(x_1, x_2, \cdots, x_n) = f_{x_1}(x_1)\, f_{x_2}(x_2)\cdots f_{x_n}(x_n). \tag{1}$$

$\therefore\ E\{g_1(X_1)\}E\{g_2(X_2)\}\cdots E\{g_n(X_n)\}$

$$=\int_{-\infty}^{\infty} g_1(x_1) f_{x_1}(x_1)\, dx_1 \int_{-\infty}^{\infty} g_2(x_2) f_{x_2}(x_2)\, dx_2 \cdots \int_{-\infty}^{\infty} g_n(x_n) f_{x_n}(x_n)\, dx_n$$

$$=\int_{-\infty}^{\infty}\int_{-\infty}^{\infty}\cdots\int_{-\infty}^{\infty} g_1(x_1) g_2(x_2)\cdots g_n(x_n) f_{x_1}(x_1) f_{x_2}(x_2)\cdots f_{x_n}(x_n)\, dx_1\, dx_2\cdots dx_n,$$

as the integrals converge absolutely

$$=\int_{-\infty}^{\infty}\int_{-\infty}^{\infty}\cdots\int_{-\infty}^{\infty} g_1(x_1) g_2(x_2)\cdots g_n(x_n) f_{x_1,x_2,\cdots,x_n}(x_1, x_2, \cdots, x_n)\, dx_1\, dx_2\cdots dx_n \ \text{[by (1)]}$$

$$= E\{g_1(X_1)\, g_2(X_2)\cdots g_n(X_n)\}.$$

Discrete case. As $X_1, x_2, \cdots, X_n$ are mutually independent. $\therefore\ f_{i_1i_2\cdots i_n} = f_{1i_1}f_{2i_2}\cdots f_{ni_n}$. (2)

$$\therefore\ E\{g_1(X_1)\}E\{g_2(X_2)\}\cdots E\{g_n(X_n)\} = \sum_{i_1=-\infty}^{\infty} g_1(x_{1i_1})\, f_{1i_1}\sum_{i_2=\infty}^{\infty} g_2(x_{2i_2})\, f_{2i_2}\cdots\sum_{i_n=\infty}^{\infty} g_n(x_{ni_n})\, f_{ni_n}$$

$$=\sum_{i_1=-\infty}^{\infty}\sum_{i_2=-\infty}^{\infty}\cdots\sum_{i_n=-\infty}^{\infty} g_1(x_{1i_1})\, g_2(x_{2i_2})\cdots g_n(x_{ni_n})\, f_{1i_1}f_{2i_2}\cdots f_{ni_n},$$

such a rearrangement is possible as the individual series converges absolutely

$$=\sum_{i_1=-\infty}^{\infty}\sum_{i_2=-\infty}^{\infty}\cdots\sum_{i_n=-\infty}^{\infty} g_1(x_{1i_1})\, g_2(x_{2i_2})\cdots g_n(x_{ni_n})\, f_{1i_12i_2\cdots i_n}, \ \text{[by (2)]}$$

$$= E\{g_1(X_1)\, g_2(X_2)\cdots g_n(X_n)\}.$$

7.9 Reproductive property

Sometimes it may so happen that each member of a set of independent RVs follows the same distribution with different parameters, then their sum also follows that distribution with some suitable parameters. Such a property of reproducing the same distribution is called the *reproductive property* of that distribution.

7.9.1 Reproductive property of Poisson distribution

Statement. If $X_1, X_2, \cdots, X_n$ be n-mutually independent Poisson variates with parameters $\mu_1, \mu_2, \cdots, \mu_n$, then the RV S_n of their sum, i.e., $S_n = X_1 + X_2 + \cdots + X_n$ is also a Poisson variate with the parameter $\mu = \mu_1 + \mu_2 + \cdots + \mu_n$. [$S_n$ is a Poisson variate which means S_n follows Poisson distribution.]
Proof. The characteristic function χ_k of the RV X_k following Poisson (μ_k) distribution is given by

$$\begin{aligned}\chi_k(t) &= E\left(e^{itX_k}\right),\ i=\sqrt{-1}\\ &= \sum_{j=0}^{\infty} e^{itj}\, e^{-\mu_k} \frac{\mu_k^j}{j!} \text{ [Spectrum is } X_k = x_{kj} = j, j = 0,1,2,\cdots]\\ &= e^{-\mu_k} \sum_{j=0}^{\infty} \frac{(\mu_k e^{it})^j}{j!} = e^{-\mu_k} e^{\mu_k e^{it}} = e^{\mu_k(e^{it}-1)},\ k=1,2,\cdots,n.\end{aligned}$$

As $X_1, X_2, \cdots, X_n$ are independent, so the characteristic function $\chi(t)$ of their sum S_n is given by

$$\begin{aligned}\chi(t) &= \chi_1(t)\chi_2(t)\cdots\chi_n(t) = e^{\mu_1(e^{it}-1)} e^{\mu_2(e^{it}-1)} \cdots e^{\mu_n(e^{it}-1)}\\ &= e^{(\mu_1+\mu_2+\cdots+\mu_n)\left(e^{it}-1\right)}\\ &= e^{\mu(e^{2t}-1)} \text{ as } \mu = \sum_{j=1}^{n} \mu_j,\end{aligned}$$

which is the characteristic function of a Poisson (μ) distribution. Uniqueness of characteristic function of a distribution implies that S_n is a Poisson distribution with parameter μ. Hence, proved.

7.9.2 Reproductive property of binomial distribution

Statement. If $X_1, X_2, \cdots, X_n$ be n-mutually independent binomial variates with parameters $(\lambda_1, p), (\lambda_2, p)$, $\cdots$ (λ_n, p), respectively, then the RV S_n of their sum, i.e., $S_n = X_1 + X_2 + \cdots + X_n$ is also a binomial variate with parameters (λ, p), where $\lambda = \lambda_1 + \lambda_2 + \cdots + \lambda_n$.
Proof. The characteristic function χ_k of X_k is given by

$$\chi_k(t) = E\left(e^{itX_k}\right) = \sum_{j=0}^{\lambda_k} e^{itj} \binom{\lambda_k}{j} p^j q^{\lambda_k - j} = \sum_{j=0}^{\lambda_k} \binom{\lambda_k}{j} \left(pe^{it}\right)^j q^{\lambda_k - j} = \left(pe^{it}+q\right)^{\lambda_k},\ k = 1(1)n.$$

$$\begin{aligned}\therefore\ \chi_{s_n}(t) &= E\left(e^{itS_n}\right) = E\left(e^{it(X_1+X_2+\cdots+X_n)}\right) = E\left\{e^{itX_1} e^{itX_2} \cdots e^{itX_n}\right\}\\ &= E\left(e^{itX_1}\right) E\left(e^{itX_2}\right) \cdots E\left(e^{itX_n}\right), \text{ as } X_1, X_2, \cdots, X_n \text{ are independent}\\ &= \left(pe^{it}+q\right)^{\lambda_1} \left(pe^{it}+q\right)^{\lambda_2} \cdots \left(pe^{it}+q\right)^{\lambda_n} = \left(pe^{it}+q\right)^{\lambda_1+\lambda_2+\cdots+\lambda_n}\\ &= \left(pe^{it}+q\right)^{\lambda}, \text{ where } \lambda = \lambda_1 + \lambda_2 + \cdots + \lambda_n.\end{aligned}$$

This is the characteristic function of a binomial (λ, p) distribution. Hence, the one-to-one correspondence of a distribution with its characteristic function implies that S_n follows binomial (λ, p) distribution.

7.9.3 Reproductive property of normal distribution

Statement. If $X_1, X_2, \cdots, X_n$ are n mutually independent normal variates with parameters $(m_1,\sigma_1),(m_2,\sigma_2),\cdots,(m_n,\sigma_n)$ respectively, then the RV S_n of their sum, i.e., $S_n = X_1+X_2+\cdots+X_n$ is also a normal variate with parameters (m,σ), when $m = m_1+m_2+\cdots+m_n$ and $\sigma=\sqrt{\sigma_1^2+\sigma_2^2+\cdots+\sigma_n^2}$.

Proof. The characteristic function χ_k of X_k is given by

$$\chi_k(t) = E\left(e^{itX_k}\right) = \frac{1}{\sqrt{2\pi}\sigma_k}\int_{-\infty}^{\infty} e^{itx_k} e^{-\frac{(x_k-m_k)^2}{2\sigma_k^2}}dx_k = \frac{1}{\sqrt{2\pi}\sigma_k}\int_{-\infty}^{\infty} e^{itx} e^{-\frac{(x-m_k)^2}{2\sigma_k^2}}dx$$

[by the property of definite integral]

$$= \frac{1}{\sqrt{2\pi}\sigma_k}\int_{-\infty}^{\infty} e^{-\frac{(x-m_k)^2-2it\sigma_k^2 x}{2\sigma_k^2}}dx = \frac{1}{\sqrt{2\pi}\sigma_k}\int_{-\infty}^{\infty} e^{\frac{-\left(x-m_k-it\sigma_k^2\right)^2-2itm_k\sigma_k^2+t^2\sigma_k^4}{2\sigma_k^2}}dx$$

$$= e^{im_kt-\frac{1}{2}\sigma_k^2t^2}\frac{1}{\sqrt{2\pi}\sigma_k}\int_{-\infty}^{\infty} e^{-\frac{(x-u)^2}{2\sigma_k^2}}dx, \text{ when } u = m_k+it\sigma_k^2$$

$$= e^{im_kt-\frac{1}{2}\sigma_k^2t^2}, \because \frac{1}{\sqrt{2\pi}\sigma_k}\int_{-\infty}^{\infty} e^{-\frac{(x-u)^2}{2\sigma_k^2}}dx = 1,\ k = 1,2,\cdots,n.$$

$$\therefore \chi_{S_n}(t) = E\left(e^{it(X_1+X_2+\cdots+X_n)}\right) = E\left(e^{itX_2}\right) = E\left(e^{itX_1}.e^{itX_2}\ldots e^{itX_n}\right)$$

$$= E\left(e^{itX_1}\right)E\left(e^{itX_2}\right)\cdots E\left(e^{itX_n}\right), \text{ as } X_1,X_2,\cdots,X_n \text{ are mutually independent}$$

$$= \chi_1(t)\chi_2(t)\cdots\chi_n(t) = e^{im_1t-\frac{1}{2}\sigma_1^2t^2}.e^{im_2t-\frac{1}{2}\sigma_2^2t^2}\ldots e^{im_nt-\frac{1}{2}\sigma_n^2t^2}$$

$$= e^{i(m_1+m_2+\cdots+m_n)t-\frac{1}{2}\left(\sigma_1^2+\sigma_2^2+\cdots+\sigma_n^2\right)t^2}$$

$$= e^{imt-\frac{1}{2}\sigma^2t^2}, \text{ where } m = m_1+m_2+\cdots+m_n \text{ and } \sigma = \sqrt{\sigma_1^2+\sigma_2^2+\cdots+\sigma_n^2}.$$

This is the characteristic function of a normal (m,σ) distribution. Hence, by the one-one correspondence between a distribution with its characteristic function, we conclude that S_n follows normal (m,σ) distribution.

Theorem 7.9.4 If X_1, X_2, $\cdots$, X_n are n mutually independent normal variates with parameter (m_1,σ_1), (m_2,σ_2), $\cdots$, (m_n,σ_n) respectively, then the RV S_n, the linear combination of them, i.e., $S_n = a_1X_1 + a_2X_2+\cdots+a_nX_n$, $a_i\in\mathbb{R}$, $\forall\ i = 1,2,\cdots,n$, is also a normal variate with parameters (m,σ), where $m = a_1m_1+a_2m_2+\cdots+a_nm_n$ and $\sigma = \sqrt{a_1^2\sigma_1^2+a_2^2\sigma_2^2+\cdots+a_n^2\sigma_n^2}$.

Proof. The characteristic function χ_k of X_k is given by $\chi_k(t) = e^{im_kt-\frac{1}{2}\sigma_k^2t^2}$.

$$\therefore \chi_{S_n}(t) = E\left(e^{itS_n}\right) = E\left\{e^{it(a_1X_1+a_2X_2+\cdots+a_nX_n)}\right\} = E\left(e^{ita_1X_1}.e^{ita_2X_2}\ldots e^{ita_nX_n}\right)$$

$$= E\left(e^{ita_1X_1}\right)E\left(e^{ita_2X_2}\right)\cdots E\left(e^{ita_nX_n}\right), \text{ as } X_1,X_2,\cdots,X_n \text{ are mutually independent}$$

$$= \chi_1(a_1t)\chi_2(a_2t)\cdots\chi_n(a_nt) = e^{ia_1m_1t-\frac{1}{2}\sigma_1^2a_1^2t^2}.e^{ia_2m_2t-\frac{1}{2}\sigma_2^2a_2^2t^2}\ldots e^{ia_nm_nt-\frac{1}{2}\sigma_n^2a_n^2t^2}$$

$$= e^{i(a_1m_1+a_2m_2+\cdots+a_nm_n)t-\frac{1}{2}\left(a_1^2\sigma_1^2+a_2^2\sigma_2^2+\cdots+a_n^2\sigma_n^2\right)t^2} = e^{imt-\frac{1}{2}\sigma^2t^2},$$

where $m = a_1m_1+a_2m_2+\cdots+a_nm_n$ and $\sigma = \sqrt{a_1^2\sigma_1^2+a_2^2\sigma_2^2+\cdots+a_n^2\sigma_n^2}$.

This is the characteristic function of a normal (m,σ) distribution. Hence, by the uniqueness of characteristic function for a probability distribution, we conclude that S_n is a normal variate with parameters (m,σ).

7.9.5 Reproductive property of gamma distribution

Statement. If the RVs $X_1, X_2, \cdots, X_n$ are mutually independent gamma variates with parameters $l_1, l_2, \cdots, l_n$ respectively, then the RV S_n of their sum, i.e., $S_n = X_1+X_2+\cdots+X_n$ is also a gamma variate with parameter l, where $l = l_1 + l_2 + \cdots + l_n$.

Proof. The characteristic function χ_k of the RV X_k is given by

$$\chi_k(t) = E\left(e^{itX_k}\right) = \int_0^\infty e^{itx_k} \frac{e^{-x_k} x_k^{l_k-1}}{\Gamma(l_k)} dx_k, \ l_k > 0$$

$$= \int_0^\infty e^{itx} \frac{e^{-x} x^{l_k-1}}{\Gamma(l_k)} dx, \text{ by the property of definite integral. (Here } x_k \text{ is a dummy variable)}$$

$$= \frac{1}{\Gamma(l_k)} \int_0^\infty e^{-(1-it)x} x^{l_k-1} dx = \frac{1}{\Gamma(l_k)} \lim_{\substack{R\to\infty \\ \epsilon\to 0+}} \int_\epsilon^R e^{-(1-it)x} x^{l_k-1} dx$$

Put $(1-it)x = u$. $\therefore$ $(1-it)dx = du$

$$= \frac{1}{\Gamma(l_k)} \lim_{\substack{R\to\infty \\ \epsilon\to 0+}} \int_{(1-it)\epsilon}^{(1-it)R} e^{-u} u^{l_k-1} (1-it)^{-l_k+1} \frac{du}{1-it} = (1-it)^{-l_k} \cdot \frac{1}{\Gamma(l_k)} \int_0^\infty e^{-u} u^{l_k-1} du$$

[from the theory of complex analysis]

$$= (1-it)^{-l_k} \cdot \frac{1}{\Gamma(l_k)} \cdot \Gamma(l_k) = (1-it)^{-l_k}, \ k = 1, 2, \cdots, n.$$

$$\therefore \chi_{s_n}(t) = E\left(e^{itS_n}\right) = E\left(e^{it(X_1+X_2+\cdots+X_n)}\right) = E\left(e^{itX_1} \cdot e^{itX_2} \cdots e^{itX_n}\right),$$

$$= E\left(e^{itX_1}\right) E\left(e^{itX_2}\right) \cdots E\left(e^{itX_n}\right), \text{ as } X_1, X_2, \cdots, X_n \text{ are mutually independent}$$

$$= (1-it)^{-l_1}(1-it)^{-l_2} \cdots (1-it)^{-l_n} = (1-it)^{-(l_1+l_2+\cdots+l_n)}$$

$$= (1-it)^{-l}, \text{ when } l = l_1 + l_2 + \cdots + l_n.$$

This is the characteristic function of a gamma (l) distribution. Hence, by the uniqueness of characteristic function for a probability distribution we conclude, then S_n follows gamma distribution with parameter l, where $l = l_1 + l_2 + \cdots + l_n$.

Ex. 1 If (X, Y) is a standard normal variate in two dimensions and $\rho(X, Y) = \rho$, then prove that $\rho\,(X^2, Y^2) = \rho^2$.

Solution. Since, (X, Y) is standard normal variate. $\therefore$ $E(X)=0, E(Y)=0, \text{var}(X)=1, \text{var}(Y)=1$.

We know that the MGF of a standard normal BD is given by

$$\psi_{x,y}(t,s) = \exp\left\{\tfrac{1}{2}\left(t^2+s^2+2\rho ts\right)\right\}, \text{ where } \rho = \rho(X,Y),$$

i.e., $\psi_{x,y}(t,s) = 1 + \frac{1}{2}\left(t^2+s^2+2\rho ts\right) + \frac{1}{2!}\cdot\frac{1}{4}\left(t^2+s^2+2\rho ts\right)^2 + \frac{1}{3!}\cdot\frac{1}{8}\left(t^2+s^2+2\rho ts\right)^3$

$$+\frac{1}{4!}\cdot\frac{1}{16}\left(t^2+s^2+2\rho ts\right)^4 + \cdots$$

Now, $$\rho\,(X^2, Y^2) = \frac{E\left(X^2Y^2\right) - E\left(X^2\right)E\left(Y^2\right)}{\sqrt{E\left(X^4\right) - \left\{E\left(X^2\right)\right\}^2}\sqrt{E\left(Y^4\right) - \left\{E\left(Y^2\right)\right\}^2}}.$$

Now, $E\left(X^2Y^2\right)$ = coefficient of $\frac{t^2}{2!}\cdot\frac{s^2}{2!}$ in the expansion of $\psi_{x,y}(t,s) = \frac{1}{2!}\cdot\frac{1}{4}\cdot\left(4\rho^2+2\right)\cdot 2!\cdot 2! = 2\rho^2+1$.

$E\left(X^2\right)$ = coefficient of $\frac{t^2}{2!}$ in $\psi_{x,y}(t,s) = \frac{1}{2}\cdot\sqrt{2} = 1$. Similarly, $E\left(Y^2\right)$ = coefficient of $\frac{s^2}{2!}$ in $\psi_{x,y}(t,s) = \frac{1}{2}2! = 1$.

$E\left(X^4\right)$ = coefficient of $\frac{t^4}{4!}$ in $\psi_{x,y}(t,s) = 4!\cdot\frac{1}{2!\cdot 4} = 3$. Similarly, $E\left(Y^4\right) = 3$.

$$\therefore \rho\,(X^2, Y^2) = \frac{2\rho^2+1-1}{\sqrt{3-1}\sqrt{3-1}} = \frac{2\rho^2}{2} = \rho$$

Ex. 2 For the standard normal BD, prove the following recurrence relation on ordinary moments:
$\alpha_{u,v} = (u+v-1)\rho\alpha_{u-1,v-1} + (u-1)(v-1)\{1-\rho^2)\,\alpha_{u-2,v-2}$.

Solution. The standard normal BD is defined by the following pdf:

$$f_{x,y}(x,y) = \frac{1}{2\pi\sqrt{1-\rho^2}} \exp\left\{-\frac{1}{2(1-\rho^2)}\left(x^2+y^2-2\rho xy\right)\right\},\ -\infty < x, y < \infty.$$

For such distribution we have the MGF: $\psi_{x,y}(t,s) = e^{\frac{1}{2}(t^2+s^2+2\rho ts)}$.

$$\therefore\ \frac{\partial \psi_{x,y}}{\partial t} = e^{\frac{1}{2}(t^2+s^2+2\rho ts)}(t+\rho s) = (t+\rho s)\psi_{x,y}$$

$$\frac{\partial \psi_{x,y}}{\partial s} = e^{\frac{1}{2}(t^2+s^2+2\rho ts)}(s+\rho t) = (s+\rho t)\psi_{x,y}.$$

$$\therefore\ \frac{\partial^2 \psi_{x,y}}{\partial t\partial s} = \frac{\partial}{\partial t}(s+\rho t)\psi_{x,y} = \rho\psi_{x,y} + (s+\rho t)\frac{\partial \psi_{x,y}}{\partial t} = \rho\psi_{x,y} + (s+\rho t)(t+\rho s)\psi_{x,y}.$$

$$\therefore\ \frac{\partial^2 \psi_{x,y}}{\partial t\partial s} - \rho t\frac{\partial \psi_{x,y}}{\partial t} - \rho s\frac{\partial \psi_{x,y}}{\partial s}$$
$$= \rho\psi_{x,y} + (s+\rho t)(t+\rho s)\psi_{x,y} - \rho t(t+\rho s)\psi_{x,y} - \rho s(s+\rho t)\psi_{x,y}$$
$$= \left(\rho + st + \rho s^2 + \rho t^2 + \rho^2 ts - \rho t^2 - \rho^2 st - \rho s^2 - \rho^2 st\right)\psi_{x,y}$$
$$= \left\{\rho + \left(1-\rho^2\right)st\right\}\psi_{x,y}.$$

Again, we have $\psi_{x,y}(t,s) = e^{\frac{1}{2}(t^2+s^2+2\rho ts)} = \sum_{u=0}^{\infty}\sum_{v=0}^{\infty}\frac{t^u s^v}{u!\,v!}\alpha_{u,v}$.

$$\therefore\ \sum_{u=1}^{\infty}\sum_{v=1}^{\infty}\frac{t^{u-1}s^{v-1}}{(u-1)!\,(v-1)!}\alpha_{u,v} - \rho\sum_{u=1}^{\infty}\sum_{v=0}^{\infty}\frac{t^u s^v}{(u-1)!v!}\alpha_{u,v} - \rho\sum_{u=0}^{\infty}\sum_{v=1}^{\infty}\frac{t^u s^v}{u!\,(v-1)!}\alpha_{u,v}$$
$$= \rho\sum_{u=0}^{\infty}\sum_{v=0}^{\infty}\frac{t^u s^v}{u!v!}\alpha_{u,v} + \left(1-\rho^2\right)\sum_{u=0}^{\infty}\sum_{v=0}^{\infty}\frac{t^{u+1}s^{v+1}}{u!v!}\alpha_{u,v}.$$

Equating the coefficients of $\frac{t^{u-1}s^{v-1}}{(u-1)!\,(v-1)!}$, we get

$$\alpha_{u,v} - \rho(u-1)\alpha_{u-1,v-1} - \rho(v-1)\alpha_{u-1,v-1} = \rho\alpha_{u-1,v-1} + \left(1-\rho^2\right)(u-1)(v-1)\alpha_{u-2,v-2}.$$

[How? Suppose, consider the last term. Two have the coefficient of $\frac{t^{u-1}s^{v-1}}{(u-1)!\,(v-1)!}$ we first replace u by $u-2$ and v by $v-2$.

$\therefore$ summation part becomes $\sum_u\sum_v \frac{t^{u-1}s^{v-1}}{(u-2)!\,(v-2)!}\alpha_{u-2},\ v-2,$

i.e., $\sum_u\sum_v \frac{t^{u-1}s^{v-1}}{(u-1)!\,(v-1)!}(u-1)(v-1)\,\alpha_{u-2,v-2}$.]

or, $\alpha_{u,v} = (u+v-1)\rho\alpha_{u-1,v-1} + (u-1)(v-1)\left(1-\rho^2\right)\alpha_{u-2,v-2}$.

7.10 Conditional expectation

Let us suppose (X, Y) be a bivariate RV whose probability distribution is known.

Let $g : A \times B \to \mathbb{R}$, when $(X, Y) \in A \times B$, be continuous. Then, the conditional expectation of the RV $g(X, Y)$ subject to the condition that $X = x$ is denoted by $E\{g(X, Y)|X = x\}$ and is defined by the MV of $g(X, Y)$ w.r.t. the condition distribution subject to the condition $X = x$.

Discrete case. Let (X, Y) is a bivariate discrete RV and (x_i, y_j) be a point in the spectrum such that $f_{ij} = P(X = x_i, Y = y_j)$, $\forall\ i, j = 0, \pm 1, \pm 2, \cdots$.

If we confine $X = x_i$, then we get a distribution of Y on the line $X = x_i$, known as conditional distribution of Y, given $X = x_i$ with pmf $f_{j|i} = \frac{f_{ij}}{f_{xi}}, j = 0, \pm 1, \pm 2, \cdots$ for a parametrically fixed i.

So, in this case, $E\{g(X,Y)|X = x_i\} = \sum\limits_{j=-\infty}^{\infty} g(x_i, y_j) f_{j|i} = \sum\limits_{j=-\infty}^{\infty} g(x_i, y_j) \frac{f_{ij}}{f_{xi}}, i = 0, \pm 1, \pm 2, \cdots$

provided the convergence is absolute.

$\therefore$ conditional mean of Y, given $X = x_i$, is given by

$$m_{y|i} = E\left\{Y \middle| X = x_i\right\} = \sum_{j=-\infty}^{\infty} y_j f_{j|i} = \frac{\sum\limits_{j=-\infty}^{\infty} y_j f_{ij}}{f_{xi}}$$

and similarly,

$$m_{x|j} = E\left\{X | Y = y_j\right\} = \sum_{i=-\infty}^{\infty} x_i f_{i|j} = \frac{\sum\limits_{i=-\infty}^{\infty} x_i f_{ij}}{f_{yj}}.$$

Conditional variance of Y, given $X = x_i$, is given by

$$\sigma^2_{y|i} = E\left\{\left(Y - m_{y|i}\right)^2 \middle| X = x_i\right\} = \frac{\sum\limits_{j=-\infty}^{\infty} \left(y_j - m_{y|i}\right)^2 f_{ij}}{f_{xi}}, \quad i = 0, \pm 1, \pm 2, \cdots$$

Conditional variance of X, given $Y = y_j$, is given by

$$\sigma^2_{x|j} = E\left\{\left(X - m_{x|j}\right)^2 \middle| Y = y_j\right\} = \frac{\sum\limits_{i=-\infty}^{\infty} \left(x_i - m_{x|j}\right)^2 f_{ij}}{f_{yj}}, \quad j = 0, \pm 1, \pm 2, \cdots$$

Conditional central moment of order $k (k \in \mathbb{N} \cup \{0\})$ of X subject to the condition $Y = y_j$, is given by

$$\mu_{xk|j} = E\left\{\left(X - m_{x|j}\right)^k \middle| Y = y_j\right\} = \frac{\sum\limits_{i=-\infty}^{\infty} \left(x_i - m_{x|j}\right)^k f_{ij}}{f_{yj}}, \quad j = 0, \pm 1, \pm 2, \cdots$$

Conditional central moment of order $k (k \in \mathbb{N} \cup \{0\})$ of Y subject to the condition $X = x_i$, is given by

$$\mu_{yk|i} = E\left\{\left(Y - m_{y|i}\right)^k \middle| X = x_i\right\} = \frac{\sum\limits_{j=\infty}^{\infty} \left(y_j - m_{y|i}\right)^k f_{ij}}{f_{xi}}, \quad i = 0, \pm 1, \pm 2, \cdots$$

Similar ordinary moments of order k are given by

$$\alpha_{xk|j} = E\left(X^k \middle| Y = y_j\right) = \sum_{i=-\infty}^{\infty} x_i^k f_{ij} / f_{yj}, \ j = 0, \pm 1, \pm 2, \cdots$$

$$\text{and } \alpha_{yk|i} = E\left(Y^k \middle| X = x_i\right) = \sum_{j=-\infty}^{\infty} y_j^k f_{ij} / f_{xi}, \ i = 0, \pm 1, \pm 2, \cdots.$$

In case when X and Y are independent, $f_{i|j} = f_{xi}$ and $f_{j|i} = f_{yj}$ and so, $m_{x|j} = m_x$, $m_{y|i} = m_y$, $\sigma_{x|j} = \sigma_x$, $\mu_{xk|j} = \mu_{xk}$, etc.

Continuous case. Let (X, Y) be a continuous bivariate RV whose probability distribution is given by the pdf: $f_{x,y}(x,y), -\infty < x, y < \infty$.

The conditional distribution of the continuous RV $g(X, Y)$ subject to the condition $X = x$ is given by

$$E\{g(X,Y)|X=x\} = \int_{-\infty}^{\infty} g(x,y) f_y(y|x), dy = \frac{\int_{-\infty}^{\infty} g(x,y) f_{x,y}(x,y) dy}{f_x(x)}, \ x \in \mathbb{R}.$$

The conditional distribution of the continuous RV $g(X, Y)$ subject to the condition $Y = y$ is given by

$$E\{g(X,Y)|Y=y\} = \int_{-\infty}^{\infty} g(x,y) f_x(x|y) dx = \frac{\int_{-\infty}^{\infty} g(x,y) f_{x,y}(x,y) dx}{f_y(y)}, \ y \in \mathbb{R}.$$

The conditional mean, variance, ordinary moment of order k, central moment of order k (non-negative integer) of the RV X subject to condition $Y = y$ are respectively given by

$$m_x(y) = E(X|Y=y) = \frac{\int_{-\infty}^{\infty} x f_{x,y}(x,y) dx}{f_y(y)}$$

$$\sigma_x^2(y) = E\left\{(X - m_x(y))^2 \middle| Y = y\right\} = \frac{\int_{-\infty}^{\infty} (x - m_x(y))^2 f_{x,y}(x,y) dx}{f_y(y)}$$

$$\alpha_{xk}(y) = E\left\{X^k \middle| Y = y\right\} = \frac{\int_{-\infty}^{\infty} x^k f_{x,y}(x,y) dx}{f_y(y)}$$

$$\mu_{xk}(y) = E\left\{(X - m_x(y))^k \middle| Y = y\right\} = \frac{\int_{-\infty}^{\infty} (x - m_x(y))^k f_{x,y}(x,y) dx}{f_y(y)}, \ y \in \mathbb{R},$$

provided all the integrals converge absolutely.

The conditional mean, variance, ordinary moment of order k, central moment of order k (non-negative integer) of Y subject to the condition $X = x$ are respectively given by

$$m_y(x) = E(Y|X=x) = \frac{\int_{-\infty}^{\infty} y f_{x,y}(x,y) dy}{f_x(x)}$$

$$\sigma_y^2(x) = E\left\{(Y - m_y(x))^2 \middle| X = x\right\} = \frac{\int_{-\infty}^{\infty} (y - m_y(x))^2 f_{x,y}(x,y) dy}{f_x(x)}$$

$$\alpha_{yk}(x) = E\left\{Y^k \middle| X = x\right\} = \frac{\int_{-\infty}^{\infty} y^k f_{x,y}(x,y) dy}{f_x(x)}$$

$$\text{and } \mu_{yk}(x) = E\left\{\left(Y - m_y(x)\right)^k \middle| X = x\right\} \frac{\int_{-\infty}^{\infty} \left(y - m_y(x)\right)^k f_{x,y}(x,y)\,dy}{f_x(x)}, \quad x \in \mathbb{R},$$

provided the integrals converge absolutely.

In case when X and Y are independent, $m_x(y) = m_x, m_y(x) = m_y, \sigma_x^2(y) = \sigma_x^2$, etc.

To make the idea of conditional expectation follow the following example:

Suppose, there are 100 students in a class. The marks for mathematics and physics in a class test of them are displayed in the notice board. If X and Y denote respectively the marks of mathematics and physics, then we have a discrete distribution of the bivariate RV (X, Y).

The conditional distribution of X, the marks in mathematics, subject to the condition $Y = 50$ means a one-dimensional distribution of the marks of mathematics for those students who secured 50 in physics. $E\{X|Y = 50\} = m_{x|50}$ = the mean of the one-dimensional conditional distribution, etc.

Theorem 7.10.1 If (X, Y) be a continuous bivariate RV and h_1, h_2 be two continuous functions on $\mathbb{R}$, then $h_1(X), h_2(X), h_1(Y), h_2(Y)$ are all continuous RVs. Show that

(i) $E\left[\{h_1(X) + h_2(X)\} \middle| Y = y\right] = E\left\{h_1(X) \middle| Y = y\right\} + E\left\{h_2(X) \middle| Y = y\right\}$,

(ii) $E\left[h_1(X) h_2(T) \middle| Y = y\right] = h_2(y) E\left\{h_1(X) \middle| Y = y\right\}$,

provided the conditional expectations of the RHS exist.

Proof. (i) The existence of $E\left\{h_1(X) \middle| Y = y\right\}$ and $E\left\{h_2(X) \middle| Y = y\right\}$ implies that the infinite integrals

$$\int_{-\infty}^{\infty} h_1(x) f_x(x|y)\,dx \text{ and } \int_{-\infty}^{\infty} h_2(x) f_x(x|y)\,dx \text{ converge absolutely,}$$

where $f_x(x|y) = \frac{f_{x,y}(x,y)}{f_y(y)}$ and $f_{x,y}(x,y)$ = pdf of (X, Y), $f_y(y)$ = marginal pdf of Y.

$$\therefore\ E\left[\{h_1(X) + h_2(X)\} | Y = y\right] = \int_{-\infty}^{\infty} \{h_1(x) + h_2(x)\} f_x(x|y)\,dx = \frac{\int_{-\infty}^{\infty} \{h_1(x) + h_2(x)\} f_{x,y}(x,y)\,dx}{f_y(y)}$$

$$= \frac{\int_{-\infty}^{\infty} h_1(x) f_{x,y}(x,y)\,dx}{f_y(y)} + \frac{\int_{-\infty}^{\infty} h_2(x) f_{x,y}(x,y)\,dx}{f_y(y)}$$

$$= \int_{-\infty}^{\infty} h_1(x) \frac{f_{x,y}(x,y)}{f_y(y)}\,dx + \int_{-\infty}^{\infty} h_2(x) \frac{f_{x,y}(x,y)}{f_y(y)}\,dx$$

$$= \int_{-\infty}^{\infty} h_1(x) f_x(x|y)\,dx + \int_{-\infty}^{\infty} h_2(x) f_x(x|y)\,dx$$

$$= E\{h_1(X) | Y = y\} + E\{h_2(X) | Y = y\}.$$

$$\text{(ii) } E\left[h_1(X) h_2(Y) | Y = y\right] = \int_{-\infty}^{\infty} h_1(x) h_2(y) f_x(x|y)\,dx - \int_{-\infty}^{\infty} h_1(x) h_2(y) \frac{f_{x,y}(x,y)}{f_y(y)}\,dx$$

$$= h_2(y) \int_{-\infty}^{\infty} h_1(x) f_x(x|y)\,dx = h_2(y) E\{h_1(X) | Y = y\}.$$

Theorem 7.10.2 If (X, Y) be a discrete bivariate RV and h_1, h_2 be two continuous functions on $\mathbb{R}$, then $h_1(X), h_2(X), h_1(Y), h_2(Y)$ are all discrete RVs. Show that

(i) $E\left[\left\{h_1(X) + h_2(X) | Y = y_j\right\}\right] = E\left\{h_1(X) | Y = y_j\right\} + E\left\{h_2(X) | Y = y_j\right\}$

(ii) $E\left[h_1(X)h_2(Y)|Y=y_j\right]=h_2\left(y_j\right)E\left\{h_1(X)|Y=y_j\right\}$.

Proof. (i) The existence of $E\left\{h_1(X)|Y=y_j\right\}$ and $E\left\{h_2(X)|Y=y_j\right\}$ implies that the series $\sum\limits_{i=-\infty}^{\infty} h_1(x_i)\, f_{i|j}$ and $\sum\limits_{i=-\infty}^{\infty} h_2(x_i)\, f_{i|j}$ converge absolutely, where $f_{i|j}=\frac{f_{ij}}{f_{yj}}$ and $f_{ij}=P\left(X=x_i, Y=y_j\right), f_{yj}=$ marginal pmf of Y.

$$\therefore\ E\left[\{h_1(X)+h_2(X)\}\,|Y=y_j\right]=\sum_{i=-\infty}^{\infty}\{h_1(x_i)+h_2(x_i)\}\, f_{i|j}=\sum_{i=-\infty}^{\infty}\{h_1(x_i)+h_2(x_i)\}\frac{f_{ij}}{f_{yj}}$$

$$=\sum_{i=-\infty}^{\infty} h_i(x_i)\frac{f_{ij}}{f_{yj}}+\sum_{i=-\infty}^{\infty} h_2(x_i)\frac{f_{ij}}{f_{yj}}=\sum_{i=-\infty}^{\infty} h_1(x_i)\, f_{i|j}+\sum_{i=-\infty}^{\infty} h_2(x_i)\, f_{i|j}$$

$$=E\left\{h_1(X)|Y=y_j\right\}+E\left\{h_2(X)|Y=y_j\right\}.$$

$$\text{(ii) } E\left[h_1(X)h_2(Y)|Y=y_j\right]=\sum_{i=-\infty}^{\infty} h_1(x_i)\, h_2\left(y_j\right) f_{i|j}=\sum_{i=-\infty}^{\infty} h_1(x_i)\, h_2\left(y_j\right)\frac{f_{ij}}{f_{yj}}$$

$$=h_2\left(y_j\right)\sum_{i=-\infty}^{\infty} h_1(x_i)\, f_{i|j}=h_2\left(y_j\right)E\left\{h_1(X)|Y=y_j\right\}.$$

7.11 Regression curves

It is interesting to note that $m_y(x)$ must depend on x. As x varies continuously within a prescribed domain $m_y(x)$ functions as a function of x and the curve $y=m_y(x)$, the locus of the conditional mean, is called the regression curve of Y and X, whereas $m_y(x)$ is called the *regression function* of Y and X.

Sometimes $y=m_y(x)$ is called the *regression curve* for the mean of Y.

The distribution of the RV $m_y(X)$ has the mean

$$E\left\{m_y(X)\right\}=\int_{-\infty}^{\infty} m_y(x)f_x(x)\,dx=\int_{-\infty}^{\infty} E(Y|X=x)f_x(x)\,dx=\int_{-\infty}^{\infty}\left(\int_{-\infty}^{\infty} y f_y(y\,|\,x)\,dy\right)f_x(x)\,dx$$

$$=\int_{-\infty}^{\infty}\int_{-\infty}^{\infty} y\frac{f_{x,y}(x,y)}{f_x(x)}f_x(x)\,dx\,dy=\int_{-\infty}^{\infty}\int_{-\infty}^{\infty} y f_{x,y}(x,y)\,dx\,dy=m_y,$$

which is the y-coordinate of centre of mass of the bivariate distribution of (X, Y).

Similarly, the regression function of X on Y and the regression curve of X on Y are respectively given by $m_x(y)$ and $x=m_x(y)$.

The mean of the distribution of the RV $m_x(Y)$ is given by

$$E\{m_x(Y)\}=\int_{-\infty}^{\infty} m_x(y)f_y(y)\,dy=\int_{-\infty}^{\infty} E\{X|Y=y\}\, f_y(y)\,dy=\int_{-\infty}^{\infty}\left(\int_{-\infty}^{\infty} x f_x(x\,|\,y)\,dx\right)f_y(y)\,dy$$

$$=\int_{-\infty}^{\infty}\int_{-\infty}^{\infty} x\frac{f_{x,y}(x,y)}{f_y(y)}f_y(y)\,dx\,dy=\int_{-\infty}^{\infty}\int_{-\infty}^{\infty} x f_{x,y}(x,y)\,dx\,dy=m_x,$$

which is the x-coordinate of the centre of mass of the BD of (X, Y).

Hence, the centre of mass of the BD of (X, Y) is $\left(m_x, m_y\right)$.

Note: If a regression curve is a straight line the corresponding regression is said to be linear.

Now, we see that though $E\left(m_y(X)\right) = m_y$ and $E(m_x(Y)) = m_y, E\left\{\sigma_y^2(X)\right\} \neq \sigma_y^2$ and $E\left\{\sigma_x^2(Y)\right\} \neq \sigma_x^2$.

$$E\left\{\sigma_y^2(X)\right\} = \int_{-\infty}^{\infty} \sigma_y^2(x) f_x(x)\,dx = \int_{-\infty}^{\infty}\int_{-\infty}^{\infty} \{y - m_y(x)\}^2 f_y(y|x) f_x(x)\,dy\,dx$$

$$= \int_{-\infty}^{\infty} E\left[\{Y - m_y(x)\}^2 | X = x\right] f_x(x)\,dx = \int_{-\infty}^{\infty}\int_{-\infty}^{\infty} \{y - m_y(x)\}^2 f_y(y|x) f_x(x)\,dy\,dx$$

$$= \int_{-\infty}^{\infty}\int_{-\infty}^{\infty} \{y - m_y(x)\}^2 f_{x,y}(x,y)\,dx\,dy = E\left[\{Y - m_y(X)\}^2\right] \neq \sigma_y^2.$$

Similarly, $$E\left\{\sigma_x^2(Y)\right\} = \int_{-\infty}^{\infty} \sigma_x^2(y) f_y(y)\,dy = \int_{-\infty}^{\infty} E\left[\{X - m_x(y)\}^2 | Y = y\right] f_y(y)\,dy$$

$$= \int_{-\infty}^{\infty}\int_{-\infty}^{\infty} \{x - m_x(y)\}^2 f_x(x|y) f_y(y)\,dx\,dy$$

$$= \int_{-\infty}^{\infty}\int_{-\infty}^{\infty} \{x - m_x(y)\}^2 f_{x,y}(x,y)\,dx\,dy = E\left[\{X - m_x(Y)\}^2\right] \neq \sigma_x^2.$$

Theorem 7.11.1 Let (X, Y) be a bivariate continuous RV and $g : \mathbb{R} \to \mathbb{R}$ be continuous. Then, the MV $E\left[\{Y - g(X)\}^2\right]$ is minimum, when $g(x) = m_y(x)$.

Proof.

$$E\left[\{Y - g(X)\}^2\right] = \int_{-\infty}^{\infty}\int_{-\infty}^{\infty} \{y - g(x)\}^2 f_{x,y}(x,y)\,dx\,dy, \text{whenever the integral converges absolutely}$$

$$= \int_{-\infty}^{\infty}\left[\int_{-\infty}^{\infty} \{y - g(x)\}^2 f_y(y|x)\,dy\right] f_x(x)\,dx = \int_{-\infty}^{\infty} \left[E\left\{(Y - g(X))^2 | X = x\right\}\right] f_x(x)\,dx.$$

Now, $E\left\{(Y - g(X))^2 | X = x\right\}$ = second order moment of the one-dimensional conditional distribution of Y, when $X = x$ about the point $g(x)$.

Now, we know that such a moment will be minimum, when taken about the corresponding mean (conditional) of the distribution in one dimension, i.e., $E\left\{(Y - g(X))^2 | X = x\right\} \geq E\left\{\left(Y - m_y(x)\right)^2 | X = x\right\}$.

Hence, $E\left[\{Y - g(X)\}^2\right]$ is minimum, when $g(x) = m_y(x)$ and this is true for all $x \in \mathbb{R}$ and for all curves $y = g(x)$.

Hence, among all the curves the MV of the square deviations of the BD of (X, Y) parallel to the direction of y-axis is the minimum from the regression curve $y = m_y(x)$.

Hence, $y = m_y(x)$ is the best fitting curve of the BD and $Y = m_y(X)$ is the best possible relationship between the RVs X and Y, i.e., $m_y(X)$ is the best representation of Y in terms of X. The RV $U = Y - m_y(X)$ is the residue of Y, when it is represented by X. It is interesting to note that the average of such a residue vanishes, i.e., $E(U) = 0$.

Theorem 7.11.2 Let (X, Y) be a bivariate continuous RV and $h : \mathbb{R} \to \mathbb{R}$ be continuous. Then, the MV $E\left[\{X - h(Y)\}^2\right]$ is minimum, when $h(y) = m_x(y)$.

Proof.

$$E\left[\{X - h(Y)\}^2\right] = \int_{-\infty}^{\infty}\int_{-\infty}^{\infty} \{x - h(y)\}^2 f_x(x,y)\,dx\,dy, \text{ provided the integral converges absolutely}$$

$$= \int_{-\infty}^{\infty}\left[\int_{-\infty}^{\infty} \{x - h(y)\}^2 f_x(x|y)\,dx\right] f_y(y)\,dy = \int_{-\infty}^{\infty} \left[E\left\{(X - h(Y))^2 | Y = y\right\}\right] f_y(y)\,dy.$$

Now, $E\{(X-h(y))^2|Y=y\}$ = second order moment of the one-dimensional conditional distribution of X, when $Y=y$ about the point $h(y)$.

Also we know that such a moment will be minimum, when taken about the corresponding mean (conditional) of the distribution in one dimension, i.e., $E\{(X-h(y))^2|Y=y\} \geq E\{(X-m_x(y))^2|Y=y\}$.

Hence, $E\left\{(X-h(Y))^2\,|\,Y=y\right\}$ is minimum, when $h(y)=m_x(y)$ and this is true $\forall\ y \in \mathbb{R}$ and $\forall$ curves $x = h(y)$. Hence, $x = m_x(y)$ is the best fitting curve of the BD and $X = m_x(Y)$ is the best relationship between X and Y, i.e., $m_x(Y)$ is the best representation of X in terms of Y. The RV $V=X-m_x(Y)$ is the residue of X after taking its best representation in terms of Y. Note that such a residue has its mean zero, i.e., $E(V)=0$. Follow the following figures:

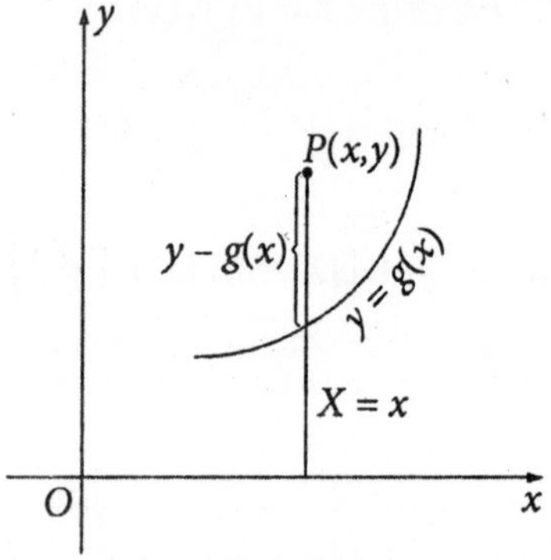

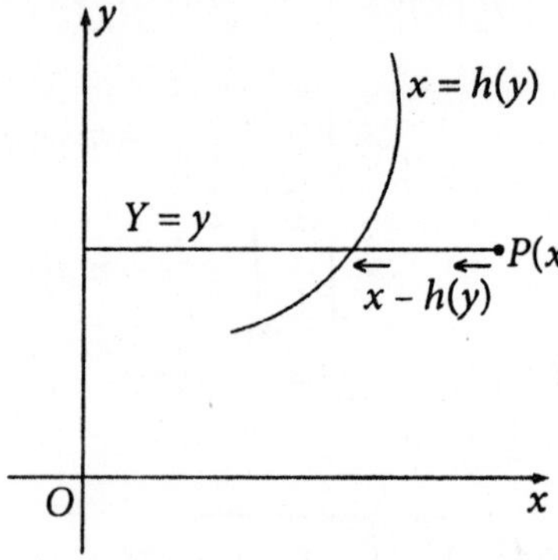

Note: When the RVs X and Y are independent, both the regression curves are linear and are parallel to the axes.

Verification. Consider the regression curve $x = m_x(y)$. When X and Y are independent, the curve becomes

$$
\begin{aligned}
x &= m_x(y) \\
&= E(X|Y=y) \\
&= \int_{-\infty}^{\infty} x f_x(x|y)\,dx \\
&= \int_{-\infty}^{\infty} x f_x(x)\,dx, \text{ as } X, Y \text{ are independent} \\
&= m_x,
\end{aligned}
$$

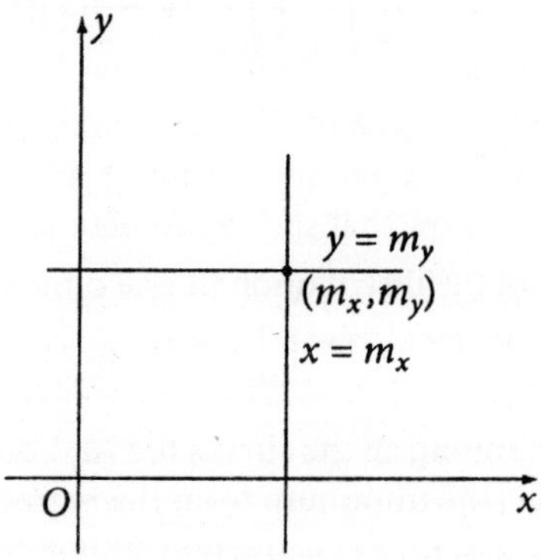

i.e., $x = m_x$ is linear equation and hence a straight line and it is parallel to y-axis.

Similarly, the other regression curve reduces to $y = m_y$, which is also linear equation and hence a straight line and it is parallel to x-axis. Their point of intersection is (m_x, m_y), which is the centre of mass of the BD of the RV (X, Y).

Theorem 7.11.3 The regression curves of any bivariate normal distribution are both linear but not parallel to the coordinate axes unless they are uncorrelated.

Proof. It is enough to prove the theorem for a standard normal BD as any bivariate normal distribution with parameters $m_x, m_y, \sigma_x, \sigma_y$ and ρ may be converted to a standard normal BD by the simple linear transforms: $x^* = \frac{x-m_x}{\sigma_x}, y^* = \frac{y-m_y}{\sigma_y}$.

Let us now consider the bivariate standard normal distribution defined by the pdf:

$$f_{x,y}(x,y)=\frac{1}{2\pi\sqrt{1-\rho^2}}\exp\left\{-\frac{1}{2(1-\rho^2)}\left(x^2+y^2-2\rho xy\right)\right\},\ -\infty<x,y<\infty$$

$$m_x(y)=E\{X\mid Y=y\}=\int_{-\infty}^{\infty}x f_x(x\mid y)\,dx=\int_{-\infty}^{\infty}x\frac{f_{x,y}(x,y)}{f_y(y)}\,dx$$

$$=\int_{-\infty}^{\infty}x\cdot\frac{\frac{1}{2\pi\sqrt{1-\rho^2}}\exp\left\{-\frac{1}{2(1-\rho^2)}(x^2+y^2-2\rho xy)\right\}}{\frac{1}{\sqrt{2\pi}}\exp\left(-\frac{y^2}{2}\right)}\,dx$$

$$=\frac{1}{\sqrt{2\pi}\sqrt{1-\rho^2}}\int_{-\infty}^{\infty}x\exp\left\{-\frac{1}{2(1-\rho^2)}(x^2+y^2-2\rho xy)+\frac{y^2}{2}\right\}dx$$

$$=\frac{1}{\sqrt{2\pi}\sqrt{1-\rho^2}}\int_{-\infty}^{\infty}x\exp\left\{-\frac{1}{2(1-\rho^2)}\left(x^2+y^2-2\rho xy-y^2+\rho^2y^2\right)\right\}dx$$

$$=\frac{1}{\sqrt{2\pi}\sqrt{1-\rho^2}}\int_{-\infty}^{\infty}x\exp\left\{-\frac{(x-\rho y)^2}{2(1-\rho^2)}\right\}dx.$$

Let $\frac{x-\rho y}{\sqrt{1-\rho^2}}=u$. $\therefore\ dx=\sqrt{1-\rho^2}\,du$.

As $x\to\infty, u\to\infty$ and $x\to-\infty, u\to-\infty$ as $-1<\rho<1$ and $\sqrt{1-\rho^2}>0$ [y is a parameter]

$$=\frac{1}{\sqrt{2\pi}}\int_{-\infty}^{\infty}\left(\rho y+u\sqrt{1-\rho^2}\right)e^{-u^2/2}\,du=\rho y\cdot\frac{1}{\sqrt{2\pi}}\int_{-\infty}^{\infty}e^{-u^2/2}\,du+\frac{\sqrt{1-\rho^2}}{\sqrt{2\pi}}\int_{-\infty}^{\infty}ue^{-u^2/2}\,du$$

$=\rho y,\ \because\ \frac{1}{\sqrt{2\pi}}\int_{-\infty}^{\infty}e^{-u^2/2}\,du=1$ and $\int_{-\infty}^{\infty}ue^{-u^2/2}\,du=0$ as integrand is an odd function.

$\therefore$ the regression curve of X on Y is $x=m_x(y)=\rho y$, i.e., $x=\rho y$, which is linear in x and y and hence a straight line and is not parallel to y-axis unless $\rho=0$.

Similarly, the other regression curve of Y on X is $y=\rho x$, which is also a straight line and is not parallel to x-axis unless $\rho=0$.

Note: The regression lines of the bivariate normal distribution of (X, Y) with parameters $m_x, m_y, \sigma_x, \sigma_y$ and ρ are obtained by replacing x by $\frac{x-m_x}{\sigma_x}$ and y by $\frac{y-m_y}{\sigma_y}$ as:

Regression line of X on Y: $\frac{x-m_x}{\sigma_x}=\rho\frac{y-m_y}{\sigma_y}$, i.e., $x=m_x+\rho\frac{\sigma_x}{\sigma_y}\left(y-m_y\right)$.

Regression line of Y on X: $\frac{y-m_y}{\sigma_y}=\rho\frac{x-m_x}{\sigma_x}$, i.e., $y=m_y+\rho\frac{\sigma_y}{\sigma_x}(x-m_x)$.

Here $\rho\frac{\sigma_x}{\sigma_y}$ is called regression coefficient of X on Y and is denoted by β_{xy}

and $\rho\frac{\sigma_y}{\sigma_x}$ is called regression coefficient of Y on X and is denoted by β_{yx},

i.e., $\beta_{xy}=\rho\frac{\sigma_x}{\sigma_y}$ = regression coefficient of X and Y and $\beta_{yx}=\rho\frac{\sigma_y}{\sigma_x}$ = regression coefficient of Y and X.

Ex. 7.11.4 Given the regression lines $x-y+1=0$ and $3x-2y+1=0$. Find which one is the regression line of Y on X. Find m_x, m_y and $\rho(X, Y)$.

Solution. We know that the two regression lines meet at the centre of mass $\left(m_x, m_y\right)$. Solving, we get the point $(1,2)$. $\therefore\ m_x=1$ and $m_y=2$.

$\therefore$ the lines may be written as $y-2=x-1$ and $2(y-2)=3(x-1)$. We distinguish them by trial and error process:

If $y=2+(x-1)$ be the regression line of Y on X and $x=1+\frac{2}{3}(y-2)$ be the regression line of X on Y, then $\rho\frac{\sigma_y}{\sigma_x}=1$ and $\rho\frac{\sigma_x}{\sigma_y}=\frac{2}{3}$. Multiplying, $\rho^2=\frac{2}{3}<1$ (possible).

Hence, our supposition is correct and $\rho=\sqrt{2/3}$.

$\therefore$ $x-y+1=0$ is the regression line of Y on X and $3x-2y+1=0$ is the regression line of X on Y.

Note: The reverse consideration gives us $\rho^2=\frac{3}{2}\not<1$. Hence, not possible.

7.12 Least square regression lines

Here we apply the least square method to find the regression lines.

Regression line of Y on X.

Consider a family of straight lines $y=a+bx$, where a,b are parameters. Among the members of the family we find that one, from which the mean square deviation of the distribution in the direction of y-axis is minimum possible. Such a line is the regression line of Y on X.

Let $S=E\left\{(Y-a-bX)^2\right\}$, (X,Y) is a random point of the BD. The necessary conditions for S to be minimum are $\frac{\partial S}{\partial a}=0$ and $\frac{\partial S}{\partial b}=0$.

These are called the *normal equations.* The first normal equation gives

$$E\{-2(Y-a-bX)\}=0 \text{ or, } m_y=a+bm_x. \tag{1}$$

The second normal equation gives

$$E\{-2X(Y-a-bX)\}=0 \text{ or, } \alpha_{11}=am_x+b\alpha_{x2}. \tag{2}$$

(1) and (2) $\Rightarrow \alpha_{11}-m_xm_y=b\left(\alpha_{x2}-m_x^2\right)$ or, $b=\dfrac{\mu_{11}}{\mu_{x2}}=\dfrac{\rho\sigma_x\sigma_y}{\sigma_x^2}=\rho\dfrac{\sigma_y}{\sigma_x}=b^*$, say.

$\therefore$ $a=m_y-\rho\dfrac{\sigma_y}{\sigma_x}m_x=a^*$, say.

Now, we show that S is minimum, when $a=a^*$ and $b=b^*$. As $a=a^*, b=b^*$ are the solutions of $E(Y-a-bX)=0$ and $E\{X(Y-a-bX)\}=0$,

$$\therefore\ E\left(Y-a^*-b^*X\right)=0 \tag{3}$$

$$\text{and } E\left\{X\left(Y-a^*-b^*X\right)\right\}=0. \tag{4}$$

Let $S=S^*$, when $a=a^*$ and $b=b^*$, i.e., $S^*=E\left\{(Y-a^*-b^*X)^2\right\}$.

$$\begin{aligned}\therefore\ S&=E\left\{(Y-a-bX)^2\right\}=E\left[\left\{Y-a^*-b^*X+a^*+b^*X-a-bX\right\}^2\right]\\&=E\left[\left\{\left(Y-a^*-b^*X\right)+\left(b^*-b\right)X+\left(a^*-a\right)\right\}^2\right]\\&=E\left\{\left(Y-a^*-b^*X\right)^2\right\}+\left(b^*-b\right)^2E\left(X^2\right)+\left(a^*-a\right)^2+2\left(b^*-b\right)E\left\{X\left(Y-a^*-b^*X\right)\right\}\\&\quad+2\left(a^*-a\right)\left(b^*-b\right)E(X)+2\left(a^*-a\right)E\left\{\left(Y-a^*-b^*X\right)\right\}\\&=S^*+\left(b^*-b\right)^2E\left(X^2\right)+\left(a^*-a\right)^2+2\left(a^*-a\right)\left(b^*-b\right)E(X), \text{ [using (3) and (4)]}.\end{aligned}$$

$$\begin{aligned}\therefore\ S-S^*&=\left(b^*-b\right)^2\left(\sigma_x^2+m_x^2\right)+2\left(a^*-a\right)\left(b^*-b\right)m_x+\left(a^*-a\right)^2,\ \because\ E\left(X^2\right)=\alpha_{x2}=\mu_{x2}+m_x^2\\&=\left(b^*-b\right)^2\sigma_x^2+\left\{\left(b^*-b\right)m_x+\left(a^*-a\right)\right\}^2\geq 0.\end{aligned}$$

$\therefore$ $S^*\leq S$, $\forall\, a,b\in\mathbb{R}$.

$\therefore$ S is minimum, when $a = a^*$ and $b = b^*$ and

$$
\begin{aligned}
S_{\min} = S^* &= E\left\{(Y - a^* - b^* X)^2\right\} = E\left[\left\{Y - \left(m_y - \rho\frac{\sigma_y}{\sigma_x}m_x\right) - \rho\frac{\sigma_y}{\sigma_x}X\right\}^2\right] \\
&= E\left[\left\{\left(Y - m_y\right) - \rho\frac{\sigma_y}{\sigma_x}(X - m_x)\right\}^2\right] \\
&= E\left\{\left(Y - m_y\right)^2\right\} - 2\rho\frac{\sigma_y}{\sigma_x}E\left\{(X - m_x)(Y - m_y)\right\} + \left(\rho\frac{\sigma_y}{\sigma_x}\right)^2 E\left\{(X - m_x)^2\right\} \\
&= \sigma_y^2 - 2\rho\frac{\sigma_y}{\sigma_x}\mu_{11} + \rho^2\sigma_y^2, \because\ E\left\{(X - m_x)^2\right\} = \sigma_x^2 \\
&= \sigma_y^2 - 2\rho\frac{\sigma_y}{\sigma_x}\cdot\sigma_x\sigma_y\rho + \rho^2\sigma_y^2, \because\ \rho = \frac{\operatorname{cov}(X, Y)}{\sigma_x\sigma_y} = \frac{\mu_{11}}{\sigma_x\sigma_y} \\
&= \sigma_y^2\left(1 - \rho^2\right), \text{ where } \rho = \rho(X, Y).
\end{aligned}
$$

This is the inverse measure of goodness of fit of the regression line $y = a^* + b^* x$, to the distribution, because when $\sigma_y^2\,(1 - \rho^2)$ is small the goodness of fit is high.

The regression line of Y on X is $y = a^* + b^* x$, or, $y = m_y - \rho\frac{\sigma_y}{\sigma_x}m_x + \rho\frac{\sigma_y}{\sigma_x}x$, or, $y - m_y = \rho\frac{\sigma_y}{\sigma_x}(x - m_x)$.

The quantity $\rho\frac{\sigma_y}{\sigma_x}$ is called the regression coefficient of Y on X and is sometimes denoted by β_{yx}. Thus, $\beta_{yx} = \rho\frac{\sigma_y}{\sigma_x}$ = regression coefficient of Y on X.

7.12.1 Regression line of *X* on *Y* [by LSM]

Consider the family of straight line $x = c + dy, c, d$ are parameters. Among the members of the family we want to find that member from which mean square deviation of the BD from the line $x = c + dy$ in the direction of x-axis is minimum possible. That member is the best fitting line to the distribution among the family of lines $x = c + dy$ and so is the regression line of X on Y.

Our sole purpose is to find c and d so that the mean square deviation $S = E\left\{(X - c - dY)^2\right\}$ is minimum for having the regression line of X on Y.

For optimal values of S we have the normal equations $\frac{\partial S}{\partial c} = 0$ and $\frac{\partial S}{\partial d} = 0$.

$\therefore$ the normal equations are

$$E\{-2(X - c - dY)\} = 0, \text{ i.e., } m_x = c + d\,m_y \tag{1}$$

$$\text{and } E\{-2Y(X - c - dY)\} = 0, \text{ i.e., } \alpha_{11} - c\,m_y + d\,\alpha_{y2}. \tag{2}$$

(1) and (2) $\Rightarrow d = \rho\frac{\sigma_x}{\sigma_y} = d^*$, say, and $c = m_x - \rho\frac{\sigma_x}{\sigma_y}m_y = c^*$, say.

Arguing like previous article we conclude that S is minimum, when $c = c^*$ and $d = d^*$ giving the regression line of X on Y:

$$x = c^* + d^* y = m_x - \rho\frac{\sigma_x}{\sigma_y}m_y + \rho\frac{\sigma_x}{\sigma_y}y \text{ or, } x - m_x = \rho\frac{\sigma_x}{\sigma_y}\left(y - m_y\right).$$

The quantity $\rho\frac{\sigma_x}{\sigma_y}$ to be denoted by β_{xy} is called the *regression coefficient* of X and Y.

$$S_{\min} = S^* = E\left\{(X - c^* - d^* Y)^2\right\} = \left(1 - \rho^2\right)\sigma_x^2,\ \rho = \rho(X, Y),$$

which is the inverse measure of goodness of fit of the regression line $x = c^* + d^* y$ to the distribution.

7.12.2 Regression parabola

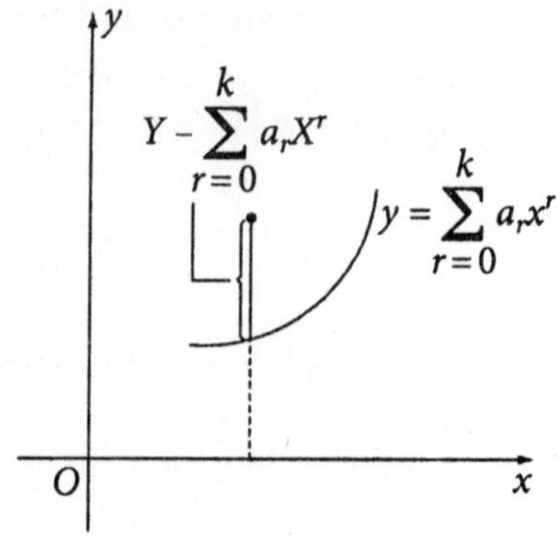

Like least square regression line we now want to find the least square regression parabola of degree k from the family of parabolas: $y=a_0+a_1x+a_2x^2+\cdots+a_kx^k$ $(a_i\in\mathbb{R},\ \forall\ i=1,2,\cdots,k)$. (1)

To have the least square regression parabola we proceed to minimize the mean square deviations of the distribution from each member of the family in some direction, say, y-axis.

The mean square deviation of the distribution from any member of the parabolas $y=a_0+a_1x+a_2x^2+\cdots+a_kx^k$ in the direction of y-axis is
$S=E\left[\{Y-(a_0+a_1X+a_2X^2+\cdots+a_kX^k)\}^2\right]$. (2)

To minimize S the normal equations are

$$\frac{\partial S}{\partial a_0}=0,\ \frac{\partial S}{\partial a_1}=0,\ \frac{\partial S}{\partial a_2}=0,\ \cdots,\ \frac{\partial S}{\partial a_k}=0,$$

i.e.,

$$\begin{aligned}
&E\left[-2\left\{Y-\left(a_0+a_1X+a_2X^2+\cdots+a_kX^k\right)\right\}\right]=0,\\
&E\left[-2X\left\{Y-\left(a_0+a_1X+a_2X^2+\cdots+a_kX^k\right)\right\}\right]=0,\\
&E\left[-2X^2\left\{Y-\left(a_0+a_1X+a_2X^2+\cdots+a_kX^k\right)\right\}\right]=0\\
&\cdots\cdots\cdots\cdots\cdots\cdots\cdots\cdots\cdots\cdots\cdots\cdots\\
&\cdots\cdots\cdots\cdots\cdots\cdots\cdots\cdots\cdots\cdots\cdots\cdots\\
&E\left[-2X^k\left\{Y-\left(a_0+a_1X+a_2X^2+\cdots+a_kX^k\right)\right\}\right]=0
\end{aligned}$$

i.e.,

$$\left.\begin{aligned}
&a_0\alpha_{00}+a_1\alpha_{10}+a_2\alpha_{20}+\cdots+a_k\alpha_{k0}=\alpha_{01}\\
&a_0\alpha_{10}+a_1\alpha_{20}+a_2\alpha_{30}+\cdots+a_k\alpha_{k+1,0}=\alpha_{11}\\
&a_0\alpha_{20}+a_1\alpha_{30}+a_2\alpha_{40}+\cdots+a_k\alpha_{k+2,0}=\alpha_{21}\\
&\cdots\cdots\cdots\cdots\cdots\cdots\cdots\cdots\cdots\cdots\\
&\cdots\cdots\cdots\cdots\cdots\cdots\cdots\cdots\cdots\cdots\\
&a_0\alpha_{k0}+a_1\alpha_{k+1,0}+a_2\alpha_{k+2,0}+\cdots+a_k\alpha_{2k,0}=\alpha_{k1}
\end{aligned}\right\}\qquad(3)$$

To have the unique solution of the system of eqs. (3) we assume that

$$\begin{vmatrix}
\alpha_{00} & \alpha_{10} & \alpha_{20} & \cdots & \alpha_{k0}\\
\alpha_{10} & \alpha_{20} & \alpha_{30} & \cdots & \alpha_{k+1,0}\\
\alpha_{20} & \alpha_{30} & \alpha_{40} & \cdots & \alpha_{k+2,0}\\
\cdots & \cdots & \cdots & \cdots & \cdots\\
\cdots & \cdots & \cdots & \cdots & \cdots\\
\alpha_{k0} & \alpha_{k+1,0} & \alpha_{k+2,0} & \cdots & \alpha_{2k,0}
\end{vmatrix}\neq 0$$

Let the unique solution be $a_0=a_0^*, a_1=a_1^*,\cdots,a_k=a_k^*$. It can be shown that S is minimum for such values of $a_0,a_1,\cdots,a_k$.

$\therefore$ the best fitting member of the family (1), i.e., the least square regression parabola is $y=a_0^*+a_1^*x+a_2^*x^2+\cdots+a_k^*x^k$ and $S_{\min}=\left[\left\{Y-\left(a_0^*+a_1^*X+a_2^*X^2+\cdots+a_k^*X^k\right)\right\}^2\right]$.

To have it, let us consider $Q=E\left[\{tY-(a_0+a_1X+a_2X^2+\cdots+a_kX^k)\}^2\right]$, which is a homogeneous function of degree 2 of the parameters $t,a_0,a_1,\cdots,a_k$.

$\therefore$ by Euler's theorem on homogeneous functions: $t\dfrac{\partial Q}{\partial t}+a_0\dfrac{\partial Q}{\partial a_0}+a_1\dfrac{\partial Q}{\partial a_1}+\cdots+a_k\dfrac{\partial Q}{\partial a_k}=2Q.$

Now, taking $t=1, a_0=a_0^*, a_1=a_1^*, \cdots, a_k=a_k^*$, we get

$$2S_{\min}=\left[t\frac{\partial Q}{\partial t}+a_0\frac{\partial Q}{\partial a_0}+a_1\frac{\partial Q}{\partial a_1}+\cdots+a_k\frac{\partial Q}{\partial a_k}\right]_{t=1,a_0=a_0^*,a_1=a_1^*,\cdots,a_k=a_k^*}$$
$$=E\left[2Y\left\{Y-\left(a_0^*+a_1^*X+a_2^*X^2+\cdots+a_k^*X^k\right)\right\}\right],$$

since $\frac{\partial Q}{\partial a_0}, \frac{\partial Q}{\partial a_1}, \cdots, \frac{\partial Q}{\partial a_k}$ all vanish by virtue of the normal eq. (3), when $t=1$, $a_0=a_0^*$, $a_1=a_1^*, \cdots, a_k=a_k^*$.

$\therefore$ $S_{\min}=\alpha_{02}-a_0^*\alpha_{01}-a_1^*\alpha_{11}-\cdots-a_k^*\alpha_{k1}$, which is the inverse measure of goodness of fit of the least square regression parabola $y=a_0^*+a_1^*x+a_2^*x^2+\cdots+a_k^*x^k$ to the BD of (X,Y).

This is an inverse measure because, as $S_{\min}$ is small, the goodness of fit is high.

7.12.3 Fitting a curve of type $y=ax^b$ (Geometric curve)

The regression curve of Y on X of the type $y=ax^b$ is obtained by determining a,b so that the mean square deviation $S=E\left(Y-aX^b\right)$ in the direction of y-axis of the set of points $(x_i,y_i), i=1,2,\cdots,n$ from the curve $y=ax^b$ is minimum possible.

Now, $y=ax^b \Leftrightarrow \log y=\log a+b\log x$ or, $y^*=A+bx^*$, where $y^*=\log y, A=\log a$ and $x^*=\log x$.

In terms of the new variables x^* and y^*, the given type becomes linear, i.e., in new variables we are to fit a straight line, which has been discussed earlier.

Note: Fitting a curve of the type $y=ab^x$ or, $y=ae^{bx}$ is obtained in similar procedure.

Ex. 7.12.4 Fit a curve of the type $y=ax^b$, which best fits the following set of points $(4,8)$, $(5,12.5)$, $(6,18)$, $(7,24.5)$ and $(8,32)$.

Solution. $y=ax^b \Leftrightarrow \log_{10}y=\log_{10}a+b\log_{10}x$ or, $y^*=A+bx^*$, where $y^*=\log_{10}y$, etc.

The normal equations are

$$\Sigma y^*=5A+b\Sigma x^* \text{ and } \Sigma y^*x^*=A\Sigma x^*+b\Sigma x^{*2}.$$

x	y	x^*	y^*	x^*y^*	x^{*2}
4	8	0.60	0.90	0.5400	0.3600
5	12.5	0.70	1.10	0.7700	0.4900
6	18	0.78	1.26	0.9828	0.6084
7	24.5	0.85	1.39	1.1815	0.7225
8	32	0.90	1.51	1.3590	0.8100
Total	—	3.83	6.16	4.8333	2.9909

$\therefore$ the normal equations are

$$6.16=5A+3.83b, 4.8333=3.83A+2.9909b.$$

Solving, $A=-0.3, b=2$. $\therefore$ $a=\text{antilog}(-0.3)=0.5$.

$\therefore$ best fitting curve w.r.t. the given points within the family is $y=0.5x^2$.

WORKED-OUT EXAMPLES

Ex. 1. The RV X and Y are both standard normal and are mutually independent. Find the expectation of $\max\{|X|,|Y|\}$. [C.H. 2000]

Solution. As X and Y are independent standard normal variables. $\therefore$ the pdf $f_{x,y}$ of (X,Y) is given by:

$$f_{x,y}(x,y)=f_x(x)f_y(y)=\frac{1}{2\pi}e^{-\frac{1}{2}(x^2+y^2)}, \; -\infty<x<y<\infty.$$

$$\therefore\ E[\max\{|X|,|Y|\}] = \int_{-\infty}^{\infty}\int_{-\infty}^{\infty} \max\{|x|,|y|\}\cdot\frac{1}{2\pi}e^{-\frac{1}{2}(x^2+y^2)}\,dx\,dy$$

$$= \frac{1}{2\pi}\underset{|x|>|y|}{\int_{-\infty}^{\infty}\int_{-\infty}^{\infty}} |x|e^{-\frac{1}{2}(x^2+y^2)}\,dx\,dy + \frac{1}{2\pi}\underset{|y|>|x|}{\int_{-\infty}^{\infty}\int_{-\infty}^{\infty}} |y|e^{-\frac{1}{2}(x^2+y^2)}\,dx\,dy \qquad (1)$$

$$\text{First integral} = \frac{1}{2\pi}\int_{-\infty}^{\infty}\int_{|y|}^{\infty} |x|e^{-\frac{1}{2}(x^2+y^2)}\,dx\,dy$$

$$= \frac{1}{2\pi}\int_{-\infty}^{\infty} e^{-\frac{1}{2}y^2}\left[\int_{|y|}^{\infty} xe^{-\frac{1}{2}x^2}\,dx + \int_{-\infty}^{-|y|}(-x)e^{-\frac{1}{2}x^2}\,dx\right]dy$$

$$= \frac{1}{\pi}\int_{-\infty}^{\infty} e^{-\frac{1}{2}y^2}\left(\int_{|y|}^{\infty} xe^{-\frac{1}{2}x^2}\,dx\right)dy,$$

as putting $-x = u$ in second integral within bracket, $dx = -du$ and limits $\infty, |y|$.

$$\therefore\ \int_{-\infty}^{-|y|}(-x)e^{-\frac{1}{2}x^2}\,dx = -\int_{\infty}^{|y|} ue^{-\frac{u^2}{2}}\,du = \int_{|y|}^{\infty} xe^{-\frac{1}{2}x^2}\,dx.$$

Now, to evaluate integral within bracket, put $\frac{1}{2}x^2 = v$. $\therefore\ x\,dx = dv$.

$$= \frac{1}{\pi}\int_{-\infty}^{\infty} e^{-\frac{1}{2}y^2}\left(\int_{\frac{1}{2}y^2}^{\infty} e^{-v}\,dv\right)dy = \frac{1}{\pi}\int_{-\infty}^{\infty} e^{-\frac{1}{2}y^2}\left[-e^{-v}\right]_{\frac{1}{2}y^2}^{\infty}dy$$

$$= \frac{1}{\pi}\int_{-\infty}^{\infty} e^{-y^2}\,dy = \frac{2}{\pi}\int_{0}^{\infty} e^{-y^2}\,dy = \frac{2}{\pi}\cdot\frac{\sqrt{\pi}}{2} = \frac{1}{\sqrt{\pi}}.$$

Due to symmetry, second integral of (1) is also $\frac{1}{\sqrt{\pi}}$.

$$\therefore\ E[\{|X|,|Y|\}] = \frac{1}{\sqrt{\pi}} + \frac{1}{\sqrt{\pi}} = \frac{2}{\sqrt{\pi}}.$$

Ex. 2. The pdf of (X, Y) is

$$f_{x,y}(x,y) = \frac{1}{\pi\sqrt{3}}\exp\left\{-2\left(x^2 - xy + \frac{y^2}{3}\right)\right\},\ -\infty < x < \infty,\ -\infty < y < \infty.$$

Find the regression curve of Y on X. [C.H. 1995]

Solution. The regression curve of Y on X is given by $y = m_y(x)$.

$$m_y(x) = E(Y|X=x) = \int_{-\infty}^{\infty} y f_y(y|x)\,dy = \frac{\int_{-\infty}^{\infty} y f_{x,y}(x,y)\,dy}{f_x(x)}$$

Now,

$$\int_{-\infty}^{\infty} y f_{x,y}(x,y)\,dy = \int_{-\infty}^{\infty} y\cdot\frac{1}{\sqrt{3}\pi}e^{-2\left(x^2-xy+\frac{y^2}{3}\right)}dy = \frac{1}{\pi\sqrt{3}}e^{-2x^2}\int_{-\infty}^{\infty} y e^{-\frac{2}{3}(y^2-3xy)}\,dy$$

$$= \frac{1}{\pi\sqrt{3}}e^{-2x^2}\int_{-\infty}^{\infty} y e^{-\frac{2}{3}\left\{\left(y-\frac{3}{2}x\right)^2-\frac{9}{4}x^2\right\}}\,dy$$

$$= \frac{1}{\pi\sqrt{3}}e^{-2x^2+\frac{3}{2}x^2}\left\{\int_{-\infty}^{\infty}\left(y-\frac{3}{2}x\right)e^{-\frac{2}{3}\left(y-\frac{3}{2}x\right)^2}dy + \frac{3}{2}x\int_{-\infty}^{\infty} e^{-\frac{2}{3}\left(y-\frac{3}{2}x\right)^2}dy\right\}.$$

Put $\sqrt{\frac{2}{3}}\left(y-\frac{3}{2}x\right)=u$. $\therefore$ $\sqrt{\frac{2}{3}}\,dy=du$. Also, as $y\to-\infty, u\to-\infty$ and $y\to\infty, u\to\infty$.

$$= \frac{1}{\pi\sqrt{3}}e^{-\frac{x^2}{2}}\left\{\int_{-\infty}^{\infty}\sqrt{\frac{3}{2}}\,u e^{-u^2}\sqrt{\frac{3}{2}}\,du + \frac{3}{2}x\int_{-\infty}^{\infty} e^{-u^2}\cdot\sqrt{\frac{3}{2}}\,du\right\}$$

$$= \frac{1}{\pi\sqrt{3}}e^{-\frac{x^2}{2}}\cdot\frac{3}{2}x\sqrt{\frac{3}{2}}\int_{-\infty}^{\infty} e^{-u^2}\,du,$$ first integral vanishes as the integrand is an odd function

$$= \frac{3xe^{-x^2/2}}{2\sqrt{2}\sqrt{\pi}} \quad \because \int_{-\infty}^{\infty} e^{-u^2}\,du=\sqrt{\pi}.$$

Again,

$$f_x(x)=\int_{-\infty}^{\infty} f(x,y)\,dy = \int_{-\infty}^{\infty}\frac{1}{\sqrt{3}\pi}e^{-2\left(x^2-xy+\frac{y^2}{3}\right)}dy = \frac{1}{\pi\sqrt{3}}e^{-2x^2}\int_{-\infty}^{\infty} e^{-\frac{2}{3}(y^2-3xy)}\,dy$$

$$= \frac{1}{\pi\sqrt{3}}e^{-2x^2}\int_{-\infty}^{\infty} e^{-\frac{2}{3}\left\{\left(y-\frac{3}{2}x\right)^2-\frac{9}{4}x^2\right\}}dy = \frac{1}{\pi\sqrt{3}}e^{-x^2/2}\int_{-\infty}^{\infty} e^{-\frac{2}{3}\left(y-\frac{3}{2}x\right)^2}dy.$$

Put $\sqrt{\frac{2}{3}}\left(y-\frac{3}{2}x\right)=u$. $\therefore$ $\sqrt{\frac{2}{3}}\,dy=du$ and limits remain the same.

$$= \frac{1}{\pi\sqrt{3}}e^{-x^2/2}\int_{-\infty}^{\infty} e^{-u^2}\sqrt{\frac{3}{2}}\,du = \frac{1}{\pi\sqrt{2}}e^{-x^2/2}\int_{-\infty}^{\infty} e^{-u^2}\,du$$

$$= \frac{1}{\sqrt{2\pi}}e^{-x^2/2}, \because \int_{-\infty}^{\infty} e^{-u^2}\,du-\sqrt{\pi}.$$

$$\therefore\ m_y(x)=\frac{3xe^{-x^2/2}}{2\sqrt{2\pi}}\times\frac{\sqrt{2\pi}}{e^{-x^2/2}}=\frac{3x}{2}.$$

Hence, the regression line of Y on X is $y=\frac{3}{2}x$.

Ex. 3. If the RVs $X_1, X_2, \cdots, X_n$ all have the same variance s^2 and he correlation coefficient between each pair $(X_i, X_j), i\neq j$ of them is ρ, show that the correlation coefficient between $\sum_{i=1}^{n} X_i$ and $\sum_{i=n+1}^{2n} X_i$ is $\frac{np}{1+(n-1)\rho}$.

Solution. Let $E(X_i)=m_i, i=1(1)2n$.

$$\therefore \operatorname{cov}\left(\sum_{i=1}^{n}X_i, \sum_{i=n+1}^{2n}X_i\right)=E\left[\left\{\sum_{i=1}^{n}(X_i-m_i)\right\}\left\{\sum_{i=n+1}^{2n}(X_i-m_i)\right\}\right]$$

$$=E\left\{\sum_{i=1}^{n}\sum_{j=n+1}^{2n}(X_i-m_i)(X_j-m_j)\right\}=\sum_{i=1}^{n}\sum_{j=n+1}^{n}E\{(X_i-m_i)(X_j-m_j)\}$$

$$=\sum_{i=1}^{n}\sum_{j=n+1}^{n}\operatorname{cov}(X_i,X_j)$$

$$=\sum_{i=1}^{n}\sum_{j=n+1}^{2n}\text{s.s.}\rho, \text{ as } \rho(X_i,X_j)=\rho, \forall\, i,j \text{ and } \sigma(X_i)=s, \forall\, i$$

$$=n^2s^2\rho.$$

Now, SD of $\sum_{i=1}^{n}X_i=\sqrt{E\{(X_1-m_1)+(X_2-m_2)+\cdots+(X_n-m_n)\}^2}=\sqrt{ns^2+2s.s.\rho\cdot{}^nC_2}$

$$=s\sqrt{n}\sqrt{1+(n-1)\rho}.$$

Similarly, SD of $\sum_{i=n+1}^{2n}X_i=s\sqrt{n}\sqrt{1+(n-1)\rho}$.

$$\therefore \rho\left(\sum_{i=1}^{n}X_i, \sum_{i=n+1}^{2n}X_i\right)=\frac{n^2s^2\rho}{s\sqrt{n}\cdot s\sqrt{n}\{1+(n-1)\rho\}}=\frac{n\rho}{1+(n-1)\rho} \quad \textbf{(Proved)}.$$

Ex. 4. Show that the acute angle θ between the least square regression lines is given by $\tan\theta=\frac{1-\rho^2}{|\rho|}\frac{\sigma_x\sigma_y}{\sigma_x^2+\alpha_y^2}$. Discuss the cases when $\rho=0$ and $\rho=\pm1$. [C.H. 2006, 08]

Solution. The regression lines are $y-m_y=\rho\frac{\sigma_y}{\sigma_x}(x-m_x)$ and $y-m_y=\frac{1}{\rho}\frac{\sigma_y}{\sigma_x}(x-m_x)$.

$\therefore$ the gradients of the regression lines are $\rho\frac{\sigma_y}{\sigma_x}$ and $\frac{1}{\rho}\frac{\sigma_y}{\sigma_x}$.

$\therefore$ if θ be the acute angle between them, then

$$\tan\theta=\left|\frac{\rho\frac{\sigma_y}{\sigma_x}-\frac{1}{\rho}\frac{\sigma_y}{\sigma_x}}{1+\frac{\sigma_y^2}{\sigma_x^2}}\right|=\frac{|\rho^2-1|}{|\rho|}\frac{\sigma_y}{\sigma_x}\times\frac{\sigma_x^2}{\sigma_x^2+\sigma_y^2}=\frac{1-\rho^2}{|\rho|}\frac{\sigma_x\sigma_y}{\sigma_x^2+\sigma_y^2} \quad \because |\rho|\le 1.$$

When $\rho=0$, the regression lines are $y=m_y$ and $x=m_x$. Angle between them is $\pi/2$ and they meet at (m_x,m_y). When $\rho=\pm1, \tan\theta=0$, i.e., $\theta=0$ and the lines coincide.

Ex. 5. If the regression lines are $x-2y+1=0$ and $2x-3y+1=0$, find the mean of X, the mean of Y and the correlation coefficient between X and Y.

Solution. The lines intersect at $(1,1)$. $\therefore$ $m_x=1$ and $m_y=1$ as the regression lines meet at (m_x,m_y).
If $x-2y+1=0$ be the regression line of Y on X, then $\rho\frac{\sigma_y}{\sigma_x}=\frac{1}{2}$
and if $2x-3y+1=0$ be the regression line of X on Y, then $\rho\frac{\sigma_x}{\sigma_y}=\frac{2}{3}$.
Multiplying, $\rho^2=\frac{1}{3}$. $\therefore$ $\rho=\pm\frac{1}{\sqrt{3}}$. But, $\rho=-\frac{1}{\sqrt{3}}$ does not satisfy the equations $\rho\frac{\sigma_y}{\sigma_x}=\frac{1}{2}$ and $\rho\frac{\sigma_x}{\sigma_y}=\frac{2}{3}$ as $\sigma_x>0, \sigma_y>0$. $\therefore$ $\rho=\frac{1}{3}=\rho(X,Y)$.

Ex. 6. (Schwartz's in equality). If X, Y be two RVs such that $E(X^2), E(Y^2), E(XY)$ exist, then $\{E(XY)\}^2\le E(X^2)\,E(Y^2)$. Hence, deduce that $-1\le\rho(X,Y)\le 1$.

Solution. If k be a real constant, then $(Y-kX)^2 \geq 0$.

$\therefore$ $E(Y-kX)^2 \geq 0$, or, $E(Y^2) - 2kE(XY) + k^2E(X^2) \geq 0$ as $E(X^2), E(Y^2)$ and $E(XY)$ all exist.

So, the roots of $k^2E(X^2) - 2kE(XY) + E(Y^2) = 0$ are either equal or complex, i.e., disc ≤ 0.

$\therefore$ $4\{E(XY)\}^2 - 4E(X^2)E(Y^2) \leq 0$ or, $\{E(XY)\}^2 \leq E(X^2)E(Y^2)$.

Applying the Schwartz's inequality for the normalised variables X^* and Y^*, then

$$\{E(X^*Y^*)\}^2 \leq E(X^{*2})\,E(Y^{*2})$$

$$\text{or, } \rho^2 \leq 1, \text{ as } E(X^*Y^*) = E\left\{\frac{X-m_x}{\sigma_x}\cdot\frac{Y-m_y}{\sigma_y}\right\} = \frac{\mu_{11}}{\sigma_x\sigma_y} = \rho(X,Y) = \rho, \text{ say}$$

$$\text{and } E(X^{*2}) = E\left(\frac{X-m_x}{\sigma_x}\right)^2 = \frac{E(X-m_x)^2}{\sigma_x^2} = \frac{\sigma_x^2}{\sigma_x^2} = 1.$$

Similarly, $E(Y^{*2}) = 1$. $\therefore$ $-1 \leq \rho \leq 1$.

Ex. 7. For the continuous distribution defined by the joint pdf f as

$$f(x,y) = 3x^2 - 8xy + 6y^2,\ 0<x<1,\ 0<y<1,$$

find the regression curves for the means and also the least square regression lines. [C.H. 1987]

Solution. Regression curve for the mean of Y is $y = m_y(x)$

and regression curve for the mean of X is $x = m_x(y)$,

$$\text{where } m_y(x) = E(Y|X=x) = \int_{-\infty}^{\infty} y f_Y(y|x)\,dy = \frac{\int_{-\infty}^{\infty} y f_{x,y}(x,y)\,dy}{f_x(x)}$$

$$\text{and } m_x(y) = \frac{\int_{-\infty}^{\infty} x f_{x,y}(x,y)\,dx}{f_y(y)}.$$

$$\text{Now, } f_x(x) = \int_{-\infty}^{\infty} f_{x,y}(x,y)\,dy = \int_0^1 (3x^2 - 8xy + 6y^2)\,dy = \left[3x^2y - 4xy^2 + 2y^3\right]_0^1$$

$$= 3x^2 - 4x + 2,\ 0<x<1$$

$$f_y(y) = \int_{-\infty}^{\infty} f_{x,y}(x,y)\,dx = \int_0^1 (3x^2 - 8xy + 6y^2)\,dx = \left[x^3 - 4xy^2 + 6y^2x\right]_0^1$$

$$= 6y^2 - 4y + 1,\ 0<y<1$$

$$\int_{-\infty}^{\infty} y f_{x,y}(x,y)\,dy = \int_0^1 (3x^2y - 8xy^2 + 6y^3)\,dy = \left[\frac{3}{2}x^2y^2 - \frac{8}{3}xy^3 + \frac{3}{2}y^4\right]_0^1$$

$$= \frac{3}{2}x^2 - \frac{8}{3}x + \frac{3}{2}$$

$$\int_{-\infty}^{\infty} x f_{x,y}(x,y)\,dx = \int_0^1 (3x^3 - 8x^2y + 6y^2x)\,dx = \left[\frac{3}{4}x^4 - \frac{8}{3}x^3y + 3y^2x^2\right]_0^1$$

$$= 3y^2 - \frac{8}{3}y + \frac{3}{4}.$$

$$\therefore\ y = m_y(x) \text{ becomes } y = \frac{\frac{3}{2}x^2 - \frac{8}{3}x + \frac{3}{2}}{3x^2 - 4x + 2} \text{ and } x = m_x(y) \text{ becomes } x = \frac{3y^2 - \frac{8}{3}y + \frac{3}{4}}{6y^2 - 4y + 1}.$$

Least square regression line of Y on X is $\dfrac{y-m_y}{\sigma_y} = \rho\dfrac{x-m_x}{\sigma_x}$

and least square regression line of X on Y is $\dfrac{y-m_y}{\sigma_y}=\dfrac{1}{\rho}\dfrac{x-m_x}{\sigma_x}$.

$m_x=$ mean for the marginal distribution of $X=\displaystyle\int_0^1 x\left(3x^2-4x+2\right)dx=\frac{5}{12}$.

$m_y=$ mean for the marginal distribution of $Y=\displaystyle\int_0^1\left(6y^2-4y+1\right)dy=\frac{2}{3}$.

$$\sigma_x^2=E\left(X^2\right)-m_x^2=\frac{67}{720},\ \sigma_y^2=E\left(Y^2\right)-m_y^2=\frac{4}{45}.$$

$$E(XY)=\int_0^1\int_0^1 xy\left(3x^2-8xy+6y^2\right)dx\,dy=\frac{17}{72}.$$

$$\therefore\ \text{cov}(X,Y)=E(XY)-m_xm_y=\frac{17}{72}-\frac{5}{18}=-\frac{1}{24}.\ \therefore\ \rho=\rho(X,Y)=\frac{\text{cov}(X,Y)}{\sigma_x\sigma_y}.$$

$\therefore$ least square regression line of Y on X is given by

$$y-\frac{2}{3}=\rho\frac{\sigma_y}{\sigma_x}\left(x-\frac{5}{12}\right)\ \text{or, } y-\frac{2}{3}=\frac{\text{cov}(X,Y)}{\sigma_x^2}\left(x-\frac{5}{12}\right)\ \text{or, } y-\frac{2}{3}=\frac{-1/24}{67/720}\left(x-\frac{5}{12}\right)$$

$$\text{or, } y-\frac{2}{3}=-\frac{30}{67}\left(x-\frac{5}{12}\right).$$

Least square regression line of X on Y is

$$x-\frac{5}{12}=\frac{\text{cov}(X,Y)}{\sigma_y^2}\left(y-\frac{2}{3}\right)\ \text{or, } x-\frac{5}{12}=\frac{-1/24}{4/45}\left(y-\frac{2}{3}\right)\ \text{or, } x-\frac{5}{12}=-\frac{15}{32}\left(y-\frac{2}{3}\right).$$

Ex. 8. The bivariate pdf of two RVs X and Y is given by $f_{x,y}(x,y)=\begin{cases}x+y, & 0<x<1, 0<y<1\\ 0, & \text{elsewhere.}\end{cases}$

Calculate the means, standard deviations, the correlation coefficient between X and Y. Find the equations of the two regression lines. [C.H. 1999, 2005]

Solution. $m_x=E(X)=\displaystyle\int_0^1\int_0^1 x(x+y)\,dx\,dy=\int_0^1\left[\frac{x^3}{3}+\frac{x^2}{2}y\right]_0^1 dy=\int_0^1\left(\frac{1}{3}+\frac{y}{2}\right)dy$

$$=\left(\frac{1}{3}y+\frac{y^2}{4}\right)_0^1=\frac{1}{3}+\frac{1}{4}=\frac{7}{12}.$$

Similarly, $m_y=E(Y)=\displaystyle\int_0^1\int_0^1 y(x+y)\,dx\,dy=\frac{7}{12}$.

$$\sigma_x^2=\int_0^1\int_0^1\left(x-\frac{7}{12}\right)^2(x+y)\,dx\,dy=\frac{11}{144}.\ \therefore\ \sigma_x=\frac{\sqrt{11}}{12}.$$

Similarly, $\sigma_y=\dfrac{\sqrt{11}}{12}$.

$$\text{cov}(X,Y)=E(XY)-E(X)E(Y)=\int_0^1\int_0^1 xy(x+y)\,dx\,dy-\frac{49}{144}=\int_0^1\int_0^1\left(x^2+xy^2\right)dx\,dy-\frac{49}{144}$$

$$=\int_0^1\left(\frac{x^2}{2}+\frac{x}{3}\right)dx-\frac{49}{144}=\frac{1}{6}+\frac{1}{6}-\frac{49}{144}=\frac{1}{3}-\frac{49}{144}=\frac{48-49}{144}=-\frac{1}{144}.$$

$$\therefore\ \rho(X,Y)=\frac{\text{cov}(X,Y)}{\sigma_x\sigma_y}=-\frac{1}{144}\times\frac{144}{11}=-\frac{1}{11}.$$

$\therefore$ regression line of Y on X is $y - m_y = \rho \frac{\sigma_y}{\sigma_x}(x - m_x)$, i.e., $y - \frac{7}{12} = -\frac{1}{11}\left(x - \frac{7}{12}\right)$

and regression line of X and Y is $y - m_y = \frac{1}{\rho}\frac{\sigma_y}{\sigma_x}(x - m_x)$, i.e., $y - \frac{7}{12} = -11\left(x - \frac{7}{12}\right)$.

Ex. 9. If two RVs X and Y have the same mean and the two regression lines are $y = ax + b, y = \alpha x + \beta$, show that $\frac{b}{\beta} = \frac{1-a}{1-\alpha}$. Find also the common mean. [C.H. 2005]

Solution. Let the common mean is m, i.e., $m_x = m_y = m$.

$\therefore$ the regression lines intersect at (m, m). $\therefore$ $m = am + b$ and $m = \alpha m \alpha m + \beta$. $\therefore$ $m = \frac{b}{1-a} = \frac{\beta}{1-\alpha}$. $\therefore$ $\frac{b}{\beta} = \frac{1-a}{1-\alpha}$ **(Proved).**

The common mean is $\frac{b}{1-a}$ or $\frac{\beta}{1-\alpha}$.

EXERCISE

1. Show that correlation coefficient between two RVs is numerically 1, if and only if they are linear.
2. Show that by a suitable rotation of coordinate axes two correlated RVs may be made uncorrelated.
3. If X and Y be two uncorrelated RVs and if $U = X + Y, V = X - Y$, find the correlation coefficient between U and V. $\left(\textit{Ans. } \rho(U,V) = \frac{\sigma_x^2 - \sigma_y^2}{\sigma_x^2 + \sigma_y^2}\right)$
4. Show that $\text{cov}(aX, bY) = ab\,\text{cov}(X, Y)$, where a, b are real constants.
5. If U and V be two RVs, prove that
   (i) $E[\min\{X, Y\}] \le \min\{E(X), E(Y)\}$ (ii) $E[\max\{X, Y\}] \ge \max\{E(X), E(Y)\}$.
   [Hints. (i) $\min\{X, Y\} = \frac{1}{2}\{X + Y - |X - Y|\}$ and $\max\{X, Y\} = \frac{1}{2}\{X + Y + |X - Y|\}$.
   $$\therefore\ E[\min\{X,Y\}] = \tfrac{1}{2}\{E(X) + E(Y) - E|X - Y|\} \le \tfrac{1}{2}\{E(X) + E(Y) - |E(X - Y)|\}, \quad \because |E(X - Y)| \le E|X - Y|$$
   $$= \tfrac{1}{2}\{E(X) + E(Y) - |E(X) - E(Y)|\}$$
   $$= \min\{E(X), E(Y)\}.$$
   (ii) Similar.]
6. Show the mean of a RV X is the mean of the conditional means of X, given $Y = y$.
   $$\text{[Hints. } E[E(X|Y=y)] = E\left[\int_{-\infty}^{\infty} x f_x(x|y)\,dx\right] = E\left[\frac{\int_{-\infty}^{\infty} x f_{x,y}(x,y)\,dx}{f_y(y)}\right]$$
   (expectation of a function of y)
   $$= \int_{-\infty}^{\infty} \frac{\int_{-\infty}^{\infty} x f_{x,y}(x,y)\,dx}{f_y(y)} f_y(y)\,dy = \int_{-\infty}^{\infty}\int_{-\infty}^{\infty} x f_{x,y}(x,y)\,dx\,dy$$
   $$= m_x(= E(X)),\ \text{i.e., } E\{m_x(y)\} = m_x.]$$
   Note: $E[E(Y|X = x)] = m_y$. Proof is similar for discrete case.
7. Show that $E\left\{\sigma_x^2(y)\right\} \ne \sigma_x^2$.
8. If the RVs X and Y are uncorrelated, then show that the RVs $a_1X + b_1, c_1Y + d_1$ (a_1, b_1, c_1, d_1 are real constants) are also uncorrelated.
9. If two RVs are independent, they must be uncorrelated. Give an example to show that the converse is not true.

[Hints. $\text{cov}(X,Y)=E\{(X-m_x)(Y-m_y)\}=E(X-m_x)E(Y-m_y)$, [when X, Y are independent]
$=0$.

$\therefore\ \rho(X,Y)=\frac{\text{cov}(X,Y)}{\sigma_x\sigma_y}=0$.

Let X be a standard normal distribution. Then, $E(X)=0=E(X^3)$.

Set $Y=X^2$, then $\text{cov}(X,Y)=E\{X(Y-m_y)\}=E(X^3-m_yX)$ [$\because Y=X^2$]
$=E(X^3)-m_yE(X)=0$. So, $\rho(X,Y)=0$.

$\therefore$ X, Y are uncorrelated but they are functionally dependent.]

10. If X and Y are independent and X is normal (m_x,σ_x), Y is normal (m_x,σ_y) and a,b are real constants, then find the characteristic function of the RV $Z=aX+bY$ and the distribution of Z. [C.H. 1983]
[Hints. See the reproductive property of normal distribution.]
11. Show that the sum of two independent Poisson variates having parameters μ_1,μ_2 is again a Poisson variate with parameter $\mu_1+\mu_2$. [C.H. 1992]
12. Prove that the sum of two binomially distributed independent RVs with parameters (n_1,p) and (n_2,p) is again a binomially distributed RV with parameters (n_1+n_2,p).
13. Show that the covariance of the RVs U and V given by
$U=X\cos\theta+Y\sin\theta, V=-X\sin\theta+Y\cos\theta$ is $-\frac{1}{2}\left(\sigma_x^2-\sigma_y^2\right)\sin 2\theta+\rho(X,Y)\sigma_x\sigma_y\cos 2\theta$.
14. If X, Y are standard normal variates with correlation coefficient ρ, then show that the correlation coefficient between X^2 and Y^2 is ρ^2.
15. If the RVs $X_1,X_2,\cdots,X_n$ be n independently and identically distributed with $E(X_i)=m$ and var $(X_i)=\sigma^2$, $\forall\, i$, then show that the RV $S_n=X_1+X_2+\cdots+X_n$ is asymptotically normal with mean $M=nm$ and variance $\Sigma^2=n\sigma^2$.
16. If the RVs $X_1,X_2,\cdots,X_n$ are independent and each follows normal (m,σ) distribution, then the RV $S_n=X_1+\cdots+X_n$ is normal (M,Σ), when $M=nm$ and $\Sigma^2=n\sigma^2$.
[Hints. Follow the reproductive property of normal distribution.]
17. If the two-dimensional distribution of the RVs X and Y is given, the pdf

$$f_{x,y}(x,y)=\begin{cases} x+y, & x+y\le 1\\ 0, & \text{elsewhere.}\end{cases}$$

Find $m_x, m_y, \sigma_x, \sigma_y, \rho(X,Y)$ and the two regression lines.

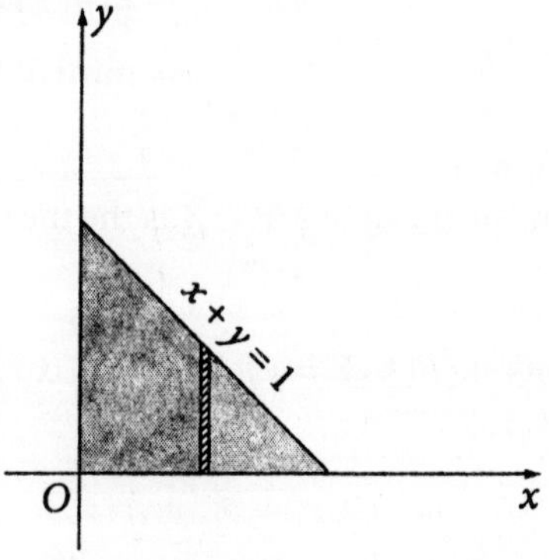

[Hints. $m_x=\displaystyle\int_0^1\int_0^{1-x} x(x+y)\,dy\,dx=\frac{1}{8}$,

$$m_y=\int_0^1\int_0^{1-y}\left(xy+y^2\right)dx\,dy=\frac{1}{8}.$$

$$\sigma_x^2=\int_0^1\int_0^{1-x}\left(x-\frac{1}{8}\right)^2(x+y)\,dy\,dx,\ \text{etc.}]$$

Chapter 8

Convergence in Probability

8.1 Introduction

In real analysis we have convergence of sequence, convergence of a series, convergence of improper integrals, etc. Here we introduce the convergence of a sequence of RV in probability and the convergence of their distributions.

Let $\{X_n\}_n$ be a sequence of RV defined on an event space S. The sequence is said to converge in probability to a RV X in S, if for any pre-assigned $\epsilon > 0$, $P(|X_n - X| \geq \epsilon)$ is smaller than any small positive quantity for large n, i.e., for an arbitrarily small $\delta > 0$ a natural number N can be found so that $P(|X_n - X| \geq \epsilon) < \delta \ \forall \ n > N$.

We express the fact as $\lim_{n\to\infty} P(|X_n - X| \geq \epsilon) = 0$ or, equivalently, $\lim_{n\to\infty} P(|X_n - X| < \epsilon) = 1$ and write

$$X_n \xrightarrow{\text{in } P} X \text{ as } n \to \infty.$$

A sequence of RV $\{X_n\}_n$ is said to converge in probability to a constant a, if for every $\epsilon > 0$,

$$\lim_{n\to\infty} P(|X_n - a| \geq \epsilon) = 0$$

$$\text{or, equivalently } \lim_{n\to\infty} P(|X_n - a| \geq \epsilon) = 1,$$

and is written as $X_n \xrightarrow{\text{in } P} a$ as $n \to \infty$.

Theorem 8.1.1 If $X_n \xrightarrow{\text{in } P} a$ and $Y_n \xrightarrow{\text{in } P} b$ as $n \to \infty$, then

(i) $X_n \pm Y_n \xrightarrow{\text{in } P} a \pm b$, as $n \to \infty$ (ii) $X_n Y_n \xrightarrow{\text{in } P} ab$, as $n \to \infty$

(iii) $\frac{X_n}{Y_n} \xrightarrow{\text{in } P} \frac{a}{b}$, as $n \to \infty$, provided $b \neq 0$.

Proof. Let us consider three events $A_n = |X_n - a| < \epsilon/2$, $B_n = |Y_n - b| < \epsilon/2$ and $C_n = |(X_n \pm Y_n) - (a \pm b)| < \epsilon$.

Now, $|(X_n \pm Y_n) - (a \pm b)| = |(X_n - a) \pm (Y_n - b)| \leq |X_n - a| + |Y_n - b|$.

$\therefore$ when A_n and B_n occur simultaneously, then $|(X_n \pm Y_n) - (a \pm b)| < \epsilon$, i.e., C_n occurs. $\therefore A_n B_n \subset C_n$.

$\therefore \overline{A_n B_n} \supset \bar{C}_n$, i.e., $\bar{C}_n \subset \overline{A_n B_n} = \bar{A}_n + \bar{B}_n$, by De Morgan's law.

$\therefore P(\bar{C}_n) \leq P(\bar{A}_n + \bar{B}_n) = P(\bar{A}_n) + P(\bar{B}_n) - P(\bar{A}_n \bar{B}_n) \leq P(\bar{A}_n) + P(\bar{B}_n) \to 0$ as $n \to \infty$,

as $P(\bar{A}_n) = P(|X_n - a| \geq \epsilon/2) \to 0$ and $P(\bar{B}_n) = P(|Y_n - b| \geq \epsilon/2) \to 0$ as $n \to \infty$,

since $X_n \xrightarrow{\text{in } P} a$ and $Y_n \xrightarrow{\text{in } P} b$ as $n \to \infty$

$\therefore \lim_{n\to\infty} P(\bar{C}_n) = 0$ or, $\lim_{n\to\infty} P[\{|X_n \pm Y_n| - (a \pm b)\} \geq \epsilon] = 0$.

$\therefore X_n \pm Y_n \xrightarrow{\text{in } P} a \pm b$ as $n \to \infty$. This proves case (i).

Note: By the principle of mathematical induction the above result may be extended to any finite number of RV.

To prove (ii), we first prove that: $X_n \xrightarrow{\text{in } P} a$ as $n \to \infty \Rightarrow cX_n \xrightarrow{\text{in } P} ca$ as $n \to \infty, c \in \mathbb{R}-\{0\}$ and $X_n^2 \xrightarrow{\text{in } P} a^2$ as $n \to \infty$.

As $X_n \xrightarrow{\text{in } P} a$ as $n \to \infty$. $\therefore$ for any pre-assigned $\epsilon > 0$, $\lim_{n\to\infty} P(|X_n - a| \geq \epsilon) = 0$.

$\therefore \lim_{n\to\infty} P(|cX_n - ca| \geq \epsilon) = \lim_{n\to\infty} \left(|X_n - a| \geq \frac{\epsilon}{|c|}\right) = 0.$

For $c = 0$, the result is trivial. $\therefore cX_n \xrightarrow{\text{in } P} ca$ as $n \to \infty$ for $c \in \mathbb{R}$.

Now, $\lim_{n\to\infty} P\left(X_n^2 \geq \epsilon\right) = \lim_{n\to\infty} P\left(|X_n| \geq \sqrt{\epsilon}\right) = 0$ if $X_n \xrightarrow{\text{in } P} 0$ as $n \to \infty$.

$\therefore$ whenever $X_n \xrightarrow{\text{in } P} 0$ as $n \to \infty$, $X_n^2 \xrightarrow{\text{in } P} 0$ as $n \to \infty$.

$X_n^2 = (X_n - a)^2 + 2a(X_n - a) + a^2 \xrightarrow{\text{in } P} a^2$ as $n \to \infty$, when $X_n \xrightarrow{\text{in } P} a$ as $n \to \infty$ [for $X_n \xrightarrow{\text{in } P} a$ as $n \to \infty$
$\Rightarrow X_n - a \xrightarrow{\text{in } P} 0$ as $n \to \infty$.
$\Rightarrow (X_n - a)^2 \xrightarrow{\text{in } P} 0$ as $n \to \infty$, also $2a(X_n - a) \xrightarrow{\text{in } P} 2a \times 0 = 0$ as $n \to \infty$.]

$\therefore X_n^2 \xrightarrow{\text{in } P} a^2$ as $n \to \infty$, whenever $X_n \xrightarrow{\text{in } P} a$ as $n \to \infty$.

$\therefore X_n Y_n = \frac{1}{4}\left\{(X_n + Y_n)^2 - (X_n - Y_n)^2\right\} \xrightarrow{\text{in } P} \frac{1}{4}\left\{(a+b)^2 - (a-b)^2\right\}$ as $n \to \infty$, whenever $X_n \xrightarrow{\text{in } P} a$ and $Y_n \xrightarrow{\text{in } P} b$ as $n \to \infty$.

$\therefore X_n Y_n \xrightarrow{\text{in } P} ab$ as $n \to \infty$, whenever $X_n \xrightarrow{\text{in } P} a$ and $Y_n \xrightarrow{\text{in } P} b$ as $n \to \infty$.

This proves (ii).

To prove (iii), we prove: When $Y_n \xrightarrow{\text{in } P} b(\neq 0)$ as $n \to \infty$, then $\frac{1}{Y_n} \xrightarrow{\text{in } P} \frac{1}{b}$ as $n \to \infty$.

$\therefore$ by (ii) it will be proved that $\frac{X_n}{Y_n} = X_n \cdot \frac{1}{Y_n} \xrightarrow{\text{in } P} a \cdot \frac{1}{b}$, i.e., $\frac{a}{b}$ as $n \to \infty$.

Consider the events $C_n : |Y_n - b| < |b|$ and $D_n : \left|\frac{1}{Y_n} - \frac{1}{b}\right| \geq \epsilon$.

$$\left|\frac{1}{Y_n} - \frac{1}{b}\right| = \frac{|Y_n - b|}{|b|\,|Y_n|} = \frac{|Y_n - b|}{|b|\,|(Y_n - b) + b|} \leq \frac{|Y_n - b|}{|b|\,||b| - |Y_n - b||} = \frac{|Y_n - b|}{|b|(|b| - |Y_n - b|)},$$

$\because |Y_n - b| < |b|$, when C_n occurs.

$$\therefore \frac{|Y_n - b|}{|b|(|b|(|b| - |Y_n - b|))} \geq \left|\frac{1}{Y_n} - \frac{1}{b}\right| \geq \epsilon, \text{ when } D_n \text{ also occurs.}$$

or, $|Y_n - b| \geq \epsilon b^2 - \epsilon|b||Y_n - b|$ as C_n occurs, i.e., $|b| - |Y_n - b| > 0$.

$$\therefore |Y_n - b| \geq \frac{\epsilon b^2}{1 + |b|\epsilon} = \epsilon_1 \text{ (say)}.$$

$\therefore$ if E_n be the event $|Y_n - b| \geq \epsilon_1$, then $C_n D_n \Rightarrow E_n$. $\therefore C_n D_n \subset E_n$. Also, $D_n \bar{C}_n \subset \bar{C}_n$.

$\therefore D_n = D_n\left(C_n + \bar{C}_n\right) = D_n C_n + D_n \bar{C}_n \subset E_n + \bar{C}_n$.

$\therefore 0 \leq P(D_n) \leq P\left(E_n + \bar{C}_n\right) \leq P(E_n) + P\left(\bar{C}_n\right)$ as $P\left(E_n \bar{C}_n\right) \geq 0$.

Since, $Y_n \xrightarrow{\text{in } P} b$ as $n \to \infty$. $\therefore \lim_{n\to\infty} P(E_n)$ and $\lim_{n\to\infty} \left(\bar{C}_n\right) = 0$.

$\therefore \lim_{n\to\infty} P(D_n) = 0$. $\therefore \frac{1}{Y_n} \xrightarrow{\text{in } P} \frac{1}{b}$ as $n \to \infty$.

8.2 Tchebycheff's inequality

STATEMENT. Let X be a RV with mean m_x and variance σ_x^2 (finite). Then, for every $\epsilon > 0$,

$$P(|X - m_x| \geq \epsilon) \leq \frac{\sigma_x^2}{\epsilon^2} \text{ or, equivalently, } P(|X - m_x| < \epsilon) \geq 1 - \frac{\sigma_x^2}{\epsilon^2}.$$

Proof. First, let X be a continuous RV. Let the distribution of X is defined by the pdf f_x.

$$\therefore\ P(|X-m_x|\geq\epsilon)=\int_{|x-m_x|\geq\epsilon} f_x(x)\,dx\leq\frac{1}{\epsilon^2}\int_{|x-m_x|\geq\epsilon}(x-m_x)^2 f_x(x)\,dx,$$

as within the range of integration $\dfrac{(x-m_x)^2}{\epsilon^2}\geq 1$.

$$\therefore\ P(|X-m_x|\geq\epsilon)\leq\frac{1}{\epsilon^2}\int_{-\infty}^{\infty}(x-m_x)^2 f_x(x)\,dx,\ \text{as the integrand is non-negative, } \forall\ x\in\mathbb{R}$$

$$\text{or, } P(|X-m_x|\geq\epsilon)\leq\frac{\sigma_x^2}{\epsilon^2},\ \because\ \int_{-\infty}^{\infty}(x-m_x)^2 f_x(x)\,dx=\sigma_x^2<\infty.$$

Secondly, let X be a discrete RV whose distribution is defined by the pmf f_{xi}.

$$\therefore\ P(|X-m_x|\geq\epsilon)=\sum_{|x_i-m_x|\geq\epsilon} f_{xi}\leq\frac{1}{\epsilon^2}\sum_{|x_i-m_x|\geq\epsilon}(x_i-m_x)^2 f_{xi},\ \text{as in the range of summation } \frac{(x_i-m_x)^2}{\epsilon^2}>1$$

$$\leq\frac{1}{\epsilon^2}\sum_{i=-\infty}^{\infty}(x_i-m_x)^2 f_{xi},\ \text{as each term of the series is non-negative}$$

$$=\frac{\sigma_x^2}{\epsilon^2},\ \text{as the series absolutely converges to } \sigma_x^2<\infty.$$

$$\therefore\ P(|X-m_x|\geq\epsilon)\leq\frac{\sigma_x^2}{\epsilon^2},\ \text{when } \sigma_x^2<\infty.$$

In both the case the probability of the complementary event

$$P(|X-m_x|<\epsilon)=1-P(|X-m_x|\geq\epsilon)\geq 1-\frac{\sigma_x^2}{\epsilon^2},\ \text{as } P(|X-m_x|\geq\epsilon)\leq\frac{\sigma_x^2}{\epsilon^2}.$$

Note: This inequality gives a great significance of σ_x. We know that, whenever the values of σ_x are smaller and smaller the spectrum of X is more and more concentrated near its mean value. The Tchebycheff's inequality gives a limit of the spectrum of X about the mean.

For example, let $m_x=0, \epsilon=0.5, \sigma_x=0.1$.

$$\therefore\ P(|X|\geq 0.5)\leq\frac{(0.1)^2}{(0.5)^2},\ \text{i.e., } P(|X|<0.5)\geq 0.96 \text{ or, } P(-0.5<X<0.5)\geq 0.96.$$

$\therefore$ when $\sigma_x=0.1$, the spectrum of X is confined in 0.5 nbd of the mean has the probability greater than or equal to 0.96.

If we take $\sigma_x=0.01$, for the same m_x and ϵ, then $P(|X|<0.5)\geq 0.9996$, i.e., the spectrum of X is confined in 0.5 nbd of the mean is almost certain.

8.2.1 Generalised Tchebycheff's inequality

Let a RV X possess a finite second order moment about some point $c\in\mathbb{R}$. Then, for each $\epsilon>0$,

$$P(|X-c|\geq\epsilon)\leq\frac{E\{(X-c)^2\}}{\epsilon^2}\ \text{ or, equivalently, } P(|X-c|<\epsilon)\geq 1-\frac{E\{(X-c)^2\}}{\epsilon^2}.$$

Proof. First, let X be a discrete RV and its distribution is given by the pmf f_{xi}.

$$\therefore\ P(|X-c|\geq\epsilon)=\sum_{|x_i-c|\geq\epsilon} f_{xi}\leq\frac{1}{\epsilon^2}\sum_{|x_i-c|\geq\epsilon}(x_i-c)^2 f_{xi},\ \text{as within the range of summation } (x_i-c)^2\geq\epsilon^2$$

$$\leq\frac{1}{\epsilon^2}\sum_{i=-\infty}^{\infty}(x_i-c)^2 f_{xi},\ \text{as each term of the series in non-negative}$$

$$=\frac{1}{\epsilon^2}E(X-c)^2,\ \text{as the series is absolutely converges to } E(X-c)^2<\infty.$$

$$\therefore\ P(|X-c|\geq\epsilon)\leq\frac{E\left\{(X-c)^2\right\}}{\epsilon^2}\ \text{when } E(X-c)^2<\infty.$$

Secondly, let X be a continuous RV and its distribution is given by the pdf f_X.

$$\therefore\ P(|X-c|\geq\epsilon)=\int_{|x-c|\geq\epsilon} f_X(x)\,dx\leq\frac{1}{\epsilon^2}\int_{|x-c|\geq\epsilon}(x-c)^2 f_X(x)\,dx,\ \text{as } (x-c)^2\geq\epsilon^2 \text{ in the range of integration}$$

$$\leq\frac{1}{\epsilon^2}\int_{-\infty}^{\infty}(x-c)^2 f_X(x)\,dx,\ \text{as the integrand is non-negative}$$

$$=\frac{E\{(X-c)^2}{\epsilon^2},\ \text{as the integral absolutely converges to } E(X-c)^2<\infty.$$

$$\therefore\ P(|X-c|\geq\epsilon)\leq\frac{E\left\{(X-c)^2\right\}}{\epsilon^2},\ \text{when } E(X-c)^2<\infty.$$

In both the cases, the probability of the complementary event

$$P(|X-c|<\epsilon)=1-P(|X-c|\geq\epsilon)\geq 1-\frac{E\{(X-c)^2\}}{\epsilon^2},\ \text{as } P(|X-c|\geq\epsilon)\leq\frac{E\{(X-c)^2\}}{\epsilon^2}.$$

8.2.2 Further generalisation of Tchebycheff's inequality

If $g:\mathbb{R}\to\mathbb{R}$ be continuous and $g(x)\geq 0,\ \forall\ x\in\mathbb{R}$, then $P\{g(X)\geq\epsilon\}\leq\frac{E\{g(X)\}}{\epsilon}$, for each $\epsilon>0$, provided $E\{g(X)\}$ exists.

Proof. (Continuous case). Let X be a continuous RV on an event space S. Then, $g(X)$ is also a RV on the same event space S and the distribution is uniquely determined by that of X.

$$\therefore\ P\{g(X)\geq\epsilon\}=\int_{g(x)\geq\epsilon} f_X(x)\,dx,\ f_X \text{ is the pdf of } X$$

$$\leq\frac{1}{\epsilon}\int_{g(x)\geq\epsilon} g(x)f_X(x)\,dx,\ \text{as } \frac{g(x)}{\epsilon}\geq 1 \text{ in the range of integration}$$

$$\leq\frac{1}{\epsilon}\int_{-\infty}^{\infty} g(x)f_X(x)\,dx,\ \text{as the integrand is non-negative}$$

$$=\frac{E\{g(X)\}}{\epsilon},\ \text{provided } E\{g(X)\} \text{ exists.}$$

Discrete case. Let X be a discrete RV on some event space S. Then, $g(X)$ is also a RV on S and the distribution of $g(X)$ is uniquely determined by that of X. Let f_{xi} be the pmf of X.

$$\therefore P\{g(X)\geq\epsilon\}=\sum_{g(x_i)\geq\epsilon} f_{xi}\leq\frac{1}{\epsilon}\sum_{g(x_i)\geq\epsilon} g(x_i)f_{xi}, \text{ as } \frac{g(x_i)}{\epsilon}\geq 1 \text{ in the range of summation}$$

$$\leq\frac{1}{\epsilon}\sum_{i=-\infty}^{\infty} g(x_i)f_{xi}, \text{ as each term in the series in non-negative}$$

$$=\frac{E\{g(X)\}}{\epsilon}, \text{ provided } E\{g(X)\} \text{ exists.}$$

8.3 Tchebycheff's theorem

Let $\{X_n\}_n$ be a sequence of RV such that the mean m_n and variance σ_n^2 of X_n for each n exists.

If $\sigma_n\to 0$ as $n\to\infty$, then $X_n-m_n\xrightarrow{\text{in } P}0$ as $n\to\infty$.

Proof. By Tchebycheff's inequality for each $\epsilon>0$, $P(|X_n-m_x|\geq\epsilon)\leq\frac{\sigma_n^2}{\epsilon^2}$. Also, $P(|X_n-m_n|\geq\epsilon)\geq 0$.

As $\lim_{n\to\infty}\sigma_n=0$, $\therefore$ $\lim_{n\to\infty}P(|X_n-m_n|\geq\epsilon)=0$ for each $\epsilon>0$, by Sandwich theorem.

$\therefore$ $X_n-m_n\xrightarrow{\text{in } P}0$ as $n\to\infty$.

Note: If further $\{m_n\}_n$ converges to m, then Tchebycheff's theorem reduces to $X_n\xrightarrow{\text{in } P}m$ as $n\to\infty$.

8.4 Bernoulli's theorem

If X_n is a binomial (n,p) variate, then $\frac{X_n}{n}\xrightarrow{\text{in } P}p$ as $n\to\infty$.

Proof. X_n is a binomial (n,p) variate. $\therefore$ $E(X_n)=np$ and var $(X_n)=np(1-p)$.

Let $Y_n=\frac{X_n}{n}$.

$\therefore$ $E(Y_n)=E\left(\frac{X_n}{n}\right)=\frac{1}{n}E(X_n)=\frac{np}{n}=p$ and var $(Y_n)=$ var $\left(\frac{X_n}{n}\right)=\frac{1}{n^2}$var $(X_n)=\frac{np(1-p)}{n^2}=\frac{p(1-p)}{n}$.

$\therefore$ for each $\epsilon>0$, the Tchebycheff's inequality gives

$$P\left(\left|Y_n-p\right|\geq\epsilon\right)\leq\frac{\text{var}(Y_n)}{\epsilon^2}, \text{ or, } P\left(\left|Y_n-p\right|\geq\epsilon\right)\leq\frac{p(1-p)}{n\epsilon^2} \text{ as var}(Y_n)=\frac{p(1-p)}{n}.$$

Also, $P\left(\left|Y_n-p\right|\geq\epsilon\right)\geq 0$. So, $0\leq P\left(\left|Y_n-p\right|\geq\epsilon\right)\leq\frac{p(1-p)}{n\epsilon^2}$.

Since, $\lim_{n\to\infty}\frac{p(1-p)}{n\epsilon^2}=0$.

$\therefore$ $\lim_{n\to\infty}P\left(\left|Y_n-p\right|\geq\epsilon\right)=0$, by Sandwich theorem.

$\therefore$ $Y_n\xrightarrow{\text{in } P}p$ as $n\to\infty$, i.e., $\frac{X_n}{n}\xrightarrow{\text{in } P}p$ as $n\to\infty$.

Alternative Method: As var $(Y_n)\to 0$ as $n\to\infty$ and $E(Y_n)=p$. $\therefore$ by Tchebycheff's theorem

$$Y_n\xrightarrow{\text{in } P}p \text{ as } n\to\infty, \text{ i.e., } \frac{X_n}{n}\xrightarrow{\text{in } P}p \text{ as } n\to\infty.$$

Note. If X_n denotes the number of success in Bernoullian sequence of n trials, then $\frac{X_n}{n}$ is the frequency ratio of success and then by the above theorem $\lim_{n\to\infty}\frac{X_n}{n}=p$, the probability of success in a single trial. This is the frequency definition of probability.

8.5 Law of large numbers

Let $\{x_n\}_n$ be a sequence of RV and the RV $S_n=X_1+X_2+\cdots+X_n$ has a finite mean M_n and finite variance Σ_n^2 for each n. Then, $\frac{S_n-M_n}{n}\xrightarrow{\text{in } P}0$ as $n\to\infty$, provided $\lim_{n\to\infty}\frac{\Sigma_n}{n}=0$.

Proof. $E\left(\frac{S_n-M_n}{n}\right)=\frac{1}{n}E(S_n-M_n)=\frac{1}{n}\{E(S_n)-M_n\}=0$.

Also, var $\left(\frac{S_n-M_n}{n}\right)=E\left\{\left(\frac{S_n-M_n}{n}-0\right)^2\right\}=\frac{1}{n^2}E(S_n-M_n)^2=\frac{\text{var}(S_n)}{n^2}=\left(\frac{\Sigma_n}{n}\right)^2$.

Let $\epsilon(>0)$ be any chosen number. $\therefore$ by Tchebycheff's inequality $P\left(\left|\frac{S_n-M_n}{n}-0\right|\geq\epsilon\right)\leq\frac{1}{\epsilon^2}\left(\frac{\Sigma_n}{n}\right)^2$.

Also, $P\left(\left|\frac{S_n-M_n}{n}-0\right|\geq\epsilon\right)\geq 0$. By the given condition, $\lim\limits_{n\to\infty}\frac{\Sigma_n}{n}=0$. $\therefore$ $\lim\limits_{n\to\infty}=\frac{1}{\epsilon}\frac{\Sigma_n}{n}=0$ as ϵ is a chosen fixed number.

$\therefore$ $\lim\limits_{n\to\infty}P\left(\left|\frac{S_n-M_n}{n}-0\right|\geq\epsilon\right)=0$, [by Sandwich theorem] $\Rightarrow \frac{S_n-M_n}{n}\xrightarrow{\text{in }P}0$ as $n\to\infty$.

If $\bar{X}=\frac{X_1+X_2+\cdots+X_n}{n}=\frac{S_n}{n}$ and $\bar{m}=\frac{m_1+m_2+\cdots+m_n}{n}=\frac{M_n}{n}$, then the above result may be written as $\bar{X}-\bar{m}\xrightarrow{\text{in }P}0$ as $n\to\infty$.

8.5.1 Law of large numbers for equal components

If the sequence of RV $\{X_n\}_n$ be such that each $X_n(n\in\mathbb{N})$ has the same distribution with finite mean m and finite SD σ and if the RV $X_1,X_2,\cdots,X_n$ are mutually independent for all $n\in\mathbb{N}$, then $\bar{X}\xrightarrow{\text{in }P}m$ as $n\to\infty$, where $\bar{X}=\frac{1}{n}(X_1+X_2+\cdots+X_n)$.

Proof. $\bar{X}=\frac{1}{n}(X_1+X_2+\cdots+X_n)$. $\therefore$ $E\left(\bar{X}\right)=\frac{1}{n}\sum\limits_{i=1}^{n}E(X_i)=\frac{1}{n}\cdot nm=m$.

$$\text{var}\left(\bar{X}\right)=E\left(\bar{X}-m\right)^2=E\left(\frac{X_1+X_2+\cdots+X_n}{n}-m\right)^2=\frac{1}{n^2}E\{(X_1-m)+(X_2-m)+\cdots+(X_n-m)\}^2$$

$$=\frac{1}{n^2}\sum_{i=1}^{n}E(X_i-m)^2,\text{ as }X_1,X_2,\cdots,X_n\text{ are mutually independent}$$

$$=\frac{1}{n^2}\cdot n\sigma^2=\frac{\sigma^2}{n}\text{ as }E(X_i-m)^2=\sigma^2\text{ for each }i.$$

Let $\epsilon(>0)$ be any given number.

$\therefore$ by Tchebycheff's inequality $P\left(\left|\bar{X}-m\right|\geq\epsilon\right)\leq\frac{\sigma^2}{n\epsilon^2}$ and $P\left(\left|\bar{X}-m\right|\geq\epsilon\right)\geq 0$, by axiom (1).

$\therefore$ $0\leq\left(\left|\bar{X}-m\right|\geq\epsilon\right)\leq\frac{\sigma^2}{n\epsilon^2}$. As $\lim\limits_{n\to\infty}\frac{\sigma^2}{n\epsilon^2}=0$ [σ is finite and ϵ is a given number].

$\therefore$ $\lim\limits_{n\to\infty}P\left(\left|\bar{X}-m\right|\geq\epsilon\right)=0$ for each $\epsilon>0$, by Sandwich theorem. $\therefore$ $\bar{X}\xrightarrow{\text{in }P}m$ as $n\to\infty$.

WORKED-OUT EXAMPLES

Ex. 1. If X be a non-negative RV with finite mean m, prove that $P(X\geq m\alpha)\leq\frac{1}{\alpha}$ for an arbitrary $\alpha>0$.

Solution. First, consider X is discrete RV

$$m=E(X)=\sum_{-\infty<x_i<\infty}x_i f_{xi}=\sum_{x_i<\alpha m}x_i f_{xi}+\sum_{x_i\geq\alpha m}x_i f_{xi},$$

$\alpha m\geq 0$ as $\alpha>0$ and $m\geq 0$, for X is non-negative.

$$\therefore\ m\geq\sum_{x_i\geq\alpha m}x_i f_{xi}\text{ as }\sum_{x_i<\alpha m}x_i f_{xi}\geq 0,\quad\begin{array}{l}\text{for }X\text{ is non-negative and }f_{xi}>0,\\ \text{i.e., }x_i\geq 0\text{ and }f_{xi}>0,\ \forall\ i\end{array}$$

$$\geq\alpha m\sum_{x_i\geq\alpha m}f_{xi}=\alpha mP(X\geq\alpha m).\ \therefore\ P(X\geq m\alpha)\leq\frac{m}{m\alpha},\text{ i.e., }P(X\geq m\alpha)\leq\frac{1}{\alpha}.$$

Secondly, consider that X is a continuous RV.

$$m = E(X) = \int_{-\infty}^{\infty} x f_x(x)\,dx = \int_{-\infty}^{m\alpha} x f_x(x)\,dx + \int_{m\alpha}^{\infty} x f_x(x)\,dx$$

$$= \int_{0}^{m\alpha} x f_x(x)\,dx + \int_{m\alpha}^{\infty} x f_x(x)\,dx, \text{ as } f_x(x) = 0 \text{ for } x < 0 \text{ [X is non-negative]}$$

$$\geq \int_{m\alpha}^{\infty} x f_x(x)\,dx, \text{ as } x f_x(x) \geq 0,\ \forall\, x \geq 0, \text{ i.e., } \int_{0}^{m\alpha} x f_x(x)\,dx \geq 0$$

$$\geq m\alpha \int_{m\alpha}^{\infty} f_x(x)\,dx \text{ as } x \geq m\alpha \text{ within the range of integration}$$

$$= m\alpha P(X \geq m\alpha).$$

$\therefore\ P(X \geq m\alpha) \leq 1/\alpha$ for any $\alpha > 0$.

Ex. 2. Let the function $h : \mathbb{R} \to \mathbb{R}$ be continuous and $h(x) \geq 0,\ \forall\, x \in \mathbb{R}$. Let X be a discrete RV and the mean of the RV $h(X)$ exists. Then, for each $\epsilon > 0, P\{h(X) \geq \epsilon\} \leq \frac{E\{h(X)\}}{\epsilon}$.

Solution. Let the distribution of the discrete RV X is defined by the pmf f_{xi}. The RV $h(X)$ is also discrete having step point $h(x_i)$, where x_i is a step point of $X, i = 0, \pm 1, \pm 2, \cdots$.

$$\therefore\ E\{h(X)\} = \sum_{h(x_i)\geq\epsilon} h(x_i) f_{xi} + \sum_{h(x_i)<\epsilon} h(x_i) f_{xi}.$$

Since, $\sum_{h(x_i)<\epsilon} h(x_i) f_{xi} \geq 0$.

$$\therefore\ \sum_{h(x_i)\geq\epsilon} h(x_i) f_{xi} + \sum_{h(x_i)<\epsilon} h(x_i) f_{xi} \geq \sum_{h(x_i)\geq\epsilon} h(x_i) f_{xi}$$

$$\therefore\ E\{h(X)\} \geq \sum_{h(x_i)\geq\epsilon} h(x_i) f_{xi} \geq \epsilon \sum_{h(x_i)\geq\epsilon} f_{xi}, \text{ as } \epsilon \leq h(x_i)$$

$$= \epsilon P\{h(X) \geq \epsilon\}.$$

$$\therefore\ P(h(X) \geq \epsilon) \leq \frac{E\{h(X)\}}{\epsilon}.$$

Note 1. The inequality is also true when X is a continuous RV.

2. This is actually a generalised form of Tchebycheff's inequality and was proved in theory part.

Ex. 3. Using Tchebycheff's inequality, show that in 1000 throws of unbiased coin the probability that the number of heads lies between 450 and 550 is at least 9/10.

Solution. If X be the RV denoting number of heads in 1000 throws of an unbiased coin, then X follows binomial (1000, 1/2) distribution. $\therefore\ E(X) = 1000 \times \frac{1}{2} = 500$ and $\sigma_x^2 = 1000 \times \frac{1}{2} \times \frac{1}{2} = 250$.

Now, $P(450 < X < 550) = P(-50 < X - 500 < 50) = P(|X - 500| < 50)$.

By Tchebycheff's inequality,

$$P(|X - 500| < 50) \geq 1 - \frac{\sigma_x^2}{50^2}, \text{ [taking } \epsilon = 50]$$

$$= 1 - \frac{250}{2500} = \frac{9}{10}. \quad \therefore\ P(450 < X < 550) \geq \frac{9}{10}.$$

Ex. 4. If X be a $\gamma(n)$ variate, then show that $P(0 < X < 2n) \geq \frac{n-1}{n}$. [C.H. 2007]

Solution.

$$E(X)=\int_0^\infty x\frac{e^{-x}x^{n-1}}{\Gamma(n)}dx=\int_0^\infty\frac{e^{-x}x^{(n+1)-1}}{\Gamma(n)}dx=\frac{1}{\Gamma(n)}\cdot\Gamma(n+1)=\frac{n\Gamma(n)}{\Gamma(n)}=n.$$

$$E\left(X^2\right)=\int_0^\infty x^2\frac{e^{-x}x^{n-1}}{\Gamma(n)}dx=\frac{1}{\Gamma(n)}\int_0^\infty e^{-x}x^{(n+2)-1}dx=\frac{\Gamma(n+2)}{\Gamma(n)}=\frac{(n+1)n\Gamma(n)}{\Gamma(n)}=n(n+1).$$

$$\therefore\ \sigma_x^2=E\left(X^2\right)-\{E(X)\}^2=n(n+1)-n^2=n.$$

$\therefore$ by Tchebycheff's inequality, $P(|X-n|\geq n)\leq\frac{\sigma_x^2}{n^2}$, as mean of $X=n$ and taking $\epsilon=n$

$$\text{or, } P(|X-n|<n)\geq 1-\frac{\sigma_x^2}{n^2}=1-\frac{n}{n^2}=\frac{n-1}{n}\quad\text{or, } P(0<X<2n)\geq\frac{n-1}{n}.$$

Ex. 5. A RV X has the pdf f_x given by $f_x(x)=\begin{cases}12x^2(1-x), & 0<x<1\\ 0, & \text{elsewhere.}\end{cases}$

Compute $P(|X-m_x|\geq 2\sigma_x)$ and compare it with the limit given by Tchebycheff's inequality, where m_x and σ_x are respectively the mean and SD of X.

Solution. $m_x=E(X)=\int_0^1 x\cdot 12x^2(1-x)dx=12\int_0^1\left(x^3-x^4\right)dx=12\times\frac{1}{20}=\frac{3}{5}.$

$$\sigma_x^2=E\left(X^2\right)-\{E(X)\}^2=\int_0^1 x^2\cdot 12x^2(1-x)dx-\frac{9}{25}=12\times\frac{1}{30}-\frac{9}{25}=\frac{2}{5}-\frac{9}{25}=\frac{10-9}{25}=\frac{1}{25}.$$

$\therefore\ \sigma_x=\frac{1}{5}.$

$$\therefore\ P\left(\left|X-\frac{3}{5}\right|\geq\frac{2}{5}\right)=1-P\left(\frac{1}{5}<X<1\right)=1-12\int_{1/5}^1 x^2(1-x)dx=1-12\cdot\left[\frac{x^3}{3}-\frac{x^4}{4}\right]_{1/5}^1$$

$$=1-12\left[\frac{1}{3}\left(1-\frac{1}{125}\right)-\frac{1}{4}\left(1-\frac{1}{625}\right)\right]=\frac{17}{625}.$$

By the Tchebycheff's inequality $P\left(\left|X-\frac{3}{5}\right|\geq\frac{2}{5}\right)\leq\frac{\sigma_x^2}{4/25}=\frac{1/25}{4/25}=\frac{1}{4}$. Obviously, $\frac{17}{625}\ll\frac{1}{4}$.

Ex. 6. The distribution of a RV X is given by $P(X=-1)=\frac{1}{8}, P(X=0)=\frac{3}{4}, P(X=1)=\frac{1}{8}$. Verify Tchebycheff's inequality for the distribution. [C.H. 1984]

Solution. Here mean $m_x=E(X)=\Sigma x_i f_{xi}=-1\cdot\frac{1}{8}+0\cdot\frac{3}{4}+1\cdot\frac{1}{8}=0.$

The variance $\sigma_x^2=E\left(X^2\right)-\{E(X)\}^2=\Sigma x_i^2 f_{xi}-0^2=(-1)^2\cdot\frac{1}{8}+0^2\cdot\frac{3}{4}+1^2\cdot\frac{1}{8}=\frac{1}{4}.$

We verify whether for each $\epsilon>0$,

$$P(|X-0|\geq\epsilon)\leq\frac{\sigma_x^2}{\epsilon^2},\ \text{i.e., } P(|X|\geq\epsilon)\leq\frac{1}{4\epsilon^2}.$$

When $\epsilon>1$, the event $|X|\geq\epsilon$ is impossible. $\therefore\ P(|X|\geq\epsilon)=0\leq\frac{1}{4\epsilon^2}$.

When $0<\epsilon\leq 1, (|X|\geq\epsilon)=P(X=-1+X=1)$

$=P(X=-1)+P(X=1)$ as the events $X=-1, X=1$ are ME

$=\frac{1}{8}+\frac{1}{8}=\frac{1}{4}.$

$\therefore\ P(|X|\geq\epsilon)=\frac{1}{4}\leq\frac{1}{4\epsilon^2}$ as $0<\epsilon\leq 1$. Hence, Tchebycheff's inequality is verified for this distribution.

Ex. 7. A unbiased die is thrown 1200 times. Find the minimum value of the probability of getting 160 to 240 sixes.

Solution. Let X be the RV denoting the number of sixes out of 1200 throws of an unbiased die.

$\therefore$ X follows binomial $(1200, \frac{1}{6})$ distribution. $\therefore$ $m_x = E(X) = 1200 \cdot \frac{1}{6} = 200$.
$\sigma_x^2 = 1200 \cdot \frac{1}{6} \cdot \frac{5}{6} = \frac{500}{3}$.
$\therefore$ $P(160 < X < 240) = P(-40 < X - 200 < 40) = P(|X - 200| < 40) \geq 1 - \frac{\sigma_x^2}{40^2}$, by Tchebycheff's inequality taking $\epsilon = 40$ or, $P(160 < X < 240) \geq 1 - \frac{500}{3 \cdot 40 \cdot 40} = 1 - \frac{5}{48} = \frac{43}{48}$.
$\therefore$ minimum value of probability of getting 160 to 240 sixes is $\frac{43}{48}$.

Ex. 8. Let the RV X_i takes up values i and $-i$ with equal probabilities, $i = 1, 2, \cdots, n$. Show that the law of large numbers cannot be applied to the mutually independent variables $X_1, X_2, \cdots, X_n$.

Solution. $P(X_i = i) = P(X_i = -i) = \frac{1}{2}, i = 1, 2, \cdots, n$. $\therefore$ $E(X_i) = i \cdot \frac{1}{2} + (-i) \cdot \frac{1}{2} = 0$.

$\therefore$ $E(X_1 + X_2 + \cdots + X_n) = 0$ (1)

and $E\left(X_i^2\right) = i^2 \cdot \frac{1}{2} + (-i)^2 \cdot \frac{1}{2} = i^2, i = 1, 2, \cdots, n$. $\therefore$ $\sigma_{x_i}^2 = E\left(X_i^2\right) - \{E(X_i)\}^2 = i^2$ [$\because$ $E(X_i) = 0$].

$$\therefore \Sigma_n^2 = \text{var}(X_1 + X_2 + \cdots + X_n) = E(X_1 + X_2 + \cdots + X_n)^2, \text{ as mean} = 0 \text{ by (1)}$$

$$= \sum_{i=1}^{n} E\left(X_i^2\right) \text{ as } X_1, X_2, \cdots, X_n \text{ are mutually independents}$$

$$= \sum_{i=1}^{n} i^2 = 1^2 + 2^2 + \cdots + n^2 = \frac{n(n+1)(2n+1)}{6}.$$

$$\therefore \lim_{n \to \infty} \frac{\Sigma_n}{n} = \lim_{n \to \infty} \sqrt{\frac{(n+1)(2n+1)}{6n}} \neq 0.$$

Hence, the law of large number cannot be applied here.

Ex. 9. Is there any distribution of a RV X such that $P\left(\mu_x - 2\sigma_x \leq X \leq \mu_x + 2\sigma_x\right) = 0.72$? [$\mu_x$ = mean of X and σ_x = SD of X.]

Solution. $P\left(\mu_x - 2\sigma_x \leq X \leq \mu_x + 2\sigma_x\right) = P\left(\left|X - \mu_x\right| \leq 2\sigma_x\right) \geq 1 - \frac{\sigma_x^2}{4\sigma_x^2} = 0.75$ (by Tchebycheff's inequality). $\therefore$ such a distribution does not exist.

Ex. 10. A special unbiased die with face 10 is thrown 1000 times. Find the lower bound of the probability that the face 10 occurs between 80 to 120 times.

Solution. Let the occurrence of 10 is a success. $\therefore$ p = probability of success in each trial is $\frac{1}{10}$, as the special die is unbiased.

$\therefore$ let the RV X denotes the number of success in 1000 trials. So, X follows binomial $(1000, \frac{1}{10})$ distribution.

$\therefore$ $m_x = np = 1000 \cdot \frac{1}{10} = 100$ and $\sigma_x = \sqrt{np(1-p)} = \sqrt{100 \cdot \frac{9}{10}} = \sqrt{90}$. $\therefore$ $\sigma_x^2 = \text{var}(X) = 90$.

By the Tchebycheff's inequality, $P\{|X - 100| < 20\} \geq 1 - \frac{90}{400}$ or, $P\{80 < X < 120\} \geq 1 - \frac{9}{40} = \frac{31}{40}$.

$\therefore$ the lower bound of the probability that the number of success lying between 80 to 120 is $\frac{31}{40}$.

Ex. 11. If X denotes the sum of faces of two unbiased die thrown simultaneously, then prove that $P\{|X - 7| < 3\} \geq \frac{19}{54}$. Also, find the actual probability.

Solution. X will take the set of values $\{2, 3, 4, \cdots, 11, 12\}$ and $P(X = 2) = \frac{1}{36}$, $P(X = 3) = \frac{2}{36}$, $P(X = 4) = \frac{3}{36}$, $P(X = 5) = \frac{4}{36}$, $P(X = 6) = \frac{5}{36}$, $P(X = 7) = \frac{6}{36}$, $P(X = 8) = \frac{5}{36}$, $P(X = 9) = \frac{4}{36}$, $P(X = 10) = \frac{3}{36}$, $P(X = 11) = \frac{2}{36}$, $P(X = 12) = \frac{1}{36}$.

$m_x = E(x) = 2 \cdot \frac{1}{36} + 3 \cdot \frac{2}{36} + 4 \cdot \frac{3}{36} + 5 \cdot \frac{4}{36} + 6 \cdot \frac{5}{36} + 7 \cdot \frac{6}{36} + 8 \cdot \frac{5}{36} + 9 \cdot \frac{4}{36} + 10 \cdot \frac{3}{36} + 11 \cdot \frac{2}{36} + 12 \cdot \frac{1}{36}$
$= \frac{1}{36}(2 + 6 + 12 + 20 + 30 + 42 + 40 + 36 + 30 + 22 + 12) = 7$.

$E\left(X^2\right) = \frac{1}{36}\left(1.2^2 + 2.3^2 + 3.4^2 + 4.5^2 + 5.6^2 + 6.7^2 + 5.8^2 + 4.9^2 + 3.10^2 + 2.11^2 + 1.12^2\right) = \frac{329}{6}$.

$\therefore$ $\sigma_x^2 = \text{var}(X) = E\left(X^2\right) - \{E(X)\}^2 = \frac{329}{6} - 49 = \frac{35}{6}$.

$E(|X - 7| < 3) \geq 1 - \frac{35}{6 \times 9} = \frac{19}{54}$.

Actual probability $= P(4 < X < 10) = P(X = 5) + P(X = 6) + P(X = 7) + P(X = 8) + P(X = 9)$
$= \frac{4}{36} + \frac{5}{36} + \frac{6}{36} + \frac{5}{36} + \frac{4}{36} = \frac{24}{36} = \frac{2}{3}$.

EXERCISE

1. Show by Tchebycheff's inequality that in 2000 throws with a fair coin the probability that the number of heads lies between 900 and 1100 is at least $\frac{19}{20}$. [C.H. 2004]
2. An unbiased die is thrown 600 times. Show by Tchebycheff's inequality that the probability of occurring face 1 between 80 to 120 is at least $\frac{19}{24}$.
3. An unbiased die is thrown 900 times. By Tchebycheff's inequality, find the minimum probability that face multiple of 3 occurs between 270 to 330. (*Ans.* 7/9)
4. If X be a RV and $E(X^2)$ is finite, prove that $P(|X| \geq a) \leq \frac{1}{a^2} E(X^2), \forall a > 0$. [C.H. 1991]
5. An urn contains 1000 white and 2000 black balls. We draw (with replacement) 300 balls. Estimate the lower limit the probability that the number m of the drawn white balls satisfies the inequality $80 < m < 120$. [C.H. 1987]
(*Ans.* 0.833)
6. Is there any distribution of a RV X such that $P(|X - m_x| \leq 2\sigma_x) = 0.718$? (*Ans.* No)
7. A special unbiased die with 8 faces is thrown 800 times. Find the lower bound of the probability that the face 5 occurs more than 80 times and less than 120 times. (*Ans.* 25/32)
8. If a RV X follows gamma distribution with parameter k, then show that $P(0 < X < 2k) \geq \frac{k-1}{k}$.
9. State and prove Tchebycheff's inequality. Use it to find the upper bound of $P\left\{\left|\frac{X_n}{n} - 3.5\right| > \epsilon\right\}, \epsilon > 0$, where the RV X_n denotes the total number obtained in n tosses of a symmetric die. [C.H. 2003]
10. Use Tchebycheff's inequality to show that for $n \geq 36$, the probability that in n throws of a fair die the number of sixes lies between $\frac{n}{6} - \sqrt{n}$ and $\frac{n}{6} + \sqrt{n}$ is at least $\frac{31}{36}$. [C.H. 1991]
[Hints. Let X = number of sixes in throws of the fair die. $\therefore$ $X \sim$ binomial $\left(n, \frac{1}{6}\right)$.
$\therefore\ E(X) = \frac{n}{6}, \text{var}(X) = n \cdot \frac{1}{6} \cdot \frac{5}{6} = \frac{5n}{36}$.
$P\left(\frac{n}{6} - \sqrt{n} < X < \frac{n}{6} + \sqrt{n}\right) = P\left(|X - \frac{n}{6}| < \sqrt{n}\right) = 1 - P\left(|X - \frac{n}{6}| \geq \sqrt{n}\right)$.
By Tchebycheff's inequality, $P\left(\left|X - \frac{n}{6}\right| \geq \sqrt{n}\right) \leq \frac{5n}{36.n} = \frac{5}{36}$.
$\therefore\ P\left(\frac{n}{6} - \sqrt{n} < X < \frac{n}{6} + \sqrt{n}\right) \geq 1 - \frac{5}{36} = \frac{31}{36}$.]
11. If $\{X_n\}_n$ be a sequence of RV such that for each $i, E(X_i) = m_i, \text{var}(X_i) = \sigma_i^2 \leq \sigma^2 < \infty$; use Tchebycheff's inequality to show that $\sum_{i=1}^{n} \frac{X_i}{n} - \sum_{i=1}^{n} \frac{m_i}{n} \xrightarrow{\text{in } P} 0$ as $n \to \infty$. [C.H. 1986]
12. If $X \sim N(m_x, \sigma_x)$, determine $P\{|X - m_x| \geq 1.96\sigma_x\}$. Compare this value with the value obtained by Tchebycheff's inequality.
Hints. $P\{|X - m_x| \geq 1.96\sigma_x\} = P\left\{\left|\frac{X - m_x}{\sigma_x}\right| \geq 1.96\right\}$
$= P(|X^*| \geq 1.96)$, when $X^* = \frac{X - m_x}{\sigma_x}$ = standard normal variate of X
$= 0.05$ (obtained from standard normal table).
By the Tchebycheff's inequality, $P\{|X - m_x| \geq 1.96\sigma_x\} \leq \frac{\sigma_x^2}{(1.96)^2 \sigma_x^2} = \frac{1}{(1.96)^2} = 0.26$.
Comparing we can say Tchebycheff's inequality gives very poor bound.
13. For the geometric distribution given by the pdf f_x, where $f_x(x) = 2^{-x}, x = 1, 2, 3, \cdots$, find the lower bound of $P(|X - 2| \leq 2)$ by Tchebycheff's inequality. Also, obtain the actual probability.
(*Ans.* 1/2, 15/16)
14. X is a continuous RV having the pdf $f_x(x) = e^{-x}, x \geq 0$. Find the upper bound of $P(|X - 1| > 2)$ by Tchebycheff's inequality. Compute the actual probability. (*Ans.* 1/4, $(e^2 - 1)/e^3$)

Chapter 9

Three Important Distributions

9.1 Introduction

In statistics three continuous distributions, viz., χ^2-distribution, t-distribution and F-distribution are much useful. So, we devote this chapter for the study of those distributions.

Note: χ^2 is a symbol and does not mean the square of some χ.

χ^2-distribution.

χ^2-distribution is a continuous distribution and is defined the following pdf:

$$f(\chi^2) = \begin{cases} \dfrac{e^{-\frac{1}{2}\chi^2}\left(\frac{1}{2}\chi^2\right)^{(n/2)-1}}{2\Gamma\left(\frac{n}{2}\right)}, & 0 < \chi^2 < \infty \\ 0, & \text{elsewhere,} \end{cases}$$

where n, a natural number, is the only parameter of the distribution and is called the *degrees of freedom of the distribution.*

Theorem 9.1.1 Let X be a RV whose distribution is a gamma distribution with parameter $\frac{n}{2} (n \in \mathbb{N})$. Then the RV $Y = 2X$ follows χ^2-distribution with n degrees of freedom.

Proof. The distribution of X is defined by the pdf f_X, where $f_X(x) = \begin{cases} \dfrac{e^{-x}.x^{(n/2)-1}}{\Gamma\left(\frac{n}{2}\right)}, & 0 < x < \infty \\ 0, & \text{elsewhere.} \end{cases}$

We have $Y = 2X$. In terms of their real values $y = 2x$. $\therefore$ $\frac{dy}{dx} = 2 > 0,\ \forall\ x \in \mathbb{R}$. So, by the law of transformation of RV the pdf f_Y of Y is given by

$$f_Y(y) = f_X(x)\left|\frac{dx}{dy}\right| = \frac{e^{-x}\,x^{(n/2)-1}}{\Gamma\left(\frac{n}{2}\right)} \cdot \frac{1}{2} = \frac{e^{-y/2}.\left(\frac{y}{2}\right)^{\frac{n}{2}-1}}{2\Gamma\left(\frac{n}{2}\right)},\ 0 < y < \infty$$

$f_Y(y) = 0$, elsewhere.

$\therefore$ Y follows a χ^2-distribution with n-degress of freedom.

Theorem 9.1.2 If a RV Y follows χ^2-distribution with n degrees of freedom, then the RV $X = \frac{Y}{2}$ is a gamma distribution with parameter $n/2$.

Proof. RV Y follows $\chi^2(n)$ distribution. So, f_Y, the pdf of Y, is given by $f_Y(y) = \begin{cases} \dfrac{e^{-y/2}\left(\frac{y}{2}\right)^{(n/2)-1}}{2\Gamma\left(\frac{n}{2}\right)}, & y > 0 \\ 0, & \text{elsewhere.} \end{cases}$

Given: $X = \frac{Y}{2}$. In term of their real values, $x = \frac{y}{2}$. $\therefore$ $\frac{dx}{dy} = \frac{1}{2} > 0,\ \forall\ y \in \mathbb{R}$.

$\therefore$ by the rule of transformation of RV, the pdf f_x of X is given by

$$f_x(x)=f_y(y)\left|\frac{dy}{dx}\right|=\frac{e^{-\frac{y}{2}}\left(\frac{y}{2}\right)^{\frac{n}{2}-1}}{2\Gamma\left(\frac{n}{2}\right)}\cdot 2=\frac{e^{-\frac{y}{2}}\left(\frac{y}{2}\right)^{\frac{n}{2}-1}}{\Gamma\left(\frac{n}{2}\right)},\ x>0$$

$f_x(x)=0$, elsewhere.

$\therefore$ X is a gamma $\left(\frac{n}{2}\right)$ variate, i.e., the distribution of X is a gamma distribution with parameter $\frac{n}{2}$.

Theorem 9.1.3 If the distribution of a RV X is a standard normal variate, then the distribution of the RV $Y=X^2$ is $\chi^2(1)$ variate.

Proof. As X is a standard normal variate, so its pdf f_x is given by

$$f_x(x)=\frac{1}{\sqrt{2\pi}}e^{-x^2/2}, -\infty<x<\infty.$$

Given: $Y=X^2$. In terms of their real values, $y=x^2$.

$\therefore$ $\frac{dy}{dx}=2x$, which does not maintain the same sign $\forall\ x\in\mathbb{R}$.

So, the law of transformation of RV is not at once applicable.

Let $x>0$. So, the event $y<Y\leq y+dy \Leftrightarrow x^2<X^2\leq(x+dx)^2$,

i.e., $y<Y\leq y+dy \Leftrightarrow -x-dx\leq X<-x+x<X\leq x+dx$, where the events on RHS are MEE.

$$\begin{aligned}\therefore\ P(y<Y\leq y+dy)&=P(-x-dx\leq X<-x+x<X\leq x+dx)\\&=P(-x-dx\leq X<-x)+P(x<X\leq x+dx),\text{ as the events are MEE}\\&=2P(x<X\leq x+dx),\text{ due to symmetry of normal distribution}\end{aligned}$$

$$P(-x-dx\leq X<-x)=P(x<X\leq x+dx).$$

$\therefore$ $f_Y(y)dy=2f_x(x)dx$.

$$\therefore\ f_Y(y)=2f_x(x)\left|\frac{dx}{dy}\right|=2\frac{1}{\sqrt{2\pi}}e^{-x^2/2}\cdot\frac{1}{2x}=\frac{e^{-y/2}\left(\frac{1}{2}y\right)^{(1/2)-1}}{2\Gamma\left(\frac{1}{2}\right)},\ \text{as }\Gamma\left(\frac{1}{2}\right)=\sqrt{\pi}, y>0.$$

$f_Y(y)=0$, elsewhere.

$\therefore$ the distribution of Y, i.e., of X^2 is $\chi^2(1)$.

9.2 Characteristic function of $\chi^2(n)$-distribution

The characteristic function is given by

$$\chi(t)=E\left(e^{itX}\right)=\int_0^\infty e^{itx}\cdot\frac{e^{-x/2}\left(\frac{x}{2}\right)^{\frac{n}{2}-1}}{2\Gamma\left(\frac{n}{2}\right)}dx=\frac{1}{2^{\frac{n}{2}}\Gamma\left(\frac{n}{2}\right)}\int_0^\infty e^{-\frac{1}{2}(1-2it)x}\cdot x^{(n/2)-1}dx$$

$$=\frac{1}{(1-2it)^{n/2}}\cdot\frac{1}{\Gamma\left(\frac{n}{2}\right)}\int_0^\infty e^{-x}x^{(n/2)-1}dx=(1-2it)^{-n/2}$$

$\therefore$ characteristic function of $\chi^2(1)$-distribution is $(1-2it)^{-1/2}$.

9.3 Reproductive property of χ^2-distribution

If $X_1, X_2, \cdots, X_n$ are mutually independent χ^2-variates with parameters $\nu_1, \nu_2, \cdots, \nu_n$ ($\nu_i\in\mathbb{N}, \forall\ i=1,2,\cdots,n$) respectively, then the RV $X=X_1+X_2+\cdots+X_n$ is a χ^2-variate with parameter $\nu=\nu_1+\nu_2+\cdots+\nu_n$.

Proof. $X_i (i=1,2,\cdots,n)$ follows a $\chi^2(\nu_i)$-distribution. So, the characteristic function χ_{X_i} of X_i is $(1-2it)^{-\nu_i/2}$.

$\therefore$ characteristic function χ_x of X is given by

$$\chi_x(t) = E\left(e^{itX}\right) = E\left(e^{it(X_1+X_2+\cdots+X_n)}\right) = E\left(e^{itX_1}\cdot e^{itX_2}\cdots e^{itX_n}\right) = E\left(e^{itX_1}\right)E\left(e^{itX_2}\right)\cdots E\left(e^{itX_n}\right),$$

as $X_1, X_2, \cdots, X_n$ are mutually independent

$$= (1-2it)^{-v_1/2}(1-2it)^{-v_2/2}\cdots(1-2it)^{-v_n/2}$$
$$= (1-2it)^{-(v_1+v_2+\cdots+v_n)/2} = (1-2it)^{-v/2}, \text{ when } v = v_1+v_2+\cdots+v_n,$$

which is the characteristic function of $\chi^2(v)$-distribution. Hence, by the uniqueness of characteristic function of RV we conclude that X is a $\chi^2(v)$-distribution.

Theorem 9.3.1 If $X_1, X_2, \cdots, X_n$ are mutually independent standard normal variates, then $X_1^2+X_2^2+\cdots+X_n^2$ is $\chi^2(n)$ variate.

Proof. In the theorem 9.1.3, we have proved that if X is standard normal variate, then X^2 is $\chi^2(1)$ variate. Hence, each of $X_1^2, X_2^2, \cdots, X_n^2$ follows χ^2-distribution with 1-degree of freedom. Since $X_1, X_2, \cdots, X_n$ are mutually independent, the RVs $X_1^2, X_2^2, \cdots, X_n^2$ are also mutually independent RVs [as we know that if X, Y are independent and g and h be two continuous functions, then the RVs $g(X), h(Y)$ are also independent].

So, by the reproductive property of χ^2-distribution $X_1^2+X_1^2+\cdots+X_n^2$ follows χ^2-distribution with the parameter $v = 1+1+\cdots n$ times $= n$.

So, $X_1^2+X_2^2+\cdots+X_n^2$ is a $\chi^2(n)$-variate.

Cor. If two RV X and Y are independent and if $X \sim \chi^2(n_1)$ and $X+Y \sim \chi^2(n_2)$, then $Y \sim \chi^2(n_2-n_1)$.

Proof. $\chi_x(t) = E\left(e^{itX}\right) = (1-2it)^{-\frac{n_1}{2}}$ and $\chi_{x+y}(t) = (1-2it)^{-\frac{n_2}{2}}$ and $\chi_{x+y}(t) = E\left(e^{it(X+Y)}\right)$
$= E\left(e^{itX}\cdot e^{itY}\right) = E\left(e^{itX}\right)E\left(e^{itY}\right)$, as X, Y are independent.

$\therefore (1-2it)^{-\frac{n_2}{2}} = (1-2it)^{-\frac{n_1}{2}}\chi_y(t)$.

$\therefore \chi_y(t) = (1-2it)^{-\frac{1}{2}(n_2-n_1)}$, which is a characteristic function of a χ^2-distribution with (n_2-n_1) degrees of freedom.

By the uniqueness of characteristic function of a distribution we may conclude that $Y \sim \chi^2(n_2-n_1)$.

Note. Similarly, if X and Y are independent and $Y \sim \chi^2(n_1)$ and $X+Y \sim \chi^2(n_2)$, then $X \sim \chi^2(n_2-n_1)$.

9.3.2 Study of χ^2-density curve

If the RV X is a $\chi^2(n)$ variate, then the pdf f_x is given by $f_x(x) = \begin{cases} \dfrac{e^{-x/2}\left(\frac{x}{2}\right)^{(n/2)-1}}{2\Gamma\left(\frac{n}{2}\right)}, & x > 0 \\ 0, & \text{elsewhere.} \end{cases}$

To have the nature of the curve, we note the following.

(i) The natural domain of f_x is the entire real line, i.e., $-\infty < x < \infty$.

(ii) Except $n = 1, 2$ the function f_x is continuous in $\mathbb{R}$.

When $n = 1, 2$, the pdf f_x has a jump discontinuity at the origin.

(iii) $f_x'(x)=\dfrac{1}{2^{n/2}\Gamma\left(\frac{n}{2}\right)}\left[-\frac{1}{2}e^{-x/2}x^{(n/2)-1}+e^{-x/2}\left(\frac{n}{2}-1\right)x^{(n/2)-2}\right]$

$$=\frac{e^{-x/2}x^{(n/2)}-2}{2^{n/2}\Gamma\left(\frac{n}{2}\right)}\left\{-\frac{1}{2}x+\left(\frac{n}{2}-1\right)\right\}=\frac{e^{-x/2}x^{(n/2)-2}}{2^{(n/2)+1}\Gamma\left(\frac{n}{2}\right)}[-x+(n+2)],\ x>0$$

and $f_x'(x)=0,\ x<0$

$$Lf_x'(0)=\lim_{x\to 0-}\frac{f(0)-f(x)}{0-x}=0$$

$$Rf_x'(0)=\lim_{x\to 0+}\frac{f(x)-f(0)}{x-0}=\lim_{x\to 0+}\frac{\frac{1}{2\Gamma\left(\frac{n}{2}\right)}e^{-x/2}\left(\frac{x}{2}\right)^{(n/2)-1}}{x}$$

$$=\frac{1}{2^{(n/2)}\Gamma\left(\frac{n}{2}\right)}\lim_{x\to 0+}e^{-x/2}x^{(n/2)-2}=\begin{cases}0, & \text{for } x\geq 5\\ 1, & \text{for } n=4\\ \infty, & \text{for } n\leq 3\end{cases}$$

$\therefore$ the pdf is smooth at the origin, only when $n\geq 5$.

When $n=4$, the tangent at the origin from right is the line $y=x$.

When $n\leq 3$, the tangent at the origin from right is the y-axis.

(iv) $f_x'(x)=0,\ \forall\ x<0$, where the function is constant and $f_x'(x)=0$ for $x>0$, when $x=n-2$.

$\therefore$ when $n>2$, $f_x'(x)$ vanishes, only when $x=n-2$.

For all other points $x(>0)$, when $n>2$ derivatives exist but do not vanish. So, f_x may have optimal value only at $x=n-2>0$.

Test of optimality: $f_x'(n-2-\delta)>0, \delta>0$ and $f_x'(n-2+\delta)<0, \delta>0$.

$\therefore$ $f_x(x)$ is increasing on the left of $n-2$ and decreasing on the right of $n-2$ and vanishes at $x=n-2$.

$\therefore$ f_x is maximum only at the point $x=n-2>0$. Note that, such such finite maximum does not exist for $n\leq 2$.

$\therefore$ $n-2$ is the only mode of the distribution when $n-2>0$.

(v) The study of concavity, convexity, point of inflexion requires the second derivative and such a study is not relevant. Only area bounded between the density curve and x-axis is required for its application in the study of statistics. So, we left such a study in general.

(vi) When $x\to\infty, f_x\to 0+$. $\therefore$ $y=0$ is the asymptote of the curve for all $n\in\mathbb{N}$.

For $n=1,2$, the nature of the curve is considerably different from the curves for $n\geq 3$.

When $n=1, f_x(x)=\frac{e^{-x/2}}{\sqrt{2\pi x}}$ and $\therefore$ $\lim_{x\to 0+}f_x(x)=\infty$.

When $n=2, f_x(x)=\frac{1}{2}e^{-x/2}$. $\therefore$ $\lim_{x\to 0+}f_x(x)=\frac{1}{2}$.

They meet at $x=\frac{2}{\pi}=0.64$. The mode for $\chi^2(n)$ is $n-2$. $\therefore$ for $\chi^2(3)$ the mode is at $1(>0.64)$.

When $n=1, x=0$, i.e., the y-axis is also an asymptote of the density curve. So, only when $n=1$, the curve has two asymptote, viz., $x=0$ and $y=0$. For all other values of n the curve has the only asymptote $y=0$.

The following is the figure showing the nature of the probability curve for different degrees of freedom n :

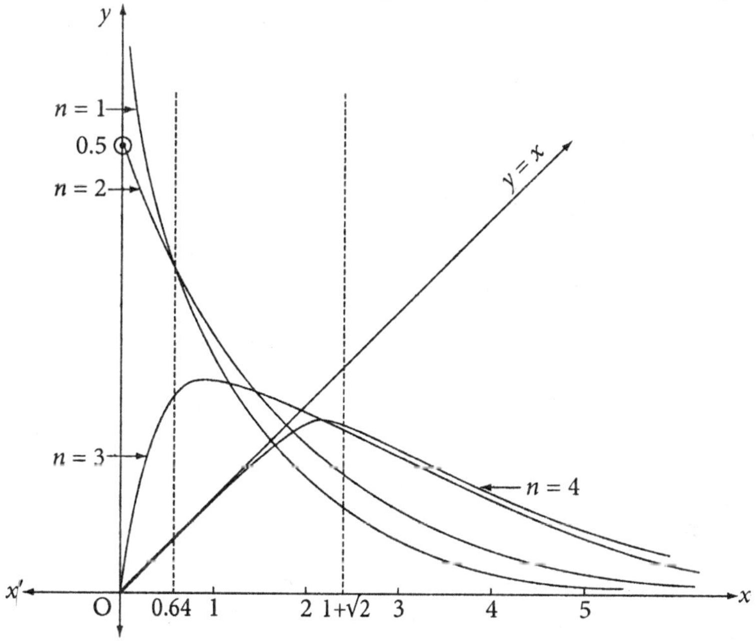

For $n=1,2, f_x'(x)<0,\ \forall\ x>0$. $\therefore\ f_x$ is decreasing for all $x>0$ and $f_x''(x)>0,\ \forall\ x>0$. $\therefore\ f_x$ is concave up for all $x>0$.

For other values of n the curve is convex up initially and concave up finally. We study the situation for $n=3$ and $n=4$. The study for higher values of n may be similarly studied.

When $n=3$:

$$f_x(x)=\frac{1}{\sqrt{2\pi}}e^{-x/2}\sqrt{x},\ f_x'(x)=\frac{e^{-x/2}}{2\sqrt{2\pi x}}(-x+1),\ f_x''(x)=\frac{1}{4\sqrt{2\pi}}\frac{e^{-x/2}}{x\sqrt{x}}\left(x-1-\sqrt{2}\right)\left(x-1+\sqrt{2}\right).$$

[$f_x'(x)=0 \Leftrightarrow x=1$, the mode of the distribution. $f_x'(1-\delta)>0$ and $f_x'(1+\delta)<0$ for $\delta>0$ and small enough.

$\therefore\ f_x$ is maximum at $x=1$, which is the mode. It is increasing on the left and decreasing on the right of 1. (Here $n-2=3-2=1$)]

In the positive half of x-axis, f_x'' vanishes at $x=1+\sqrt{2}$ and $f_x''\left(1+\sqrt{2}-\delta\right)=-\delta\left(2\sqrt{2}-\delta\right)<0$ and $f_x''\left(1+\sqrt{2}+\delta\right)=\delta\left(2\sqrt{2}+\delta\right)>0$.

$\therefore\ f_x$ has the only point of inflection when $x>0$ at $x=1+\sqrt{2}$. It is convex up for all $x>0$ in the left of $1+\sqrt{2}$ and it is concave up for all $x>1+\sqrt{2}$.

$\therefore$ the point of inflection is on the right of the mode.

When $n=4$:

$$f_x(x)=\frac{1}{4}e^{-x/2}x,\ f_x'(x)=\frac{e^{-x/2}}{8}(-x+2),\ f_x''(x)=\frac{e^{-x/2}}{16}(x-4) \text{ exists } \forall\ x>0.$$

[$f_x'(x)=0 \Leftrightarrow x=2, f_x'(2-\delta)>0$ and $f_x'(2+\delta)<0$. $\therefore\ f_x$ is maximum only at $x=2$, which is the mode. (Here $n-2=4-2=2$)]

In the positive half of x-axis f_x'' vanishes for $x=4$ and $f_x''(4-\delta)<0$ and $f_x''(4+\delta)>0$.

$\therefore$ $x=4$ is the only point of inflextion of the density curve. It is convex up for all $x \in (0,4)$ and concave up $\forall\, x>4$, etc.

Here is the table for values of f_x for different values of x and n:

n	$f_x(x)$	$f_x(0.64)$	$f_x(1)$	$f_x(2)$	$f_x(3)$	$f_x(4)$	$f_x(5)$
1	$\frac{e^{-x/2}}{\sqrt{2\pi x}}$	0.362	0.242	0.104	0.051	0.027	0.015
2	$\frac{1}{2}e^{-x/2}$	0.363	0.303	0.184	0.112	0.068	0.041
3	$\frac{1}{\sqrt{2\pi}}e^{-x/2}\sqrt{x}$	0.232	0.242	0.208	0.154	0.108	0.073
4	$\frac{1}{4}e^{-x/2}\cdot x$	0.116	0.152	0.184	0.167	0.135	0.103
5	$\frac{1}{3\sqrt{2\pi}}e^{-x/2}\cdot x\sqrt{x}$	0.049	0.081	0.138	0.154	0.144	0.122
...	...	...	...	...	...	...	...
...	...	...	...	...	...	...	...

Theorem 9.3.3 If each RV $X_r (r=1,2,\cdots,n)$ is normal (m_r,σ_r) variate and they are independent then

$$\sum_{r=1}^{n}\left(\frac{X_r-m_r}{\sigma_r}\right)^2 \text{ is } \chi^2(n) \text{ variate.}$$

Proof. X_r is normal (m_r,σ_r). $\therefore$ $\frac{X_r-m_r}{\sigma_r}$ is normal $(0,1)$ for each $r=1,2,\cdots,n$.

As the RVs X_r are mutually independent, the RVs $\frac{X_r-m_r}{\sigma_r}$ are also mutually independent.

As $\frac{X_r-m_r}{\sigma_r}$ is standard normal. $\therefore$ $\left(\frac{X_r-m_r}{\sigma_r}\right)^2$ is $\chi^2(1)$ for each r and they are mutually independent.

Hence, by the reproductive property of χ^2-distribution $\sum_{r=1}^{n}\left(\frac{X_r-m_r}{\sigma_r}\right)^2$ follows $\chi^2(n)$-distribution.

Theorem 9.3.4 If each RV $X_r (r=1,2,\cdots,n)$ is normal (m,σ) and $X_1,X_2,\cdots,X_n$ are mutually independent then the RV $\frac{nS^2}{\sigma^2}$ is $\chi^2(n-1)$ variate, where $S^2=\frac{1}{n}\sum_{r=1}^{n}\left(X_r-\bar{X}\right)^2$ and $\bar{X}=\frac{1}{n}\sum_{r=1}^{n}X_r$.

$$\textit{Proof. } \sum_{r=1}^{n}(X_r-m)^2=\sum_{r=1}^{n}\left(X_r-\bar{X}+\bar{X}-m\right)^2=\sum_{r=1}^{n}\left\{\left(X_r-\bar{X}\right)^2+2\left(\bar{X}-m\right)\left(X_r-\bar{X}\right)+\left(\bar{X}-m\right)^2\right\}$$

$$=\sum_{r=1}^{n}\left(X_r-\bar{X}\right)^2+n\left(\bar{X}-m\right)^2 \quad \left[\because \Sigma\left(X_r-\bar{X}\right)=0\right].$$

$$\therefore \sum_{r=1}^{n}\left(X_r-\bar{X}\right)^2=\sum_{r=1}^{n}(X_r-m)^2-n\left(\bar{X}-m\right)^2 \text{ or, } nS^2=\sum_{r=1}^{n}(X_r-m)^2-n\left(\bar{X}-m\right)^2.$$

$$\therefore \frac{nS^2}{\sigma^2}=\sum_{r=1}^{n}\left(\frac{X_r-m}{\sigma}\right)^2-\left(\frac{\bar{X}-m}{\sigma/\sqrt{n}}\right)^2. \tag{1}$$

As $X_1,X_2,\cdots,X_n$ are mutually independent, their continuous functions $\sum_{r=1}^{n}\left(\frac{X_r-m}{\sigma}\right)^2$ and $\left(\frac{\bar{X}-m}{\sigma/\sqrt{n}}\right)^2$ are also independent.

Now, each RV $\frac{X_r-m}{\sigma}$ is $N(0,1)$. $\therefore$ $\left(\frac{X_r-m}{\sigma}\right)^2$ is $\chi^2(1)$.

As $\left(\frac{X_1-m}{\sigma}\right)^2,\left(\frac{X_2-m}{\sigma}\right)^2,\cdots,\left(\frac{X_n-m}{\sigma}\right)^2$ are mutually independent.

$\therefore$ their sum $\sum_{r=1}^{n}\left(\frac{X_r-m}{\sigma}\right)^2$ is $\chi^2(n)$ variate by the reproductive property of χ^2-distribution.

Again, $E\left(\bar{X}\right)=\frac{1}{n}E\left(\sum_{r=1}^{n}X_r\right)=\frac{1}{n}\sum_{r=1}^{n}E(X_r)=\frac{1}{n}\sum_{r=1}^{n}m=\frac{1}{n}\cdot nm=m$ and

$$\sigma\left(\bar{X}\right)=\sqrt{E\left(\bar{X}-m\right)^2}=\sqrt{E\left\{\frac{X_1+X_2+\cdots+X_n}{n}-m\right\}^2}=\frac{1}{n}\sqrt{E\{(X_1-m)+(X_2-m)+\cdots+(X_n-m)\}^2}$$

$=\frac{1}{n}\sqrt{E(X_1-m)^2+E(X_2-m)^2+\cdots+E(X_n-m)^2}$, as $X_1, X_2\cdots$, are mutually independent

$=\frac{1}{n}\sqrt{n\sigma^2}\ \left[\because\ E(X_1-m)^2=\sigma^2 \text{ for each } i=1(1)n\right]$

$=\frac{\sigma}{\sqrt{n}}$.

Hence, by the reproductive property of normal distribution $\bar{X}\sim N\left(m,\frac{\sigma}{\sqrt{n}}\right)$.

$$\therefore\ \frac{\bar{X}-m}{\sigma/\sqrt{n}}\sim N(0,1).\ \therefore\ \left(\frac{\bar{X}-m}{\sigma/\sqrt{n}}\right)^2\sim\chi^2(1).$$

Hence, from (1) $\frac{nS^2}{\sigma^2}\sim\chi^2(n-1)$.

9.3.5 Mean and variance of $\chi^2(n)$-distribution

Let $X\sim\chi^2(n)$. $\therefore\ Y=\frac{X}{2}\sim\gamma\left(\frac{n}{2}\right)$. $\therefore$ mean of $Y=E(Y)=\frac{n}{2}$ and $\text{var}(Y)=\frac{n}{2}$, as we know that l is the mean and variance of a $\gamma(l)$-distribution.

$\therefore\ E(X)=E(2Y)=2E(Y)=2\cdot\frac{n}{2}=n.$

$$\text{var}(X)=\text{var}(2Y)=4\text{var}(Y)=4\cdot\frac{n}{2}=2n.$$

9.3.6 Moments of $\chi^2(n)$-distribution

$\alpha_k=$ ordinary moment (moment about origin) of order k

$$=E\left(X^k\right)=E\left(2^kY^k\right)=2^kE\left(Y^k\right)=2^k\cdot\frac{1}{\Gamma\left(\frac{n}{2}\right)}\int_0^\infty y^ke^{-y}y^{(n/2)-1}\,dy=\frac{2^k}{\Gamma\left(\frac{n}{2}\right)}\int_0^\infty e^{-y}y^{k+\frac{n}{2}-1}\,dy$$

$$=\frac{2^k}{\Gamma\left(\frac{n}{2}\right)}\Gamma\left(\frac{n}{2}+k\right)=2^k\left(\frac{n}{2}+k-1\right)\left(\frac{n}{2}+k-2\right)\cdots\frac{n}{2}.$$

$\mu_k=$ central moment (moment about mean) of order k

$=E\{X-E(X)\}^k=E(X-n)^k=\sum_{r=0}^{k}\binom{k}{r}\alpha_{k-r}(-1)^rn^r$, as expectation is a linear operator

$=\alpha_k-k\alpha_{k-1}n+\frac{k(k-1)}{2!}\alpha_{k-2}n^2-\cdots+(-1)^kn^k$, as $\alpha_0=1$.

9.4 t-distribution

This distribution is first introduced by the statistician W.S. Gosset. It is also known as Student's distribution according to his pseudonym *Student*. This is a continuous distribution and is very much useful in statistics.

t-distribution is defined by the following pdf:

$$f_t(t)=\frac{\Gamma\left(\frac{n+1}{2}\right)}{\sqrt{n\pi}\,\Gamma\left(\frac{n}{2}\right)}\cdot\frac{1}{\left(1+\frac{t^2}{n}\right)^{(n+1)/2}},\ \infty<t<\infty,$$

where $n\in\mathbb{N}$ is the only parameter of the distribution and is known as the degrees of freedom of the distribution.

9.4.1 Graph of $y = f_t(t)$

To have the graph we first study the nature of pdf:

(i) f_t is continuous $\forall\ t \in \mathbb{R}$.

(ii) $f_t'(t) = k\left(-\frac{n+1}{2}\right)\cdot\frac{2t}{\left(1+\frac{t^2}{n}\right)^{(n+3)/2}}$, where $k = \frac{\Gamma\left(\frac{n+1}{2}\right)}{\sqrt{n\pi}\,\Gamma\left(\frac{n}{2}\right)} > 0$

$= -k\frac{(n+1)t}{\left(1+\frac{t^2}{n}\right)^{(n+3)/2}}$ exists $\forall t$ and vanishes only when $t = 0$.

$f_t'(t) > 0$, when $t < 0$ and $f_t'(t) < 0$, when $t > 0$. $\therefore$ f_t is increasing in $(-\infty, 0)$ and decreasing in $(0, \infty)$.

$\therefore$ f_t is maximum at $t = 0$ and max. $f_t = k$.

(iii) $f_t''(t) = \frac{k(n+1)(n+2)}{n\left(1+\frac{t^2}{n}\right)^{(n+5)/2}}\left(t-\sqrt{\frac{n}{n+2}}\right)\left(t+\sqrt{\frac{n}{n+2}}\right)$. $\therefore$ $f_t''(t) = 0$ at $t = \mp\sqrt{\frac{n}{n+2}}$.

$f_t''\left(-\sqrt{\frac{n}{n+2}}-\delta\right) > 0$, $f_t''\left(-\sqrt{\frac{n}{n+2}}+\delta\right) < 0$,

$f_t''\left(\sqrt{\frac{n}{n+2}}-\delta\right) < 0$, $f_t''\left(\sqrt{\frac{n}{n+2}}+\delta\right) > 0$ for arbitrarily small $\delta > 0$.

So, there are two points of inflection at $t = \mp\sqrt{\frac{n}{n+2}}$.

Again, $f_t''(t) > 0$, $\forall\ t \in \left(-\infty, -\sqrt{\frac{n}{n+2}}\right)$, $f_t''(t) < 0$, $\forall\ t \in \left(-\sqrt{\frac{n}{n+2}}, \sqrt{\frac{n}{n+2}}\right)$

and $f_t''(t) > 0$, $\forall\ t \in \left(\sqrt{\frac{n}{n+2}}, \infty\right)$. Also, $f_t''\left(\mp\sqrt{\frac{n}{n+2}}\right) = 0$.

$\therefore$ the pdf is concave up in $\left(-\infty, -\sqrt{\frac{n}{n+2}}\right) \cup \left(-\sqrt{\frac{n}{n+2}}, \infty\right)$ and pdf is concave down in $\left(-\sqrt{\frac{n}{n+2}}, \sqrt{\frac{n}{n+2}}\right)$ and $\mp\sqrt{\frac{n}{n+2}}$ are the points of inflexion.

(iv) $f_t(t) \to 0$ as $t \to \pm\infty$. $\therefore$ $y = 0$ is the asymptote of the pdf curve. Hence, the graph of pdf curve for t-distribution is following:

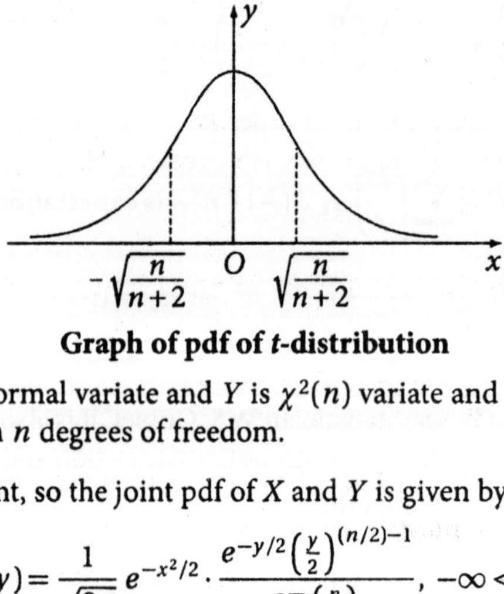

Graph of pdf of t-distribution

Theorem 9.4.2 If X is standard normal variate and Y is $\chi^2(n)$ variate and X, Y are independent, then $Z = \frac{X}{\sqrt{Y/n}}$ follows t-distribution, with n degrees of freedom.

Proof. As X and Y are independent, so the joint pdf of X and Y is given by

$$f_{X,Y}(x,y) = f_X(x)f_Y(y) = \frac{1}{\sqrt{2\pi}}\,e^{-x^2/2}\cdot\frac{e^{-y/2}\left(\frac{y}{2}\right)^{(n/2)-1}}{2\Gamma\left(\frac{n}{2}\right)},\quad -\infty < x < \infty,\ 0 < y < \infty$$

We first find the joint pdf of Z and $U = Y$ and then find the marginal distribution of Z. In terms of their real values $z = \frac{x\sqrt{n}}{\sqrt{y}}$ and $u = y$ or $x = \frac{z\sqrt{u}}{\sqrt{n}}$ and $y = u$.

$$\therefore \text{Jacobian of transformation } J = \begin{vmatrix} \frac{\partial x}{\partial z} & \frac{\partial x}{\partial u} \\ \frac{\partial y}{\partial z} & \frac{\partial y}{\partial u} \end{vmatrix} = \begin{vmatrix} \sqrt{\frac{u}{n}} & \frac{z}{2\sqrt{un}} \\ 0 & 1 \end{vmatrix} = \sqrt{\frac{u}{n}} > 0,\ \forall\, u, z.$$

$$\therefore\ f_{z,u}(z,u) = f_{x,y}(x,y)\left|\frac{\partial(x,y)}{\partial(z,u)}\right| = \frac{1}{\sqrt{2\pi}}e^{-x^2/2}\cdot\frac{e^{-y/2}\left(\frac{y}{2}\right)^{(n/2)-1}}{2\Gamma\left(\frac{n}{2}\right)}\sqrt{\frac{u}{n}}$$

$$= \frac{1}{\sqrt{2\pi}}\cdot\frac{1}{2^{n/2}\sqrt{n}\Gamma\left(\frac{n}{2}\right)}e^{-z^2u/2n}\cdot e^{-u/2}u^{(n/2)-\frac{1}{2}},\ -\infty < z < \infty, 0 < u < \infty.$$

$$\therefore\ f_z(z) = \int_{-\infty}^{\infty} f_{z,u}(z,u)\,du = \int_0^{\infty}\frac{1}{\sqrt{2\pi n}2^{n/2}\Gamma\left(\frac{n}{2}\right)}e^{-\frac{u}{2}\left(1+\frac{z^2}{n}\right)}u^{(n-1)/2}\,du.$$

Put $v = \frac{u}{2}\left(1+\frac{z^2}{n}\right)$. $\therefore\ dv = \frac{1}{2}\left(1+\frac{z^2}{n}\right)du$.

$v = 0$, when $u = 0$ and $v \to \infty$ as $u \to \infty$.

$$\therefore\ f_z(z) = \frac{1}{\sqrt{\pi n}\Gamma\left(\frac{n}{2}\right)2^{(n+1)/2}}\int_0^{\infty} e^{-v}\cdot\frac{2^{(n-1)/2}v^{(n-1)/2}}{\left(1+\frac{z^2}{n}\right)^{(n-1)/2}}\frac{2\,dv}{1+\frac{z^2}{n}}$$

$$= \frac{1}{\sqrt{\pi n}\Gamma\left(\frac{n}{2}\right)\left(1+\frac{z^2}{n}\right)^{(n+1)/2}}\int_0^{\infty} e^{-v}v^{\frac{n+1}{2}-1}\,dv$$

$$= \frac{\Gamma\left(\frac{n+1}{2}\right)}{\Gamma\left(\frac{n}{2}\right)}\cdot\frac{1}{\sqrt{\pi n}}\cdot\frac{1}{\left(1+\frac{z^2}{n}\right)^{(n+1)/2}},\ -\infty < z < \infty.$$

Hence, Z, i.e., $\frac{X}{\sqrt{Y/n}}$ follows t-distribution with n degrees of freedom when $X \sim N(0,1)$ and $Y \sim \chi^2(n)$.

9.4.3 Characteristics of t-distribution

The pdf is $f_t(t) = \frac{\Gamma\left(\frac{n+1}{2}\right)}{\sqrt{n\pi}\Gamma\left(\frac{n}{2}\right)}\cdot\frac{1}{\left(1+\frac{t^2}{n}\right)^{(n+1)/2}},\ -\infty < t < \infty, n \in \mathbb{N}$.

As f_t is an even function of t, $\therefore$ the probability density curve $y = f_t(t)$ is symmetrical about y-axis.

When $n = 1$, $f_t(t) = \frac{1}{\pi(1+t^2)}$, which is a Cauchy distribution with parameters $\lambda = 1, \mu = 0$. We know that the mean of a Cauchy distribution does not exist for the lack of absolute convergence.

$\therefore$ mean for t-distribution, when $n = 1$, does not exist.

For $n > 1$, mean = median = mode = 0 due to symmetry about y-axis.

Variance. Let $X \sim t(n), n > 2$.

$$\operatorname{var} X = E\left(X^2\right)\quad [\because\ E(X) = \text{mean} = 0]$$

$$= \int_{-\infty}^{\infty} x^2\frac{\Gamma\left(\frac{n+1}{2}\right)}{\sqrt{n\pi}\Gamma\left(\frac{n}{2}\right)}\cdot\frac{1}{\left(1+\frac{x^2}{n}\right)^{(n+1)/2}}\,dx = \frac{2\Gamma\left(\frac{n+1}{2}\right)}{\sqrt{n\pi}\Gamma\left(\frac{n}{2}\right)}\int_0^{\infty}\frac{x^2}{\left(1+\frac{x^2}{n}\right)^{(n+1)/2}}\,dx,$$

as the integrand is an even function.

Let $1+\frac{x^2}{n}=\frac{1}{y}$, i.e., $x^2=\frac{n(1-y)}{y}$. $\therefore\ 2x\,dx=-\frac{n}{y^2}\,dy$ and $y\to 1$ as $x\to 0$; $y\to 0+$ as $x\to\infty$.

$$\therefore\ \operatorname{var}X=\frac{\Gamma\left(\frac{n+1}{2}\right)}{\sqrt{n\pi}\,\Gamma\left(\frac{n}{2}\right)}\int_1^0\sqrt{\frac{n(1-y)}{y}}\cdot y^{(n+1)/2}\left(-\frac{n}{y^2}\right)dy=\frac{n\Gamma\left(\frac{n+1}{2}\right)}{\sqrt{\pi}\,\Gamma\left(\frac{n}{2}\right)}\int_0^1 y^{(n/2)-2}(1-y)^{1/2}\,dy$$

$$=\frac{n\Gamma\left(\frac{n+1}{2}\right)}{\sqrt{\pi}\,\Gamma\left(\frac{n}{2}\right)}\int_0^1 y^{\left(\frac{n}{2}-1\right)-1}(1-y)^{(3/2)-1}\,dy=\frac{n\Gamma\left(\frac{n+1}{2}\right)}{\sqrt{\pi}\,\Gamma\left(\frac{n}{2}\right)}B\left(\frac{n}{2}-1,\frac{3}{2}\right)=\frac{n\Gamma\left(\frac{n+1}{2}\right)}{\sqrt{\pi}\,\Gamma\left(\frac{n}{2}\right)}\cdot\frac{\Gamma\left(\frac{n}{2}-1\right)\Gamma\left(\frac{3}{2}\right)}{\Gamma\left(\frac{n+1}{2}\right)}$$

$$=\frac{n}{\sqrt{\pi}\left(\frac{n}{2}-1\right)}\frac{1}{2}\sqrt{\pi}=\frac{n}{n-2},\ n>2.$$

9.4.4 Moments

Let $X\sim t(n), n>2k, k\in\mathbb{N}$.

$$\therefore\ \alpha_{2k+1}=\int_{-\infty}^{\infty}x^{2k+1}f_x(x)\,dx=\int_{-\infty}^{\infty}x^{2k+1}\cdot\frac{\Gamma\left(\frac{n+1}{2}\right)}{\sqrt{n\pi}\,\Gamma\left(\frac{n}{2}\right)}\cdot\frac{1}{\left(1+\frac{x^2}{n}\right)^{(n+1)/2}}\,dx=0,$$

as the integrand is an odd function.

$$\alpha_{2k}=E\left(x^{2k}\right)=\frac{\Gamma\left(\frac{n+1}{2}\right)}{\sqrt{n\pi}\,\Gamma\left(\frac{n}{2}\right)}\int_{-\infty}^{\infty}x^{2k}\frac{1}{\left(1+\frac{x^2}{n}\right)^{(n+1)/2}}\,dx=\frac{2\Gamma\left(\frac{n+1}{2}\right)}{\sqrt{n\pi}\,\Gamma\left(\frac{n}{2}\right)}\int_0^{\infty}\frac{x^{2k}}{\left(1+\frac{x^2}{n}\right)^{(n+1)/2}}\,dx,$$

as the integrand is an even function.

Let $1+\frac{x^2}{n}=\frac{1}{y}$. $\therefore\ x^2=\frac{n(1-y)}{y}$. $\therefore\ 2x\,dx=-\frac{n}{y^2}\,dy$ and $y\to 1$ as $x\to 0$; $y\to 0+$ as $x\to\infty$.

$$\therefore\ \alpha_{2k}=\frac{\Gamma\left(\frac{n+1}{2}\right)}{\sqrt{n\pi}\,\Gamma\left(\frac{n}{2}\right)}\int_1^0\frac{\left\{\frac{n(1-y)}{y}\right\}^{(2k-1)/2}}{\left(\frac{1}{y}\right)^{(n+1)/2}}\left(-\frac{n}{y^2}\right)dy=\frac{\Gamma\left(\frac{n+1}{2}\right)\cdot n^{(2k+1)/2}}{\sqrt{2\pi}\,\Gamma\left(\frac{n}{2}\right)}\int_0^1 y^{\left(\frac{n}{2}-k\right)-1}(1-y)^{k+\frac{1}{2}-1}\,dy$$

$$=\frac{\Gamma\left(\frac{n+1}{2}\right)\cdot n^{\frac{2k+1}{2}}}{\sqrt{n\pi}\,\Gamma\left(\frac{n}{2}\right)}\cdot B\left(\frac{n}{2}-k,k+\frac{1}{2}\right),\ \text{when } \frac{n}{2}>k$$

$$=\frac{\Gamma\left(\frac{n+1}{2}\right)}{\sqrt{n\pi}\,\Gamma\left(\frac{n}{2}\right)}n^{k+\frac{1}{2}}\cdot\frac{\Gamma\left(\frac{n}{2}-k\right)\Gamma\left(k+\frac{1}{2}\right)}{\Gamma\left(\frac{n+1}{2}\right)}$$

$$=\frac{n^{k+\frac{1}{2}}}{\sqrt{n\pi}\left(\frac{n}{2}-1\right)\left(\frac{n}{2}-2\right)\cdots\left(\frac{n}{2}-k\right)}\cdot\left(k-\frac{1}{2}\right)\left(k-\frac{3}{2}\right)\cdots\frac{1}{2}\sqrt{\pi}$$

$$=n^k\cdot\frac{(2k-1)(2k-3)\cdots5\cdot3\cdot1}{(n-2)(n-4)\cdots(n-2k)},\ \text{when } n>2k, k\in\mathbb{N}.$$

As mean of the distribution is 0 for $n>1$, the ordinary moments of order k is the same as central moments of order k, when $n>k$.

$\therefore\ \mu_{2k+1}=0$ and $\mu_{2k}=\alpha_{2k}=n^k\dfrac{(2k-1)(2k-3)\cdots5\cdot3\cdot1}{(n-2)(n-4)\cdots(n-2k)}$.

When $k=1, \mu_2=\frac{n}{n-2}=$ variance (already obtained separately). $\therefore\ \sigma=\sqrt{\frac{n}{n-2}}$.

When $k=2, \mu_4=n^2\cdot\frac{3\cdot1}{(n-2)(n-4)}$, etc.

$\therefore$ coefficient of skewness $\gamma_1=\frac{\mu_3}{\sigma^3}=0$, coefficient of kurtosis $\beta_2=\frac{\mu_4}{\sigma^4}=\frac{3n^2}{n^2-6n+8}\cdot\frac{(n-2)^2}{n^2}=\frac{3(n-2)}{n-4}$ and coefficient of excess $\gamma_2=\beta_2-3=\frac{3(n-2)}{n-4}-3=\frac{6}{n-4}$.

Important remark. $\lim_{n\to\infty} \beta_2 = \lim_{n\to\infty} 3\frac{n-2}{n-4} = 3 \lim_{n\to\infty} \frac{1-\frac{2}{n}}{1-\frac{4}{n}} = 3$. $\therefore$ coefficient of excess $\gamma_2 = \beta_2 - 3 = 0$.

So, for a large value of the degree of freedom n, the t-distribution tends to coincide with normal distribution, in particular $t(n)$ tends to standard normal distribution for large n. The mean, median, mode are the same for both the distributions, i.e., for the distributions mean = median = mode = 0 [See later].

Both are symmetric about y-axis.

For a large n, the points of inflexions $\pm\sqrt{\frac{n}{n+2}}$ becomes ± 1.

For standard normal distribution:

$$f_x(x) = \frac{1}{\sqrt{2\pi}} e^{-x^2/2}, -\infty < x < \infty, f_x'(x) = \frac{1}{\sqrt{2\pi}}(-x)e^{-x^2/2} \text{ and } f_x''(x) = \frac{1}{\sqrt{2\pi}} e^{-x^2/2}\left(x^2-1\right) = 0 \Rightarrow x = \pm 1$$

and $f_x''(1-\delta)f_x''(1+\delta) < 0, f_x''(-1-\delta)f_x''(-1+\delta) < 0$ for $\delta > 0$.

$\therefore$ ± 1 are also the points of inflexion for standard normal distribution, etc.

9.4.5 Median

$$\int_{-\infty}^{\infty} \frac{\Gamma\left(\frac{n+1}{2}\right)}{\sqrt{n\pi}\,\Gamma\left(\frac{n}{2}\right)} \cdot \frac{1}{\left(1+\frac{t^2}{n}\right)^{\frac{n+1}{2}}} dt = 1 \Rightarrow 2\frac{\Gamma\left(\frac{n+1}{2}\right)}{\sqrt{n\pi}\,\Gamma\left(\frac{n}{2}\right)} \int_0^{\infty} \frac{1}{\left(1+\frac{t^2}{n}\right)^{\frac{n+1}{2}}} dt = 1$$

$$\text{or, } \frac{\Gamma\left(\frac{n+1}{2}\right)}{\sqrt{n\pi}\,\Gamma\left(\frac{n}{2}\right)} \int_0^{\infty} \frac{1}{\left(1+\frac{t^2}{n}\right)^{\frac{n+1}{2}}} dt = \frac{1}{2}. \therefore \int_{-\infty}^{0} \frac{\Gamma\left(\frac{n+1}{2}\right)}{\sqrt{n\pi}\,\Gamma\left(\frac{n}{2}\right)} \frac{1}{\left(1+\frac{t^2}{n}\right)^{\frac{n+1}{2}}} dt = 1 - \frac{1}{2} = \frac{1}{2}.$$

Hence, the median for $t(n)$ distribution is 0. We have already seen that f_t is maximum for $t = 0$. $\therefore$ the mode is at the origin.

These are due to the fact of symmetry of the distribution about origin.

9.5 *F*-distribution

Definition. A continuous RV X is said to follow F-distribution if its probability density function f_x is given by

$$f_x(x) = \begin{cases} \frac{\Gamma\left(\frac{m+n}{2}\right)}{\Gamma\left(\frac{m}{2}\right)\Gamma\left(\frac{n}{2}\right)} \cdot \left(\frac{m}{n}\right)^{m/2} \cdot \frac{x^{(m-2)/2}}{\left(1+\frac{m}{n}x\right)^{(m+n)/2}}, & 0 < x < \infty \\ 0, & \text{elsewhere.} \end{cases}$$

Here the two parameters m and n are called *degrees of freedom* of the distribution. If X follows F-distribution with parameters m and n, we write $X \sim F(m, m)$.

Note that $F(m,n) \neq F(n,m)$, when $m \neq n$.

Theorem 9.5.1 If $U \sim \chi^2(m)$, $V \sim \chi^2(n)$ and U, V are independent, then the RV X, given by $X = \frac{n}{m} \cdot \frac{U}{V}$ will follow $F(m,n)$ distribution, i.e., if $U \sim \chi^2(m)$, $V \sim \chi^2(n)$ and U, V are independent, then $\frac{n}{m}\frac{U}{V} \sim F(m,n)$.

Proof. As U, V are independent, $\therefore$ their pdf is given by

$$f_{u,v}(u,v) = f_u(u)f_v(v) = \begin{cases} \frac{e^{-u/2}\left(\frac{u}{2}\right)^{(m/2)-1}}{2\Gamma\left(\frac{m}{2}\right)} \cdot \frac{e^{-v/2}\left(\frac{v}{2}\right)^{(n/2)-1}}{2\Gamma\left(\frac{n}{2}\right)}, & 0 < u, v < \infty \\ 0, & \text{elsewhere.} \end{cases}$$

$$\text{i.e., } f_{u,v}(u,v) = \begin{cases} = \frac{1}{2^{(m+n)/2}\Gamma\left(\frac{m}{2}\right)\Gamma\left(\frac{n}{2}\right)} e^{-(u+v)/2} u^{(m/2)-1} v^{(n/2)-1}, & 0 < u, v < \infty \\ 0, & \text{elsewhere.} \end{cases}$$

Let $X = \frac{n}{m}\frac{U}{V} = \frac{U/m}{V/n}$ and $Y = V$.

In terms of their real values, $x = \frac{u/m}{v/n}, y = v$. $\therefore$ $u = \frac{m}{n}xy$, $v = y$.

The Jacobian of transformation $J = \begin{vmatrix} \dfrac{\partial u}{\partial x} & \dfrac{\partial u}{\partial y} \\ \dfrac{\partial v}{\partial x} & \dfrac{\partial v}{\partial y} \end{vmatrix} = \begin{vmatrix} \dfrac{m}{n}y & \dfrac{m}{n}x \\ 0 & 1 \end{vmatrix} = \frac{m}{n}y > 0, \ \forall\, u, v > 0.$

$\therefore$ by the law of transformation of RV

$$f_{X,Y}(x,y) = f_{U,V}(u,v)|J| = \begin{cases} \dfrac{1}{\Gamma\left(\frac{m}{2}\right)\Gamma\left(\frac{n}{2}\right)2^{(m+n)/2}} e^{-\frac{1}{2}\left(1+\frac{m}{n}x\right)y} \cdot \left(\dfrac{m}{n}\right)^{m/2} x^{(m/2)-1} y^{\frac{m+n}{2}-1}, & 0 < x, y < \infty \\ 0, & \text{elsewhere.} \end{cases}$$

$\therefore$ the marginal distribution of X is given by

$$f_X(x) = \frac{1}{\Gamma\left(\frac{m}{2}\right)\Gamma\left(\frac{n}{2}\right)2^{(m+n)/2}} \left(\frac{m}{n}\right)^{m/2} x^{(m-2)/2} \int_0^\infty e^{-\frac{1}{2}\left(1+\frac{m}{n}x\right)y} \cdot y^{(m+n-2)/2}\, dy.$$

Let $\frac{1}{2}\left(1+\frac{m}{n}x\right)y = s$. $\therefore$ $dy = \dfrac{ds}{\frac{1}{2}\left(1+\frac{m}{n}x\right)}$.

$$\therefore\ f_X(x) = \frac{1}{\Gamma\left(\frac{m}{2}\right)\Gamma\left(\frac{n}{2}\right)2^{(m+n)/2}} \cdot \left(\frac{m}{n}\right)^{m/2} x^{(m-2)/2} \int_0^\infty e^{-s} \left\{\frac{2s}{\frac{1}{2}\left(1+\frac{m}{n}x\right)}\right\}^{(m+n-2)/2} \cdot \frac{ds}{\frac{1}{2}\left(1+\frac{m}{n}x\right)}$$

$$= \frac{1}{\Gamma\left(\frac{m}{2}\right)\Gamma\left(\frac{n}{2}\right)} \cdot \left(\frac{m}{n}\right)^{m/2} \frac{x^{(m-2)/2}}{\left(1+\frac{m}{n}x\right)^{(m+n)/2}} \int_0^\infty e^{-s} s^{\frac{m+n}{2}-1}\, ds$$

$$= \begin{cases} \dfrac{\Gamma\left(\frac{m+n}{2}\right)}{\Gamma\left(\frac{m}{2}\right)\Gamma\left(\frac{n}{2}\right)} \left(\dfrac{m}{n}\right)^{m/2} \dfrac{x^{(m-2)/2}}{\left(1+\frac{m}{n}x\right)^{(m+n)/2}}, & x > 0 \\ 0, & \text{elsewhere.} \end{cases}$$

Hence, $X \sim F(m, n)$.

9.5.2 Mean

Let $X \sim F(m, n)$.

The mean of $X = E(X) = \dfrac{\Gamma\left(\frac{m+n}{2}\right)}{\Gamma\left(\frac{m}{2}\right)\Gamma\left(\frac{n}{2}\right)} \left(\dfrac{m}{n}\right)^{m/2} \displaystyle\int_0^\infty x \cdot x^{(m-2)/2} \left(1+\frac{m}{n}x\right)^{-\frac{m+n}{2}} dx.$

Let $\dfrac{\frac{m}{n}x}{1+\frac{m}{n}x} = u$. $\therefore$ $1 - u = \dfrac{1}{1+\frac{m}{n}x}$. $\therefore$ $du = \dfrac{m/n}{\left(1+\frac{m}{n}x\right)^2} dx$.

Also when $x = 0, u = 0$ and when $x \to \infty, u \to 1$.

$$\therefore\ E(X) = \frac{\Gamma\left(\frac{m+n}{2}\right)}{\Gamma\left(\frac{m}{2}\right)\Gamma\left(\frac{n}{2}\right)} \int_0^1 \left(\frac{u}{1-u}\right)^{m/2} (1-u)^{(m+n)/2} \cdot \frac{n}{m} \cdot \frac{1}{(1-u)^2}\, du$$

$$= \frac{\Gamma\left(\frac{m+n}{2}\right)}{\Gamma\left(\frac{m}{2}\right)\Gamma\left(\frac{n}{2}\right)} \cdot \frac{n}{m} \int_0^1 u^{m/2}(1-u)^{(n/2)-2}\, du, \text{ which converges absolutely, when } n > 2$$

$$= \frac{\Gamma\left(\frac{m+n}{2}\right)}{\Gamma\left(\frac{m}{2}\right)\Gamma\left(\frac{n}{2}\right)} \cdot \frac{n}{m} B\left(\frac{m}{2}+1, \frac{n}{2}-1\right) = \frac{\Gamma\left(\frac{m+n}{2}\right)}{\Gamma\left(\frac{m}{2}\right)\Gamma\left(\frac{n}{2}\right)} \cdot \frac{n}{m} \cdot \frac{\Gamma\left(\frac{m}{2}+1\right)\Gamma\left(\frac{n}{2}-1\right)}{\Gamma\left(\frac{m+n}{2}\right)}$$

$$= \frac{m/2}{\frac{n}{2}-1} \cdot \frac{n}{m} = \frac{n}{n-2},\ n > 2.$$

9.5.3 Variance

$$\operatorname{var}(X) = E\left(X^2\right) = (E(X))^2 - E\left(X^2\right) - \left(\frac{n}{n-2}\right)^2.$$

$$E\left(X^2\right) = \frac{\Gamma\left(\frac{m+n}{2}\right)}{\Gamma\left(\frac{m}{2}\right)\Gamma\left(\frac{n}{2}\right)} \cdot \left(\frac{m}{n}\right)^{m/2} \int_0^\infty x^2 \cdot x^{(m-2)/2}\left(1+\frac{m}{n}x\right)^{-\frac{m+n}{2}} dx$$

$$= \frac{\Gamma\left(\frac{m+n}{2}\right)}{\Gamma\left(\frac{m}{2}\right)\Gamma\left(\frac{n}{2}\right)} \cdot \left(\frac{n}{m}\right) \int_0^1 \left(\frac{u}{1-u}\right)^{(m+2)/2} (1-u)^{(m+n)/2} \cdot \frac{n}{m} \cdot \frac{1}{(1-u)^2}\, du,$$

when $1-u = \dfrac{1}{1+\frac{m}{n}x}$ (as before)

$$= \frac{\Gamma\left(\frac{m+n}{2}\right)}{\Gamma\left(\frac{m}{2}\right)\Gamma\left(\frac{n}{2}\right)} \cdot \left(\frac{n}{m}\right)^2 \int_0^1 u^{(m/2)+1}(1-u)^{(n/2)-3}\, du$$

$$= \frac{\Gamma\left(\frac{m+n}{2}\right)}{\Gamma\left(\frac{m}{2}\right)\Gamma\left(\frac{n}{2}\right)} \cdot \left(\frac{n}{m}\right)^2 B\left(\frac{m}{2}+2, \frac{n}{2}-2\right), \text{ when } n > 4$$

$$= \frac{\Gamma\left(\frac{m+n}{2}\right)}{\Gamma\left(\frac{m}{2}\right)\Gamma\left(\frac{n}{2}\right)} \cdot \frac{n^2}{m^2} \cdot \frac{\Gamma\left(\frac{m}{2}+2\right)\Gamma\left(\frac{n}{2}-2\right)}{\Gamma\left(\frac{m+n}{2}\right)}$$

$$= \frac{n^2\left(\frac{m}{2}+1\right)\frac{m}{2}}{m^2\left(\frac{n}{2}-1\right)\left(\frac{n}{2}-2\right)} = \frac{n^2}{m^2} \cdot \frac{m(m+2)}{(n-2)(n-4)}.$$

$$\therefore\ \operatorname{var}(X) = \frac{n^2}{m} \frac{m+2}{(n-2)(n-4)} - \frac{n^2}{(n-2)^2} = \frac{2n^2(m+n-2)}{m(n-2)^2(n-4)}, \text{ when } n > 4.$$

Theorem 9.5.4 If Y is $\beta_1\left(\frac{n}{2}, \frac{m}{2}\right)$ variate, then $X = \frac{n}{m}\left(\frac{1}{Y}-1\right)$ is $F(m, n)$ variate.

Proof. Let $X = \frac{n}{m}\left(\frac{1}{Y}-1\right)$. In terms of their real variables, $x = \frac{n}{m}\left(\frac{1}{y}-1\right)$.

$$\therefore\ \frac{dx}{dy} = -\frac{n}{m} \cdot \frac{1}{y^2} < 0 \text{ and } y = \frac{1}{1+\frac{m}{n}x}, 1-y = \frac{\frac{m}{n}x}{1+\frac{m}{n}x}.$$

$\therefore$ by the law of transformation of RV, the pdf of X is given by

$$f_Y(y) = f_X(x)\left|\frac{dy}{dx}\right| = \frac{y^{(n/2)-1}(1-y)^{(m/2)-1}}{B\left(\frac{n}{2},\frac{m}{2}\right)}\cdot\frac{m}{n}y^2 = \frac{\Gamma\left(\frac{m+n}{2}\right)}{\Gamma\left(\frac{n}{2}\right)\Gamma\left(\frac{m}{2}\right)}\cdot\frac{m}{n}\frac{1}{\left(1+\frac{m}{n}x\right)^{(m+n)/2}}\cdot\left(\frac{m}{n}x\right)^{(m/2)-1}$$

$$= \frac{\Gamma\left(\frac{m+n}{2}\right)}{\Gamma\left(\frac{m}{2}\right)\Gamma\left(\frac{n}{2}\right)}\cdot\left(\frac{m}{n}\right)^{m/2}\frac{x^{(m-2)/2}}{\left(1+\frac{m}{n}x\right)^{(m+n)/2}},\ 0<x<\infty.$$

$\therefore$ X is $F(m,n)$ variate.

Theorem 9.5.5 If X follows $F(m,n)$ distribution, then prove that $\frac{1}{X}$ follows $F(n,m)$ distribution.

Proof. Left for the students.

EXERCISE

1. Find the characteristic function of χ^2-distribution with n degrees of freedom. $\left(Ans.\ (1-2it)^{-\frac{n}{2}}\right)$
2. Find the mode of the $\chi^2(n)$-distribution. $(Ans.\ n-2)$
3. Find the mean and variance of $\chi^2(n)$-distribution. $(Ans.\ \text{mean} = n,\ \text{variance} = 2n)$
4. Find the points of inflexion of the t-distribution with degrees of freedom n. $\left(Ans.\ \pm\sqrt{\frac{n}{n+2}}\right)$
5. Find the position of the mode of t-distribution with the degrees of freedom n. $(Ans.\ t=0)$
6. Find the variance of the t-distribution with the degree of freedom $n(>2)$. $\left(Ans.\ \frac{n}{n-2}\right)$
7. Find the median of t-distribution with degrees of freedom n. $(Ans.\ 0)$
8. Find the mean and variance of the $F(m,n)$-distribution. $\left(Ans.\ \text{mean} = \frac{n}{n-2}, n>2;\ \text{variance} = \frac{2n^2(m+n-2)}{m(n-2)^2(n-4)}, n>4\right)$

Chapter 10

Some Important Limit Theorems

10.1 Asymptotically normal

If $\{X_n\}_n$ be a sequences of RV and $\{a_n\}_n, \{b_n\}_n$ two sequence of real numbers such that the distribution function of the RV $\frac{X_n - a_n}{b_n}$ tends to the standard normal distribution function Φ as $n \to \infty$, then we say that X_n is asymptotically normal (a_n, b_n). [a_n, b_n may not be the mean and SD of X_n].

10.2 Central limit theorem

If $\{X_n\}_n$ be a sequence of independent RV with finite variances (this means that their means are also finite) and the RV corresponding to the nth partial sum $S_n = X_1 + X_2 + \cdots + X_n$ with finite mean M_n and SD. Σ_n will follow asymptotically normal (M_n, Σ_n) distribution, iff the following Lindeberg's condition is satisfied:

For a pre-assigned $\epsilon > 0$,

$$\lim_{n\to\infty} \sum_{k=1}^{n} \left\{ \int_{|x-m_k| \geq \epsilon\Sigma_n} (x - m_k)^2 f_k(x)\,dx \right\} = 0, \text{ for a continuous case}$$

$$\text{or, } \lim_{n\to\infty} \sum_{k=1}^{n} \left\{ \sum_{|x_k-m_k| \geq \epsilon\Sigma_n} (x_k - m_k)^2 f_{xk} \right\} = 0, \text{ for a discrete case.}$$

10.2.1 We have seen that a binomial (n, p) distribution will tend to become Poisson-μ distribution when n is tending to ∞, p is tending to zero but $np(=\mu)$ is of moderate magnitude. In case p is not small but $n \to \infty$, the binomial (n, p) will tend to normal distribution as stated in the following theorem.

Theorem 10.2.2 De Moivre-Laplace limit theorem

If X_n be a binomial (n, p) variate for every n (p is given), then the sequence $\left\{\frac{X_n - np}{\sqrt{npq}}\right\}_n$ is convergent in distribution to a standard normal variate as $n \to \infty$, where $q = 1 - p$.

Proof. If the RVs $Y_1, Y_2, \cdots, Y_n$ are independent binomial $(1, p)$ variate, then by the reproductive property of binomial distribution, we get that the RV $X_n = Y_1 + Y_2 + \cdots + Y_n$ is binomial (n, p) variate.

Each Y_i $(i = 1, 2, \cdots, n)$ has the same distribution with mean p and SD $\sqrt{pq}$, as $n = 1$ and $Y_1, Y_2, \cdots, Y_n$ are mutually independent for all n. So, by central limit theorem for equal components (given in next theorem) the RV

$$\frac{\bar{Y} - p}{\frac{\sqrt{pq}}{\sqrt{n}}}, \text{ i.e., } \frac{\frac{Y_1 + Y_2 + \cdots + Y_n}{n} - p}{\frac{\sqrt{pq}}{\sqrt{n}}}$$

is asymptotically normal $(0, 1)$, i.e., $\frac{X_n - np}{\sqrt{npq}}$ is asymptotically standard normal.

So, $\left\{\frac{X_n - np}{\sqrt{npq}}\right\}_n$ is convergent in distribution to a standard normal variate as $n \to \infty$,

$$\text{i.e., } \lim_{n\to\infty} F_n(x) = \frac{1}{\sqrt{2\pi}} \int_{-\infty}^{x} e^{-x^2/2} dx, \ \forall\, x \in \mathbb{R},$$

where F_n is the probability distribution function of the RVs $\frac{X_n - np}{\sqrt{npq}}$.

Central limit theorem for Equal components

Theorem 10.2.3 If $\{X_n\}_n$ is a sequence of independent RV such that each X_i has the same distribution with mean m and SD σ, then $\frac{\bar{X}-m}{\sigma/\sqrt{n}}$ is asymptotically normal (0, 1), i.e., $\bar{X} \sim N\left(m, \frac{\sigma}{\sqrt{n}}\right)$, where $\bar{X} = \frac{1}{n}(X_1 + X_2 \cdots + X_n)$.

Proof. Each X_i has same distribution with mean m and SD σ.

$\therefore$ each $\frac{X_i - m}{\sigma}$ has the same distribution with mean 0 and SD 1, $i = 1(1)n$. (1)

As $\{X_n\}_n$ be a sequence of independent RVs.

$\therefore \left\{\frac{X_n - m}{\sigma}\right\}_n$ is a sequence of independent RVs.

$$\text{Let } Y_n = \frac{\bar{X} - m}{\sigma/\sqrt{n}} = \frac{\frac{X_1 - m}{\sigma} + \frac{X_2 - m}{\sigma} + \cdots + \frac{X_n - m}{\sigma}}{\sigma}.$$

Characteristic function of Y_n is given by

$$\chi_n(t) = E\left\{e^{\frac{it}{\sqrt{n}}\left(\frac{X_1 - m}{\sigma}\right)}\right\} E\left\{e^{\frac{it}{\sqrt{n}}\left(\frac{X_2 - m}{\sigma}\right)}\right\} \cdots E\left\{e^{\frac{it}{\sqrt{n}}\left(\frac{X_n - m}{\sigma}\right)}\right\} = \left\{\chi\left(\frac{t}{\sqrt{n}}\right)\right\}^n;$$

where $\chi(t)$ is the character function of $\frac{X_i - m}{\sigma}$, $i = 1(1)n$.

[As the distribution of $\frac{X_i - m}{\sigma}$, $i = 1, 2, \cdots, n$ is the same, so their characteristic functions are also the same]

$$= \left\{1 - \frac{t^2}{2n} + O\left(\frac{t^2}{n}\right)\right\}^n, \text{ by Taylor's theorem,}$$

$$\text{as } E\left(e^{\frac{it}{\sqrt{n}}\frac{(X_1 - m)}{\sigma}}\right) = E\left\{1 + \frac{i}{\sqrt{n}}\left(\frac{X_1 - m}{\sigma}\right) + \frac{t^2}{2!}\frac{i^2}{n}\left(\frac{X_1 - m}{\sigma}\right)^2 + \cdots\right\}$$

$$= 1 - \frac{t^2}{2n} + 0\left(\frac{t^2}{n}\right) \because E\frac{(X_1 - m)}{\sigma} = 0,\ E\left(\frac{X_1 - m}{\sigma}\right)^2 = 1 \text{ [by (1)]}.$$

$$\therefore \lim_{n\to\infty} \chi_n(t) = \lim_{n\to\infty}\left\{1 - \frac{t^2}{2n} + 0\left(\frac{t^2}{n}\right)\right\}^n = e^{-t^2/2}.$$

which is the characteristic function of a standard normal variate.

So, by the limit theorem of characteristic function, we get Y_n, i.e., $\frac{\bar{X}-m}{\sigma/\sqrt{n}}$ converges in distribution to a standard normal variate as $n \to \infty$. So, $\bar{X}$ is asymptotically normal $\left(m, \frac{\sigma}{\sqrt{n}}\right)$.

Theorem 10.3 Central limit theorem for equal components, implies the law of large numbers for equal components.

Proof. From the central limit theorem for equal components we have:

If $\{X_n\}_n$ be a sequence of mutually independent RVs such that each X_i has the same distribution with mean m and SD σ (both finite), then $Y_n = \frac{\bar{X}-m}{\sigma/\sqrt{n}}$ is asymptotically normal (0, 1), where $\bar{X} = \frac{1}{n}\sum_{i=1}^{n} X_i$.

That is, the distribution function F_n of Y_n tends to ϕ, the distribution function for standard normal variate as $n \to \infty$.

We are to show that $X \xrightarrow{\text{in } P} m$ as $n \to \infty$, i.e., $\lim_{n\to\infty} P\left(\left|\bar{X}-m\right| \geq \epsilon\right)=0,\ \forall \epsilon > 0$.

Now,

$$P\left(\left|\bar{X}-m\right| \geq \epsilon\right)=P\left\{\frac{\left|\bar{X}-m\right|}{\sigma/\sqrt{n}} \geq \sqrt{n}\frac{\epsilon}{\sigma}\right\}=P\left\{|Y_n| \geq \sqrt{n}\frac{\epsilon}{\sigma}\right\}. \tag{1}$$

$|Y_n| \geq \sqrt{n}\frac{\epsilon}{\sigma} \Rightarrow |Y_n| > b$, if $0 < b < \sqrt{n}\frac{\epsilon}{\sigma}$.

$$\begin{aligned}
\therefore\ P\left\{|Y_n| \geq \sqrt{n}\frac{\epsilon}{\sigma}\right\} &\leq P(|Y_n| > b), \text{ whenever } 0 < b < \sqrt{n}\frac{\epsilon}{\sigma}, \text{ i.e., } n > \frac{b^2\sigma^2}{\epsilon^2} \\
&= 1-P(|Y_n| \leq b) = 1-P(-b \leq Y_n \leq b) \\
&= 1-\{F_n(b)-F_n(-b)\}, \because F_n \text{ is the distribution function of } Y_n \text{ which is } N(0,1) \\
&= 1-\{F_n(b)-(1-F_n(b))\} = 2-2F_n(b) = 2(1-F_n(b)). \qquad (2)
\end{aligned}$$

$\therefore$ by (1) and (2), we get $0 \leq P\left(\left|\bar{X}-m\right| \geq \epsilon\right) \leq 2(1-F_n(b))$.

Taking limit as $n \to \infty$, $0 \leq P\left(\left|\bar{X}-m\right| \geq \epsilon\right) \leq 2(1-\Phi(b))$, by central limit theorem for equal components where $\Phi =$ distribution function of standard normal distribution.

This is true for all $b > 0$. Taking limit as $b \to \infty$, we get $P\left(\left|\bar{X}-m\right| \geq \epsilon\right)=0$.

Hence $\bar{X} \xrightarrow{\text{in } P} m$ as $n \to \infty$, which is the law of large numbers for equal components.

10.3 Lindeberg-Levy theorem

If $X_1, X_2, \cdots, X_n$ are independently and identically distributed (iid) RVs with $E(X_i)=\mu$ and var $(X_i)=\sigma^2$, $\forall\ i=1(1)n$, then their sum $S_n = X_1+X_2+\cdots+X_n$ is asymptotically normal with mean $\mu^* = n\mu$ and variance $\sigma^{*2}=n\sigma^2$.

Proof. Standardised variate of S_n is $S_n^* = \frac{S_n-\mu^*}{\sigma^*}$, i.e., $S_n^* = \frac{X_1+X_2+\cdots+X_n-n\mu}{\sigma^*} = \sum_{i=1}^{n}\frac{X_i-\mu}{\sigma^*}$.

Since $X_1, X_2, \cdots, X_n$ are mutually independent,

$\therefore\ \frac{X_1-\mu}{\sigma^*}, \frac{X_2-\mu}{\sigma^*}, \cdots, \frac{X_n-\mu}{\sigma^*}$ are also mutually independent.

$\therefore$ moment generating function S^* is given by

$$\begin{aligned}
M_{S_n^*}(t) &= E\left(e^{tS_n^*}\right) = E\left(e^{t\sum_{i=1}^{n}\frac{x_i-\mu}{\sigma^*}}\right) = E\left(e^{t\frac{x_1-\mu}{\sigma^*}}\right)\cdot E\left(e^{t\frac{x_2-\mu}{\sigma^*}}\right)\cdots E\left(e^{t\frac{x_n-\mu}{\sigma^*}}\right) \\
&= \left\{E\left(e^{t\frac{x_1-\mu}{\sigma^*}}\right)\right\}^n \text{ as } X_i\text{'s are iid} \\
&= \left[M_1\left(\frac{t}{\sigma^*}\right)\right]^n = \left[M_1\left(\frac{t}{\sqrt{n}\sigma}\right)\right]^n, \text{ where } M_1 = \text{MGF of } X_1 \\
&= \left[1+\frac{t^2}{2n}+0\left(t^3\right)\right]^n
\end{aligned}$$

$\therefore\ \lim_{n\to\infty} M_{S_n^*}(t)=e^{t^2/2}$, which is the MGF of a standard normal variate.

Hence by the uniqueness of MGF of standard variate.

$S_n^* = \frac{S_n-\mu^*}{\sigma^*}$ is asymptotically normal $(0,1)$.

$\therefore\ S_n$ is asymptotically normal (μ^*, σ^*), where $\mu^* = n\mu$ and $\sigma^* = \sqrt{n}\sigma$.

Chapter 11

Miscellaneous Examples

Ex. 1. Two dice are thrown n times in succession. What is the probability of obtaining double six at least once? Also find the least number of throws so that the probability of obtaining at least one double six will be greater than $\frac{1}{2}$. **[C.H. 2009, 2010]**

Solution. In a single throw with two dice, the event space is $S=\{(1,1),(1,2),(1,3),\cdots,(6,5),(6,6)\}$. $n(S)=36$. Out of 36 elementary events only one is in favour of *double six*.

$\therefore$ probability of obtaining *double six* in a single throw is $\frac{1}{36}$

and probability of not obtaining *double six* in a single throw is $\frac{35}{36}$.

As the throws are independent, $\therefore$ the probability of not obtaining a double six in n throws is $\left(\frac{35}{36}\right)^n$.

$\therefore$ probability of obtaining at least one double six is $1-\left(\frac{35}{36}\right)^n$.

According to the condition,

$$1-\left(\frac{35}{36}\right)^n > \frac{1}{2} \text{ or, } \left(\frac{35}{36}\right)^n < \frac{1}{2} \text{ or, } n\{\log_{10}35-\log_{10}36\} < \log_{10}1-\log_{10}2$$

$$\text{or, } n > \frac{\log_{10}2}{\log_{10}36-\log_{10}35} \text{ or, } n > 24.61 \text{ (approx.)}$$

$\therefore$ minimum number of throw is 25.

Ex. 2. If A and B be any two events, then show that $P\left(\bar{A}+B\right)=1-P(A)+P(AB)$. **[B.H. 2010]**

Solution. LHS $= P\left(\bar{A}\right)+P(B)-P\left(\bar{A}B\right)$. (1)

We have $A+\bar{A}=S$ and $A\bar{A}=0$.

$\therefore P\left(A+\bar{A}\right)=P(S)=1$ or, $P(A)+P\left(\bar{A}\right)=1$.

$\therefore P\left(\bar{A}\right)=1-P(A)$. (2)

Again, $B=SB=\left(A+\bar{A}\right)B=AB+\bar{A}B$, by distributive law.

$\therefore P(B)=P(AB)+P\left(\bar{A}B\right)$, $\because AB\bar{A}B=A\bar{A}B=0$

or, $P(B)-P\left(\bar{A}B\right)=P(AB)$. (3)

Hence, (1) with the help of (2) and (3) becomes

LHS $=1-P(A)+P(AB)=$ RHS **(Proved).**

Ex. 3. If A and B be two mutually exclusive events and $P(A+B)\neq 0$, prove that $P(A|(A+B))=\frac{P(A)}{P(A)+P(B)}$. **[B.H. 2010]**

Solution. Given $AB=0$. $\therefore$ $P(A+B)=P(A)+P(B)$.

$$\therefore P(A|(A+B))=\frac{P(A(A+B))}{P(A+B)}=\frac{P(A+AB)}{P(A)+P(B)}=\frac{P(A)}{P(A)+P(B)}, \because AB=0.$$

Ex. 4. The digits 1, 2, 3, 4, 5, 6, 7 are written in random order to form a 7-digit number. What is the probability that it is divisible by 4. [B.H. 2010]

Solution. The total number of 7-digit numbers by the digits 1, 2, 3, 4, 5, 6, 7 is the number of their permutations taking all at a time is 7!

Among these 7! numbers, those will be divisible by 4 for which the number formed by the last two figures are divisible by 4. The numbers formed by the last two figures and divisible by 4 are 12, 16, 24, 32, 36, 44, 52, 56, 64, 72, 76.

So, total number of favourable cases $= 11 \times 5!$.

By classical definition required probability $= \frac{11\times 5!}{7!} = \frac{11}{7\times 6} = \frac{11}{42}$.

Ex. 5. An urn contains a red balls and b white balls. A ball is drawn at random and the ball together with c balls of the same colour are returned to the urn. The same operation is done once again. Then a ball is drawn from the urn. Find the probability that the balls drawn are red for all the three times.

Solution. Let R_i be the event that the ith draw is a red ball, $i = 1,2,3$.

$$\therefore\ P(R_1) = \frac{a}{a+b},\ P(R_2 \mid R_1) = \frac{a+c}{a+b+c},\ P(R_3 \mid R_1 R_2) = \frac{a+2c}{a+b+2c}.$$

$$\therefore \text{required probability} = P(R_1 R_2 R_3) = P(R_1)P(R_2 \mid R_1)P(R_3 \mid R_1 R_2) = \frac{a}{a+b}\cdot\frac{a+c}{a+b+c}\cdot\frac{a+2c}{a+b+2c}.$$

Ex. 6. For what value of k, f represents the pdf of the continuous RVs X and Y, where

$$f(x,y) = \begin{cases} k(4-2x+y), & 0<x<3, 2<y<4 \\ 0, & \text{elsewhere,} \end{cases}$$

and find $P(X<2 \mid Y<3)$.

Solution. For any pdf, we have

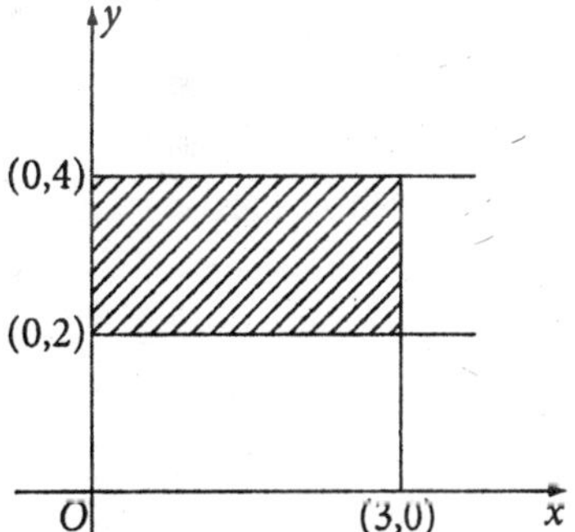

$$\int_{-\infty}^{\infty}\int_{-\infty}^{\infty} f(x,y)\,dx\,dy = 1.$$

$$\therefore k\int_2^4\int_0^3 (4-2x+y)\,dx\,dy = 1 \Rightarrow \frac{1}{k} = \int_2^4 \left[4x - x^2 + yx\right]_0^3 dy$$

$$\text{or, } \frac{1}{k} = \int_2^4 (3+3y)\,dy = \left[3y + \frac{3}{2}y^2\right]_2^4 = 6 + 18 = 24.$$

$\therefore\ k = \frac{1}{24}$.

$$P(X<2 \mid Y<3) = \frac{P(X<2, Y<3)}{P(Y<3)} = \frac{\frac{1}{24}\int_0^2\int_2^3 (4-2x+y)\,dy\,dx}{\frac{1}{24}\int_0^3\int_2^3 (4-2x+y)\,dy\,dx} = \frac{\int_0^2 \left[4y - 2xy + \frac{y^2}{2}\right]_2^3 dx}{\int_0^3 \left[4y - 2xy + \frac{y^2}{2}\right]_2^3 dx}$$

$$= \frac{\int_0^2 \left(4 - 2x + \frac{5}{2}\right) dx}{\int_0^3 \left(4 - 2x + \frac{5}{2}\right) dx} = \frac{\left[\frac{13}{2}x - x^2\right]_0^2}{\left[\frac{13}{2}x - x^2\right]_0^3} = \frac{13-4}{21/2} = \frac{6}{7}.$$

Ex. 7. The joint pdf of two RVs X and Y is $f(x,y)=\begin{cases}8xy, & 0\le x<y, 0\le y\le 1\\ 0, & \text{otherwise.}\end{cases}$

Examine whether X and Y are independent. Also compute $\operatorname{var}(X)$ and $\operatorname{var}(Y)$. [C.H. 2010]

Solution.

$$f_X(x)=\begin{cases}8\displaystyle\int_x^1 xy\,dy=8x\left[\frac{y^2}{2}\right]_x^1=4x\left(1-x^2\right), & 0\le x\le 1\\ 0, & \text{otherwise.}\end{cases}$$

$$f_Y(y)=\begin{cases}8\displaystyle\int_0^y xy\,dx=4y^3, & 0\le y\le 1\\ 0, & \text{otherwise.}\end{cases}$$

$f(x,y)\neq f_X(x)f_Y(y)$. Hence, X and Y are not independent.

$$E(X)=\int_0^1 4x^2\left(1-x^2\right)dx=4\left(\frac{1}{3}-\frac{1}{5}\right)=\frac{8}{15},\ E\left(X^2\right)=\int_0^1 4x^3\left(1-x^2\right)dx=4\left(\frac{1}{4}-\frac{1}{6}\right)=\frac{1}{3}.$$

$$\therefore\ \operatorname{var}(X)=E\left(X^2\right)-\{E(X)\}^2=\frac{1}{3}-\frac{64}{225}=\frac{11}{225}.$$

$$E(Y)=4\int_0^1 y^4\,dy=\frac{4}{5},\ E\left(Y^2\right)=4\int_0^1 y^5\,dy=\frac{2}{3}.$$

$$\therefore\ \operatorname{var}(Y)=E\left(Y^2\right)-\{E(Y)\}^2=\frac{2}{3}-\frac{16}{25}=\frac{2}{75}.$$

Ex. 8. Show that the mean deviation about the mean of a normal (m,σ) distribution is $\sqrt{\frac{2}{\pi}}\sigma$. [C.H. 2007, 2010]

Solution. Mean deviation about the mean is

$$E(|X-m|)=\frac{1}{\sqrt{2\pi}\sigma}\int_{-\infty}^{\infty}|x-m|e^{-\frac{(x-m)^2}{2\sigma^2}}dx=\frac{1}{\sqrt{2\pi}\sigma}\int_{-\infty}^{\infty}|u|e^{-\frac{u^2}{2\sigma^2}}du,\ \text{when } u=x-m$$

$$=\frac{2}{\sqrt{2\pi}\sigma}\int_0^{\infty}|u|e^{-\frac{u^2}{2\sigma^2}}du,\ \text{as the integrand is an even function}$$

$$=\frac{2}{\sqrt{2\pi}\sigma}\int_0^{\infty}ue^{-\frac{u^2}{2\sigma^2}}du,\ \text{as } u\ge 0.$$

[Let $\frac{u^2}{2\sigma^2}=v$. $\therefore$ $u\,du=\sigma^2\,dv$. Now $v=0$ when $u=0$ and $v\to\infty$ as $u\to\infty$]

$$=\sqrt{\frac{2}{\pi}}\sigma\int_0^{\infty}e^{-v}\,du=\sqrt{\frac{2}{\pi}}\sigma\lim_{R\to\infty}\int_0^R e^{-v}\,dv$$

$$=\sqrt{\frac{2}{\pi}}\sigma\lim_{R\to\infty}\left[e^{-v}\right]_R^0=\sqrt{\frac{2}{\pi}}\sigma\lim_{R\to\infty}\left(1-e^{-R}\right)=\sqrt{\frac{2}{\pi}}\sigma\ \text{(Proved).}$$

Ex. 9. If a RV X has the probability density function $f(x)=\frac{1}{2}e^{-|x|}$, $-\infty<x<\infty$, show that the moment generating function ϕ of X is given by $\phi(t)=\frac{1}{1-t^2}$. Hence, find $\operatorname{var}(X)$. [C.H. 2010]

Solution.

$$\phi(t)=E\left(e^{tX}\right)=\int_{-\infty}^{\infty} e^{tx}\cdot\frac{1}{2}e^{-|x|}\,dx=\frac{1}{2}\int_{-\infty}^{0} e^{(t+1)x}\,dx+\frac{1}{2}\int_{0}^{\infty} e^{(t-1)x}\,dx$$

$$=\frac{1}{2}\int_{-\infty}^{0} e^{(1+t)x}\,dx+\frac{1}{2}\int_{0}^{\infty} e^{-(1-t)x}\,dx=\frac{1}{2}\lim_{R\to-\infty}\left[\frac{e^{(1+t)x}}{1+t}\right]_R^0+\frac{1}{2}\lim_{S\to\infty}\left[\frac{e^{-(1-t)x}}{-(1-t)}\right]_0^S$$

$$=\frac{1}{2}\lim_{R\to-\infty}\left[\frac{1}{1+t}-\frac{e^{(1+t)R}}{1+t}\right]+\frac{1}{2}\lim_{S\to\infty}\left[\frac{e^{-(1-t)S}}{-(1-t)}+\frac{1}{1-t}\right]$$

$$=\frac{1}{2}\cdot\frac{1}{1+t}+\frac{1}{2}\cdot\frac{1}{1-t},\text{ provided } |t|<1$$

$$=\frac{1}{1-t^2}.$$

$$\therefore\ \phi(t)=\sum_{k=0}^{\infty}\frac{\alpha_k(X)}{k!}t^k=\frac{1}{1-t^2}=1+t^2+t^4+\cdots\infty$$

$$\therefore\ \frac{\alpha_2(X)}{2!}=1,\ \frac{\alpha_1(X)}{1!}=0.\ \therefore\ \alpha_2(X)=2,\ \alpha_1(X)=0.$$

$\therefore$ $\operatorname{var}(X)=E\left(X^2\right)-\{E(X)\}^2=\alpha_2(X)-\alpha^2(X)=2.$

Ex. 10. What is the probability that x^2-y^2 is divisible by 3, if x,y are chosen at random from the set $\{1,2,3,\cdots,3n\}$.

Solution. $\{1,2,3,\cdots,3n\}=\{3k:1\le k\le n\}\cup\{3k-1:1\le k\le n\}\cup\{3k-2:1\le k\le n\}$, $k\in\mathbb{N}$; be a partition of the given set.

$x^2-y^2=(x-y)(x+y)$ and 3 is prime, so $3|x^2-y^2\Leftrightarrow 3|x-y$ or $3|x+y$.

If we take x,y from any one class of the partition, then $3|x-y$. The number of ways of choosing two elements from the same partition $=3\times{}^nC_2=\frac{3n(n-1)}{2}$.

The condition $3|x+y$ is satisfied if one of x and y is chosen from the partition $\{3k-1:1\le k\le n\}$ and the other from $\{3k-2:1\le k\le n\}$ or x,y from first partition (which is already taken). The number of ways of such a selection is ${}^nC_1\times{}^nC_1=n^2$.

$\therefore$ the total number of favourable ways $=\frac{3n(n-1)}{2}+n^2=\frac{5n^2-3n}{2}$.

The total number of selection of x and y from the given set $={}^{3n}C_2=\frac{3n(3n-1)}{2}$.

$\therefore$ required probability $=\frac{5n^2-3n}{3n(3n-1)}=\frac{5n-3}{3(3n-1)}$.

Ex. 11. Let A,B be two events such that $P(A)=P(A|B)=\frac{1}{3}$ and $P(B|A)=\frac{1}{4}$. Check, whether the following statements are true or not.

(a) A and B are independent, (b) A and B are mutually exclusive, (c) $P(A'|B)=\frac{2}{3}$, (d) $P(B'|A')=\frac{3}{4}$.

Solution. $P(AB)=P(B)P(A|B)=P(B)P(A)$ [as $P(A|B)=P(A)$, given].

Hence, A and B are independent. $\therefore$ statement (a) is true.

As two proper events [events with nonzero probability] cannot be simultaneously mutually exclusive and independent.

Hence, statement (b) is not correct.

The independence of A and B implies that A', B are independent; also A', B' are independent.

$\therefore$ $P(A'|B)=P(A')=1-P(A)=1-\frac{1}{3}=\frac{2}{3}$

and $P(B'|A')=P(B')=1-P(B)$

$=1-P(B|A)$, as A,B are independent

$=1-\frac{1}{4}=\frac{3}{4}$

Hence, (c) and (d) are both correct.

Ex. 12. *A speaks the truth in 75% cases and B in 80% cases. What is the probability that they will contradict each other?* **[C.H. 2008]**

Solution. Let $A_1 =$ the event that A speaks the truth.

$B_1 =$ the event that B speaks the truth.

$\therefore\ P(A_1) = 0.75,\ P(B_1) = 0.8.\ \therefore\ P(\bar{A}_1) = 0.25, P(\bar{B}_1) = 0.2.$

The event that they contradict each other $= A_1\bar{B}_1 + \bar{A}_1 B_1$ (as the event occurs only when one of them speaks the truth and the other false)

$$\begin{aligned} P(A_1\bar{B}_1 + \bar{A}_1 B_1) &= P(A_1\bar{B}_1) + P(\bar{A}_1 B_1), \text{ as } A_1\bar{B}_1\bar{A}_1 B_1 = 0 \\ &= P(A_1)P(\bar{B}_1) + P(\bar{A}_1)P(B_1), \text{ as } A_1, B_1 \text{ are independent} \\ &= 0.75\times 0.2 + 0.25\times 0.8 = 0.35. \end{aligned}$$

in 35% cases they will contradict.

Ex. 13. *If n objects are distributed among a men and b women $(b < a)$, show that the probability that the women get an odd number of objects is $\frac{1}{2}\left[\frac{(a+b)^n-(a-b)^n}{(a+b)^n}\right]$.* **[C.H. 2008]**

Solution. Let the things are distributed one by one at random. The probability that a man receives it is $\frac{a}{a+b} = p$ (say) and the probability than an woman receives it is $\frac{b}{a+b} = q$ (say).

That the women get an odd number of objects has the probability

$$\begin{aligned} \binom{n}{1}qp^{n-1} + \binom{n}{3}q^3p^{n-5} + \binom{n}{5}q^5p^{n-5} + \cdots &= \frac{1}{2}[(p+q)^n - (p-q)^n] \\ &= \frac{1}{2}\left[1 - \left(\frac{a-b}{a+b}\right)^n\right] = \frac{1}{2}\left[\frac{(a+b)^n-(a-b)^n}{(a+b)^n}\right]. \end{aligned}$$

Ex. 14. *Find the moment generating function of the continuous probability distribution whose density function is $\frac{1}{2}x^2e^{-x}$, $0 < x < \infty$ and deduce the values of mean and variance.* **[K.H. 2009]**

Solution. The MGF is

$$M_x(t) = \int_0^\infty e^{tx}\frac{1}{2}x^2e^{-x}\,dx = \frac{1}{2}\int_0^\infty e^{(t-1)x}x^2\,dx = \frac{1}{2}\int_0^\infty e^{-(1-t)x}x^2\,dx.$$

The integral will be convergent absolutely provided $|t| < 1$.

Let $(1-t)x = u$. $\therefore (1-t)dx = du$. When $x = 0, u = 0$ and when $x \to \infty, u \to \infty$.

$$\begin{aligned} \therefore\ M_x(t) &= \frac{1}{2}\int_0^\infty e^{-u}\frac{u^2}{(1-t)^2}\cdot\frac{du}{1-t} = \frac{1}{2}\cdot\frac{1}{(1-t)^3}\int_0^\infty e^{-u}u^{3-1}\,du \\ &= \frac{1}{2}\cdot\frac{1}{(1-t)^3}\cdot\Gamma(3) = \frac{2!}{2(1-t)^3} = \frac{1}{(1-t)^3}. \end{aligned}$$

$$\therefore\ M_x(t) = (1-t)^{-3} = 1 + 3t + 6t^2 + \cdots = \sum_{k=0}^{\infty}\frac{\alpha_k(x)}{k!}t^k.$$

$$\therefore\ \alpha_1(X) = 3,\ \frac{\alpha_2(X)}{2!} = 6.\ \therefore\ \alpha_2(X) = 12.$$

$\therefore$ mean $= E(X) = \alpha_1(X) = 3.$

Variance $= E(X^2) - \{E(X)\}^2 = \alpha_2(X) - \{\alpha_1(X)\}^2 = 12 - 9 = 3.$

Ex. 15. Find the characteristic function $\chi_n(t)$ of a continuous RV X with pdf:

$$f_x(x)=\begin{cases}1-|x|, & \text{if } |x|<1\\ 0, & \text{elsewhere.}\end{cases}$$

[C.H. 2009]

Solution.

$$\chi_x(t)=E\left(e^{itX}\right)=\int_{-1}^{1}e^{itx}(1-|x|)dx=\int_{-1}^{1}e^{itx}-\int_{-1}^{1}e^{itx}|x|dx$$

$$=\frac{e^{it}-e^{-it}}{it}+\int_{-1}^{0}e^{itx}x\,dx-\int_{0}^{1}e^{itx}x\,dx$$

$$=\frac{2}{t}\sin t+\left[\frac{e^{itx}}{it}x\right]_{-1}^{0}-\int_{-1}^{0}\frac{e^{itx}}{it}dx-\left[\frac{e^{itx}}{it}\cdot x\right]_{0}^{1}+\int_{0}^{1}\frac{e^{itx}}{it}dx$$

$$=\frac{2}{t}\sin t+\frac{e^{-it}}{it}+\frac{1}{t^2}\left(1-e^{-it}\right)-\frac{e^{it}}{it}-\frac{1}{t^2}(e^{it}-1)$$

$$=\frac{2}{t}\sin t-\frac{2}{t}\sin t+\frac{1}{t^2}(1-e^{-it}-e^{it}+1)=\frac{2}{t^2}-\frac{1}{t^2}\left(e^{it}+e^{-it}\right)$$

$$=\frac{2}{t^2}-\frac{2\cos t}{t^2}=\frac{2}{t^2}(1-\cos t).$$

Ex. 16. A point is chosen at random within the ellipse $\frac{x^2}{a^2}+\frac{y^2}{b^2}=1$ $(a<b)$. What is the chance that its distance from the centre exceeds a? [B.H. 2006]

Solution. Since the point is chosen at random within the ellipse, $\therefore$ the RV (X,Y) follows uniform distribution. The shaded region is in favour of the event that the point (x,y) is at a distance greater than a and within the ellipse.

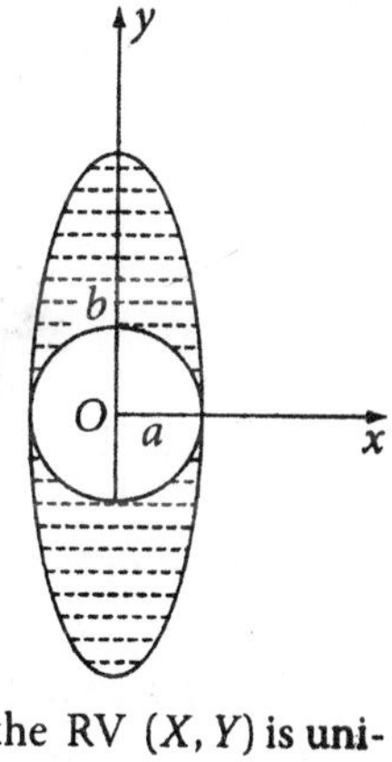

$$\therefore \text{ required probability}=\frac{\text{area in favour of the event}}{\text{total area within the ellipse}}$$

$$=\frac{\pi ab-\pi a^2}{\pi ab}=\frac{b-a}{b}$$

$$=1-\frac{a}{b}.$$

Alternative method: Since the point is chosen at random within the ellipse, $\therefore$ the RV (X,Y) is uniformly distributed within it. So, the pdf $f_{x,y}$ of (X,Y) is given by

$$f_{x,y}(x,y)=\begin{cases}\frac{1}{\pi ab}, & \text{when } (x,y) \text{ is a point within ellipse}\\ 0, & \text{elsewhere.}\end{cases}$$

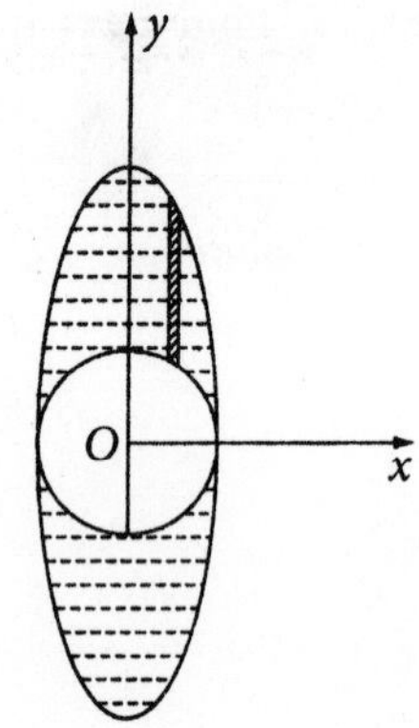

$$\therefore \text{ required probability} = 2\int_{-a}^{a}\int_{\sqrt{a^2-x^2}}^{\frac{b}{a}\sqrt{a^2-x^2}} \frac{1}{\pi ab}\,dy\,dx$$

$$= \frac{2}{\pi ab}\int_{-a}^{a}\left(\frac{b}{a}-1\right)\sqrt{a^2-x^2}\,dx$$

$$= \frac{2}{\pi ab}\cdot\frac{b-a}{a}\cdot 2\int_{0}^{a}\sqrt{a^2-x^2}\,dx,$$

as the integrand is an even function

$$= \frac{4(b-a)}{\pi a^2 b}\left[\frac{x\sqrt{a^2-x^2}}{2}+\frac{a^2}{2}\sin^{-1}\frac{x}{a}\right]_0^a$$

$$= \frac{4(b-a)}{\pi a^2 b}\cdot\frac{a^2}{2}\cdot\frac{\pi}{2} = 1-\frac{a}{b}.$$

Ex. 17. Let A and B be two events such that $P(A)=\frac{3}{4}$ and $P(B)=\frac{5}{8}$. Show that $\frac{3}{8}\le P(A\cap B)\le\frac{5}{8}$. [B.H. 2006]

Solution. We have $P(A\cup B)\le 1$.

$\therefore\ P(A)+P(B)-P(A\cap B)\le 1$ or, $\frac{3}{4}+\frac{5}{8}-P(A\cap B)\le 1$ or, $P(A\cap B)\ge\frac{11}{8}-1$, i.e., $\frac{3}{8}\le P(A\cap B)$. (1)

Again, $A\cup B\subseteq B$. $\therefore\ P(A\cap B)\le P(B)$.

$$\therefore\ P(A\cap B)\le\tfrac{5}{8}. \tag{2}$$

$\therefore$ by (1) and (2) we get $\frac{3}{8}\le P(A\cap B)\le\frac{5}{8}$.

Ex. 18. Let X_1, X_2, X_3 be three independent $N(\mu,1)$ variables. Prove that $P[X_1-2X_2+X_3>0]=\frac{1}{2}$. [B.H. 2006]

Solution. Let $X=X_1-2X_2+X_3$.

$\therefore\ m_x=E(X)=E(X_1)-2E(X_2)+E(X_3)=\mu-2\mu+\mu=0.$

$\sigma_x^2=\sigma_{x_1}^2+4\sigma_{x_2}^2+\sigma_{x_3}^2$, as X_1, X_2, X_3 are independent

$=1+4+1=6.$

$\therefore\ X\sim N\left(0,\sqrt{6}\right).$

$$\therefore\ P(X>0)=\frac{1}{\sqrt{2\pi}\cdot\sqrt{6}}\int_0^\infty e^{-\frac{x^2}{12}}\,dx=\frac{1}{\sqrt{12\pi}}\int_0^\infty e^{-u^2}\sqrt{12}\,du,\ \text{when } u=\frac{x}{\sqrt{12}}$$

$$=\frac{1}{\sqrt{\pi}}\cdot\frac{\sqrt{\pi}}{2}=\frac{1}{2}.$$

Ex. 19. The pdf of a RV X is

$$f_x(x)=\begin{cases}\frac{1}{\theta}e^{-\frac{x}{\theta}}, & x>0, \theta>0\\ 0, & \text{otherwise,}\end{cases}$$

find $E\left(X^2\right)$. [B.H. 2005]

Solution. $E(X^2)=\dfrac{1}{\theta}\displaystyle\int_0^{\infty} x^2 e^{-\frac{x}{\theta}}\,dx$, provided the integral converges absolutely.

$$\therefore\ \theta E\left(X^2\right)=\lim_{R\to\infty}\int_0^R x^2 e^{-\frac{x}{\theta}}\,dx=\lim_{R\to\infty}\left[-\theta x^2 e^{-\frac{x}{\theta}}\Big|_0^R+2\theta\int_0^R x e^{-\frac{x}{\theta}}\,dx\right]$$

$$=2\theta\lim_{R\to\infty}\left[-\theta x e^{-\frac{x}{\theta}}\Big|_0^R+\theta\int_0^R e^{-\frac{x}{\theta}}\,dx\right]=2\theta^2\lim_{R\to\infty}\left[-\theta e^{-\frac{x}{\theta}}\right]_0^R=2\theta^2\lim_{R\to\infty}\left[-\theta e^{-\frac{R}{\theta}}+\theta\right]$$

$$=2\theta^3.\ \therefore\ E\left(X^2\right)=2\theta^2.$$

Ex. 20. The bivariate RV (X,Y) jointly follows the pdf

$$f_{x,y}(x,y)=\begin{cases}kx^2(8-y), & x<y<2x,\ 0\le x\le 2\\ 0, & \text{elsewhere.}\end{cases}$$

Find (i) k and (ii) the marginal pdf of X and Y (iii) the conditional pdf $f_{x|y}(x\,|\,y)$ and $f_{y|x}(y\,|\,x)$.

[B.H. 2006]

Solution. We have

$$\int_{-\infty}^{\infty}\int_{-\infty}^{\infty} f_{x,y}(x,y)\,dx\,dy=1 \text{ or, } \int_0^3\int_x^{2x} kx^2(8-y)\,dy\,dx=1$$

$$\text{or, } \frac{1}{k}=\int_0^2\left[8x^2y-\frac{x^2y^2}{2}\right]_x^{2x}dx=\int_0^2\left(8x^3-\frac{3}{2}x^4\right)dx=\left[2x^4-\frac{3}{10}x^5\right]_0^2$$

$$=32-\frac{3}{10}\cdot 32=\frac{7}{10}\cdot 32=\frac{112}{5}.$$

$\therefore\ k=\frac{5}{112}$.

$$f_x(x)=\frac{112}{5}\int_x^{2x} x^2(8-y)\,dy=\frac{112}{5}\left[8x^2y-\frac{x^2y^2}{2}\right]_x^{2x}=\frac{112}{5}\left(8x^3-\frac{3}{2}x^4\right)$$

$$=\frac{56}{5}x^3(16-3x),\ 0\le x\le 2.$$

$f_x(x)=0$, elsewhere.

$$f_Y(y)=\frac{112}{5}\int_{y/2}^{y} x^2(8-y)\,dx=\frac{112}{5}\left[\frac{8}{3}x^3-\frac{y}{3}x^3\right]_{y/2}^{y}=\frac{112}{5}\cdot\frac{7}{24}y^3(8-y)$$

$$=\frac{98}{15}y^3(8-y),\ 0<y<4.$$

$f_Y(y)=0$, elsewhere.

$$f_{x|y}(x\,|\,y)=\frac{f_{x,y}(x,y)}{f_Y(y)}=\frac{\frac{112}{5}x^2(8-y)}{\frac{98}{15}y^3(8-y)}=\frac{24}{7}\cdot\frac{x^2}{y^3},\ 0\le x\le 2, \text{ for some fixed } y.$$

$$f_{y|x}(y\,|\,x)=\frac{f_{x,y}(x,y)}{f_x(x)}=\frac{\frac{112}{5}x^2(8-y)}{\frac{56}{5}x^3(16-3x)}=\frac{2(8-y)}{x(16-3x)},\ 0<y<4, \text{ for some fixed } x.$$

Ex. 21. Find $\rho(X,Y)$ of X and Y having the joint pdf

$$f_{X,Y}(x,y)=\begin{cases}\frac{1}{2}\sin(x+y), & 0\le x\le\frac{\pi}{2},\ 0\le y\le\frac{\pi}{2}\\ 0, & \text{otherwise.}\end{cases}$$ [B.H. 2006]

Solution.

$$f_X(x)=\frac{1}{2}\int_0^{\pi/2}\sin(x+y)\,dy=\frac{1}{2}\Big[\cos(x+y)\Big]_{\pi/2}^{0}=\frac{1}{2}(\cos x+\sin x),\ 0\le x\le\frac{\pi}{2}.$$

$$f_Y(y)=\frac{1}{2}\int_0^{\pi/2}\sin(x+y)\,dx=\frac{1}{2}\Big[\cos(x+y)\Big]_{\pi/2}^{0}=\frac{1}{2}(\cos y+\sin y),\ 0\le y\le\frac{\pi}{2}.$$

$$m_x=E(X)=\frac{1}{2}\int_0^{\pi/2}x(\cos x+\sin x)\,dx=\frac{1}{2}\int_0^{\pi/2}\left(\frac{\pi}{2}-x\right)(\sin x+\cos x)\,dx$$

$$\therefore\ 2m_x=E(X)=\frac{1}{2}\int_0^{\pi/2}\frac{\pi}{2}(\sin x+\cos x)\,dx=\frac{\pi}{4}\Big[\sin x-\cos x\Big]_0^{\pi/2}=\frac{\pi}{2}.$$

$$\therefore\ m_x=\frac{\pi}{4}.$$

Similarly, $m_y=\frac{\pi}{4}$.

$$\sigma_x^2=E\left(X-\frac{\pi}{4}\right)^2=\frac{1}{2}\int_0^{\pi/2}\left(x-\frac{\pi}{4}\right)^2(\cos x+\sin x)\,dx$$

$$2\sigma_x^2=\left[\left(x-\frac{\pi}{4}\right)^2(\sin x-\cos x)\right]_0^{\pi/2}-2\int_0^{\pi/2}\left(x-\frac{\pi}{4}\right)(\sin x-\cos x)\,dx=\frac{\pi^2}{16}-\frac{\pi^2}{16}(-1)=\frac{\pi^2}{8}.$$

$$\therefore\ \sigma_x=\frac{\pi}{4}.$$

Similarly, $\sigma_y=\frac{\pi}{4}$.

$$\alpha_{11}=E(XY)=\frac{1}{2}\int_0^{\pi/2}\int_0^{\pi/2}xy\sin(x+y)\,dx\,dy.$$

$$\therefore\ 2\alpha_{11}=\int_0^{\pi/2}y\left\{\left(\frac{\pi}{2}-1\right)\sin y+\cos y\right\}dy=\left(\frac{\pi}{2}-1\right)\int_0^{\pi/2}y\sin y\,dy+\int_0^{\pi/2}y\cos y\,dy$$

$$=\left(\frac{\pi}{2}-1\right)\left[-y\cos y\Big|_0^{\pi/2}+\int_0^{\pi/2}\cos y\,dy\right]+\left[y\sin y\Big|_0^{\pi/2}-\int_0^{\pi/2}\sin y\,dy\right]$$

$$=\left(\frac{\pi}{2}-1\right)+\left(\frac{\pi}{2}-1\right)=2\left(\frac{\pi}{2}-1\right).$$

$$\therefore\ \alpha_{11}=\frac{\pi}{2}-1.$$

$$\therefore\ \operatorname{cov}(X,Y)=\alpha_{11}-m_xm_y=\frac{\pi}{2}-1-\frac{\pi^2}{16}$$

$$\therefore\ \rho(X,Y)=\frac{\operatorname{cov}(X,Y)}{\sigma_x\sigma_y}=\frac{16}{\pi^2}\left(\frac{\pi}{2}-\frac{\pi^2}{16}-1\right)=\frac{8}{\pi}-\frac{16}{\pi^2}-1.$$

Ex. 22. Two nonnegative RVs X and Y have a joint density function $f_{X,Y}(x,y)=\frac{1}{2}x^3e^{-x(y+1)}$, $x>0, y>0$. Find their correlation coefficient. [B.H. (Old) 2006; K.H. 2009]

Solution.

$$f_X(x)=\frac{1}{2}\int_0^\infty x^3e^{-x(y+1)}\,dy=\frac{1}{2}x^2e^{-x},\ x>0$$

$$f_Y(y)=\frac{1}{2}\int_0^\infty x^3e^{-x(y+1)}\,dx=\frac{1}{2}\cdot\frac{\Gamma(4)}{(y+1)^4}=\frac{3}{(y+1)^4},\ y>0.$$

$$E(X)=\frac{1}{2}\int_0^\infty x^3e^{-x}\,dx=\frac{1}{2}\Gamma(4)=3,\ E(Y)=3\int_0^\infty\frac{y}{(y+1)^4}\,dy=3B(2,2)=\frac{1}{2}$$

$$E\left(X^2\right)=\frac{1}{2}\int_0^\infty x^4e^{-x}\,dx=\frac{1}{2}\Gamma(5)=12$$

$$E\left(Y^2\right)=3\int_0^\infty\frac{y^2}{(y+1)^4}\,dy=3B(3,1)=3\cdot\frac{\Gamma(3)\Gamma(1)}{\Gamma(4)}=1.$$

$$\sigma_x^2=E\left(X^2\right)-\{E(X)\}^2=12-9=3,\ \sigma_y^2=E\left(Y^2\right)-\{E(Y)\}^2=1-\frac{1}{4}=\frac{3}{4}.$$

$$\alpha_{11}=E(XY)=\frac{1}{2}\int_0^\infty\int_0^\infty x^4ye^{-x(y+1)}\,dx\,dy=\frac{\Gamma(5)}{2}\int_0^\infty\frac{y\,dy}{(y+1)^5}=12B(2,3)=12\frac{\Gamma(2)\Gamma(3)}{\Gamma(5)}=1$$

$$\therefore\ \rho(X,Y)=\frac{\alpha_{11}-m_xm_y}{\sigma_x\sigma_y}=\frac{1-\frac{3}{2}}{\sqrt{3}\frac{\sqrt{3}}{2}}=-\frac{1}{3}.$$

Ex. 23. The RV (X,Y) has the pdf

$$f_{x,y}(x,y)=\begin{cases}k(1-x-y), & x>0,y>0,x+y<1\\ 0, & \text{elsewhere.}\end{cases}$$

Find (i) k (ii) the marginal distributions of X and Y (iii) their covariance (iv) the regression lines. Also, test the independence of X and Y.

Solution.

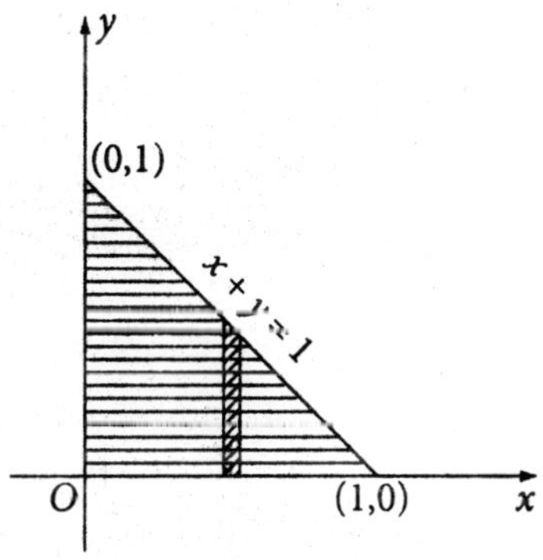

(i) $k\displaystyle\int_0^1\int_0^{1-x}(1-x-y)\,dy\,dx=1$

$$\text{or, }\frac{1}{k}=\int_0^1\left[y-xy-\frac{y^2}{2}\right]_0^{1-x}dx$$

$$=\frac{1}{2}\int_0^1(1-x)^2\,dx=\frac{1}{2}\left[x-x^2+\frac{x^3}{3}\right]_0^1=\frac{1}{6}.$$

$\therefore\ k=6.$

$$\therefore\ f_{x,y}(x,y)=\begin{cases}6(1-x-y), & x>0,y>0,x+y<1\\ 0, & \text{elsewhere.}\end{cases}$$

(ii) The pdf for the marginal distributions are

$$f_x(x)=6\int_0^{1-x}(1-x-y)\,dy=3(1-x)^2,\ 0<x<1$$

$$f_y(y)=6\int_0^{1-y}(1-x-y)\,dx=3(1-y)^2,\ 0<y<1.$$

(iii) $m_x=E(X)=3\int_0^1 x(1-x)^2\,dx=3B(2,3)=3\dfrac{\Gamma(2)\Gamma(3)}{\Gamma(5)}=\dfrac{1}{4}.$

$m_y=E(Y)=3\int_0^1 y(1-y)^2\,dx=3B(2,3)=\dfrac{1}{4}.$

$$\alpha_{11}=E(XY)=6\int_0^1\int_0^{1-x}xy(1-x-y)\,dy\,dx=6\int_0^1 x\left[(1-x)\frac{y^2}{2}-\frac{y^3}{3}\right]_0^{1-x}dx$$

$$=6\int_0^1 x\cdot\frac{1}{6}(1-x)^3\,dx=\int_0^1 x(1-x)^3\,dx$$

$$=B(2,4)=\frac{\Gamma(2)\Gamma(4)}{\Gamma(6)}=\frac{3!}{5!}=\frac{1}{20}.$$

$$\therefore\ \text{cov}(X,Y)=\alpha_{11}-m_x m_y=\frac{1}{20}-\frac{1}{4}\cdot\frac{1}{4}=-\frac{1}{80}.$$

(iv) $$E(X|Y=y)=\int_0^{1-y}x f_{x|y}(x|y)dx=\frac{\int_0^{1-y}x f_{x,y}(x,y)dx}{f_y(y)}$$

$$=\frac{2}{(1-y)^2}\int_0^{1-y}x(1-x-y)dx=\frac{2}{(1-y)^2}\cdot\frac{(1-y)^3}{6}=\frac{1-y}{3},\ \text{for a fixed } y \text{ in } (0,1).$$

Similarly, $E(Y|X=x)=\dfrac{1-x}{3}$, for a fixed x in $(0,1)$.

So, the regression line of Y on X is

$$y=m_y(x)=E(Y|X=x)=\frac{1-x}{3}\ \text{ or, } x+3y=1.$$

The regression line of X and Y is

$$x=m_x(y)=E(X|Y=y)=\frac{1-y}{3}\ \text{ or, } 3x+y=1.$$

As $m_x\neq m_x(y)$, the RVs X and Y are independent.

Ex. 24. Three coins having probabilities of heads $\frac{1}{2},\frac{2}{5},\frac{3}{7}$ respectively are thrown. Find the probability of obtaining exactly one head. [B.H. 2001]

Solution. If H_i, T_i be the events of head and tail on the ith coin, $i=1,2,3$, then the required event is $H_1T_2T_3+T_1H_2T_3+T_1T_2H_3$.

Its probability $=P(H_1)P(T_2)P(T_3)+P(T_1)P(H_2)P(T_3)+P(T_1)P(T_2)P(H_3)$,

as the events are MEE and independent.

$$=\tfrac{1}{2}\cdot\tfrac{3}{5}\cdot\tfrac{4}{7}+\tfrac{1}{2}\cdot\tfrac{2}{5}\cdot\tfrac{4}{7}+\tfrac{1}{2}\cdot\tfrac{3}{5}\cdot\tfrac{3}{7}=\tfrac{29}{70}.$$

Ex. 25. Distribution function $F(x)$ of a continuous RV X is given by

$$F(x)=\begin{cases}1-e^{-\frac{x}{\theta}}, & x\geq 0,\theta>0\\ 0, & \text{elsewhere.}\end{cases}$$

Find $P\{X\geq s+t\,|\,X\geq t\}$. [B.H. 2002]

Solution. If $f_X(x)$ be the pdf, then $f_X(x)=F'(x)=\frac{1}{\theta}e^{-\frac{x}{\theta}}$, $x\geq 0,\theta>0$.

$$\therefore\ P\{X\geq s+t\,|\,X\geq t\}=1-P\{X<s+t|X\geq t\}=1-\frac{P(t\leq X<s+t)}{P(X\geq t)}=1-\frac{\frac{1}{\theta}\int_t^{s+t}e^{-\frac{x}{\theta}}dx}{\frac{1}{\theta}\int_t^{\infty}e^{-\frac{x}{\theta}}dx}$$

$$=1-\frac{\left[-e^{-\frac{x}{\theta}}\right]_t^{s+t}}{\left[-e^{-\frac{x}{\theta}}\right]_t^{\infty}}=1-\frac{e^{-\frac{s+t}{\theta}}-e^{-\frac{t}{\theta}}}{-e^{-\frac{t}{\theta}}}=e^{-\frac{s}{\theta}}.$$

Ex. 26. The probability distribution of the RV (X,Y) is given by the following table:

$Y\downarrow$ \ $X\rightarrow$	0	1	2	Total
1	1/9	1/9	1/9	1/3
2	1/9	2/9	—	1/3
3	1/9	—	2/9	1/3
Total	1/3	1/3	1/3	1

Find $E(X)$, $E(Y)$ and $E(XY)$. Show that X and Y are correlated.

Solution. Marginal distribution of X is given by $X=0,1,2$; $P(X=0)=\frac{1}{3}$, $P(X=1)=\frac{1}{3}$, $P(X=2)=\frac{1}{3}$.

Marginal distribution of Y is given by $Y=1,2,3$; $P(Y=1)=P(Y=2)=P(Y=3)=\frac{1}{3}$.

$\therefore\ E(X)=0\cdot\frac{1}{3}+1\cdot\frac{1}{3}+2\cdot\frac{1}{3}=1, E(Y)=1\cdot\frac{1}{3}+2\cdot\frac{1}{3}+3\cdot\frac{1}{3}=2.$

$E(XY)=0\cdot1\cdot\frac{1}{9}+0\cdot2\cdot\frac{1}{9}+0\cdot3\cdot\frac{1}{9}+1\cdot1\cdot\frac{1}{9}+1\cdot2\cdot\frac{2}{9}+2\cdot1\cdot\frac{1}{9}+2\cdot3\cdot\frac{2}{9}=\frac{1}{9}+\frac{4}{9}+\frac{2}{9}+\frac{12}{9}=\frac{19}{9}.$

$\therefore\ \text{cov}(X,Y)=E(XY)-E(X)E(Y)=\frac{19}{9}-1\times2=\frac{19}{9}-2=\frac{1}{9}\neq0.$

Hence, $\rho(X,Y)\neq0$, i.e., X,Y are correlated.

Ex. 27. The regression lines of two RV X and Y are $2x+3y=5, 5x+2y=8$. Find their correlation coefficient. [B.H. 2008]

Solution. The point of intersection of the lines is $\left(\frac{14}{11},\frac{9}{11}\right)$.

$\therefore\ m_x=\frac{14}{11},\ m_y=\frac{9}{11}.$

The regression line of Y on X is $y-\frac{9}{11}=-\frac{2}{3}\left(x-\frac{14}{11}\right)$.

The regression line of X on Y is $x-\frac{14}{11}=-\frac{2}{5}\left(y-\frac{9}{11}\right)$.

$\therefore\ \rho\frac{\sigma_y}{\sigma_x}=-\frac{2}{3},\rho\frac{\sigma_x}{\sigma_y}=-\frac{2}{5}.\ \therefore\ \rho^2=\frac{4}{15}<1$, so the regression lines are classified correctly.

$\therefore\ \rho=\pm\frac{2}{\sqrt{15}}.\ \therefore\ \rho=-\frac{2}{\sqrt{15}}$, as the positive value does not satisfy $\rho\frac{\sigma_y}{\sigma_n}=-\frac{2}{3}\left[\sigma_y,\sigma_x>0\right]$.

Ex. 28. Balls are drawn one by one (without replacement) from an urn containing a white and b black balls until the first white ball appears. Determine the expected number of black balls preceeding the first white ball. [B.H. 2008]

Solution. Let $W_i=$ the event of drawing a white ball in the ith draw and

B_i = the event of drawing a black ball in the ith draw, $i = 1, 2, \cdots, a+b$.
Let X be the RV denoting the number of black balls preceding the first white ball.
$\therefore$ the distribution of X is the following: $X = 0, 1, 2, \cdots, b$.

$$P(X=0) = \frac{a}{a+b}, P(X=1) = \frac{b}{a+b} \cdot \frac{a}{a+b-1}, P(X=2) = \frac{b}{a+b} \cdot \frac{b-1}{a+b-1} \cdot \frac{a}{a+b-2} \cdots$$

$$P(X=b) = \frac{b}{a+b} \cdot \frac{b-1}{a+b-1} \cdot \frac{b-2}{a+b-2} \cdots \frac{1}{a+1} \cdot \frac{a}{a}.$$

$$E(X) = 0 \cdot \frac{a}{a+b} + 1 \cdot \frac{b}{a+b} \cdot \frac{a}{a+b-1} + 2 \cdot \frac{b}{a+b} \cdot \frac{b-1}{a+b-1} \cdot \frac{a}{a+b-2} + \cdots$$
$$+ b \cdot \frac{b}{a+b} \frac{b-1}{a+b-1} \cdot \frac{b-2}{a+b-2} \cdots \frac{1}{a+1}$$
$$= \frac{ab}{a+b} \left[\frac{1}{a+b-1} + \frac{2(b-1)}{(a+b-1)(a+b-2)} + \cdots + \frac{b(b-1)(b-2)\cdots 1}{(a+b-1)(a+b-2)\cdots(a+1)a} \right].$$

Ex. 29. A box contains n tickets numbered from 1 to n and m tickets are drawn at random at a time. Find the mathematical expectation of the sum of the numbers of the tickers drawn. [Try yourself]

$\left(\textit{Ans. } \frac{m(n+1)}{2}\right)$

»

Index